Walking in Space

Springer
London
Berlin
Heidelberg
New York
Hong Kong
Milan
Paris
Tokyo

David J. Shayler

Walking in Space

Springer

Published in association with
Praxis Publishing
Chichester, UK

PRAXIS

David J. Shayler, FBIS
Astronautical Historian
Astro Info Service Ltd
Halesowen
West Midlands
UK

SPRINGER–PRAXIS BOOKS IN ASTRONOMY AND SPACE SCIENCES
SUBJECT *ADVISORY EDITOR*: John Mason B.Sc., M.Sc., Ph.D.

ISBN 1-85233-710-9 Springer-Verlag Berlin Heidelberg New York

Springer-Verlag is a part of Springer Science + Business Media (*springeronline.com*)

British Library Cataloguing-in-Publication Data has been applied for.

ISBN 1-85233-710-9

Library of Congress Cataloging-in-Publication Data has been applied for.

Copy editing and graphics processing: R.A. Marriott
Cover design: Jim Wilkie
Typesetting: BookEns Ltd, Royston, Herts., UK

Printed in the United States of America on acid-free paper

This book is dedicated to those who dared to dream
the impossible and strived to make it happen

To those who have the courage to let go and step into the void:
the true explorers of the cosmos – the spacewalkers

To their colleagues who support them from inside the
spacecraft, and to those on the ground who build the hardware
for them, help prepare them, and monitor their every move

To their families, who can only sit, watch, wait ...
and hold their breath

Other books in the Springer–Praxis Space Science Series

By the same author:

Disasters and Accidents in Manned Spaceflight, May 2000, ISBN 1-85233-225-5
Skylab: America's Space Station, May 2001, ISBN 1-85233-407-X
Gemini: Steps to the Moon, September 2001, ISBN 1-85233-405-3
Apollo: The Lost and Forgotten Missions, June 2002, ISBN 1-85233-575-0

With Rex Hall:

The Rocket Men, April 2001, ISBN 1-85233-391-X
Soyuz: A Universal Spacecraft, May 2003, ISBN 1-85233-657-9

Table of contents

Foreword

Stepping outside your spaceship in a spacesuit is the most intimate and immediate experience you can have of being 'in space'. The view out of the Space Shuttle's windows is remarkable, but it is nothing compared to what you experience from inside the helmet of your spacesuit. It is like the difference between looking through a glass-bottomed boat and exploring the sea in a diving suit. Like the sea, space is a hostile environment, totally intolerant of mechanical failures and human error. Without a spacesuit, a human's time of useful consciousness, exposed to a vacuum, is measured in seconds. There are many hazards which can incapacitate or kill you: lack of oxygen, lack of pressure, and thermal extremes ranging over hundreds of degrees from sunlight to shadow. The initial challenge facing Alexei Leonov and Ed White in their pioneering spacewalks was to emerge into space, survive, and re-enter the spacecraft. But a spacesuit must ultimately do more than just enable you to survive; it must at the same time allow you to carry out useful work. Spacesuit engineers have done wonders in designing suit joints that maintain a reasonable degree of flexibility even when pressurised. However, a great deal of training is required in learning to move your body in a way that matches the flexibility characteristics of your spacesuit.

I have been fortunate enough to have ventured outside the Space Shuttle on four occasions, for a total of more than twenty-four hours working in the vacuum of space. On my first spaceflight in April 1985 I made the first unplanned spacewalk in NASA's history in an attempt to rescue a malfunctioning satellite. I will never forget the magnificent sight of a space sunset, which stunned me as I opened the hatch of *Discovery*'s air-lock prior to floating out into the payload bay. I felt that the Universe had hit me in the face! It was eerie to realise that between my face and my hand was nothing but *space*. One of the lessons which any successful spacewalker has to learn quickly, of course, is to not become overwhelmed by the incredible onslaught of new sensations, as the purpose of going outside is to use eyes, hands and brains near the problem area so that useful work can be accomplished. I felt justified, however, in reserving a small part of my consciousness to savour the experience, since I recognised that I had been presented with a unique opportunity.

After the launch of the Hubble Space Telescope in the spring of 1990, and the

NASA astronaut Jeffrey A. Hoffman was selected as a Mission Specialist in 1978, flew on five Shuttle missions between 1985 and 1996, and completed four EVAs during two of those missions, totalling more than 25 hours outside. These included the first unplanned Shuttle EVA in 1985 and the first Hubble Space Telescope service mission in 1993.

subsequent discovery of the optical aberration in the primary mirror, repairing the telescope became NASA's number one priority; and I was selected as part of the EVA crew that would help fix it. For me personally, as a former astrophysicist turned astronaut, the thrill of actually visiting and putting my hands on the greatest astronomical telescope ever built, while it was in its natural habitat in space, is an experience that I will carry with me all my life. The story of the Hubble Space Telescope's problems and how they were repaired – a journey from disaster to triumph – has been told many times. But I will add one personal experience – both to share the perspective of someone fortunate enough to be in the middle of the action, and also to illustrate the value of having human beings involved in complex activities in space.

While repairing the Hubble Space Telescope we had several surprises that forced us to adopt procedures completely different from those planned before the mission or suggested in real time by Mission Control: for example, in figuring how to close a warped door, which would have crippled the telescope had it been left open. It took me the best part of half an hour playing with the door to determine the cause of the problem. It required a combination of looking from many different angles, feeling how the different latches were engaging, wiggling the door in as many different ways as possible, and also – and here a precise explanation is difficult – being able to step back and appreciate the whole *gestalt* of the situation, using years of training and experience and somehow hoping for inspiration. My EVA colleague and I devised a plan to use a standard EVA tool in an unorthodox manner to latch the door. We verbally described the problem and our proposed solution to Mission Control, and sent down lots of TV pictures; but the people on the ground did not seem to understand the problem in the same way that we did. They suggested several procedures that we considered would not work – and one which we thought might even damage the telescope. On the other hand, they were afraid that *our* solution might damage the telescope. I am not sure if even the best virtual reality tools could have bridged the gap. If we had been robots waiting for reprogramming, I am afraid we would have been out of luck. Fortunately, after considerable discussion the Flight Director made a command decision that since NASA had taken great trouble to train us and send us into space, and we were there and could understand the situation better than those on the ground, they needed, in the end, to trust our judgment and let us proceed with our plan – which we did, successfully. In unplanned time-critical situations, humans are far more efficient and flexible than even our best robots.

The Hubble Space Telescope was brought back to full functionality; and moreover, the capability of EVA astronauts to accomplish complex and demanding tasks was demonstrated beyond doubt. This was critical in convincing many people that we would be able to successfully assemble the International Space Station – a task requiring more EVA time than had been performed in all the spacewalks ever accomplished throughout the entire history of the space programme.

As I write these words, the construction of the International Space Station – which so far has been accomplished brilliantly – is temporarily halted following the loss of *Columbia* and her crew. In preparing to resume flying, new EVA capabilities are being devised to allow crews to repair thermal protection tiles if damage is discovered once a Shuttle is in orbit. At the same time, astronauts are practicing EVA tasks underwater, together with newly developed EVA robots, which will some day work as assistants to (or even partners with) spacesuited astronauts. We look forward to one day sending astronauts to the Moon, to Mars, and beyond. No doubt they will work closely with robots as they explore our Solar System. The spacesuits they use will hopefully be so advanced as to make our current suits seem like old 'hard hat' diving suits in comparison to modern scuba gear. But they will have built on a rich heritage of EVA experience, and it is worth recording, as this book so eloquently does, the history of how humans have learned to survive and carry out useful work in space.

Jeffrey A. Hoffman
Cambridge, Massachusetts
NASA Mission Specialist, 1978
EVA crew-member, STS-51D, 1985
EVA crew-member, STS-61, 1993

August 2003

Author's preface

The Space Age was less than four years old when Soviet cosmonaut Yuri A. Gagarin became the first human to orbit the Earth. Barely four years later, Alexei Leonov opened the hatch of his Voskhod 2 spacecraft and floated out into the void, to become the first person to walk in space. It was a brief excursion of only a few minutes, and one that almost ended in disaster, as Leonov found it very difficult to re-enter the spacecraft. It was a daring experiment, initiating an activity that has now become almost a standard element of every manned spaceflight.

Officially designated extravehicular activity (activity outside of the spacecraft) or EVA, it is more commonly known as walking in space. It has involved a variety of different activities, including operations inside the spacecraft while wearing a pressure suit and working in a vacuum, stand-up EVAs (where only the head and shoulders are out of the vehicle), tethered and untethered activity in Earth orbit or deep space on the way back from the Moon, walking and driving across the lunar surface, and initial planning for the first excursions on the surface of Mars.

Throughout the first forty or so years of manned spaceflight fewer than 250 EVAs have been conducted by only 125 people, but with the creation and operation of the International Space Station this is expected to double over the next fifteen years. An estimated 1,500 hours of EVA operations is anticipated at the International Space Station, with each EVA taking, on average, about six hours to complete. As we approach the decision to send humans back to the Moon or on to Mars, EVA will once again focus on surface exploration, and mission planners will look back to the days of Apollo to review what was learned from those historic lunar excursions. As with most aspects of spaceflight, new advances are dependent on the learning curve of past experiences and achievements. The extensive EVA operations conducted by the Russians outside their space stations are a valuable resource for current and future International Space Station operations, and each programme of EVA has incorporated new techniques and procedures developed from earlier work.

This book does not review every EVA ever accomplished, but rather, examines the skills that were developed to approach particular objectives and complete the tasks. These objectives have been categorised as pioneering excursions, exploring the surface of the Moon, working outside space stations, and servicing satellites. The theory of leaving a vehicle in vacuum and working effectively was proposed in the early years of the twentieth century, fifty years before the first artificial satellites were launched to orbit. It was not until the mid-1960s, however, that the theory was finally put into practise, with both rewarding and surprising results.

Before any space explorer ventures out of a spacecraft, certain procedures must be followed to ensure the safety of the spacewalker and the success of the EVA. These include planning the excursions, providing the tools and equipment to achieve the goal, and training the crew to perform their task efficiently, safely and productively. Over the decades, procedures, equipment and training have evolved from past experience, and have established an expanding database from which to move forward with more ambitious spacewalks. One of the greatest challenges has been to devise techniques and procedures for EVA training in a microgravity and vacuum environment while still on Earth.

In looking to the future of EVA, it is worth reviewing past experiences. Surface exploration EVAs were the main focus of Apollo, and will be an integral element in any return to the Moon and the human exploration of Mars. Using the experiences of Apollo and recent Earth-based demonstrations, the prospect of future surface exploration can be assessed.

Another type of EVA, performed exclusively from the American Space Shuttle, has been the service and repair of satellites to increase their useful orbital lifetime, the most prominent of which is the Hubble Space Telescope. This book reviews the techniques of reaching a stranded satellite in orbit, capturing it, repairing or retrieving it, and redeploying it.

The next aspect of EVA covered is operations at large space stations, including the support, expansion, repair and servicing of the station over an extended period of time. Although the majority of work on space stations focuses on scientific research inside the vehicle, EVA has become an integral part of operations involving both the crews in space and the teams of support engineers and controllers on the ground.

The conclusion looks forward to the next step in EVA, and at what we can expect from the spacewalkers of the next four decades as we venture to Mars and possibly beyond. Can humans safely explore the moons of the distant planets, or will robots have to take the place of hands-on experience? Will the Earth, our Moon and Mars be the only places with human footprints, or are there other places in the Solar System where initial small steps could lead to giant leaps in space exploration? Only time and an adequately funded and consistent space programme will reveal the answer to these questions.

This book examines the techniques required for spacewalking, from the earliest excursions to current operations at the International Space Station, and reviews the experience and lessons learned from those pioneering years. It is an account of how to prepare on Earth for an activity in the vacuum of space, and how an EVA team who perform the spacewalk is always supported by dozens of colleagues on the ground in the planning, training and support of each activity. We are fortunate to live in a time when we can recall witnessing the initial excursions outside our atmosphere when the first spacewalks were achieved. This book records the legacy of the true explorers of the Solar System – the spacewalkers.

On 14 January 2004 (while this book was in production), President George W. Bush announced a redirection of the American manned space programme that could see a return to the Moon by 2015, leading to a possible flight to Mars by 2030. Hugely dependent on consistent political and public support and a sustained budget commitment, this had an immediate effect on current programmes and, by implication, future plans for EVA. The Shuttle was to be grounded in 2012 after completion of the International Space Station, a new Crew Exploration Vehicle was to be developed for both ISS support and for the return to the Moon, and the Hubble Space Telescope was to be abandoned in orbit, with no further service missions (although at the time of writing this is still being hotly debated).

This announcement also had huge implications for international ventures on the ISS and proposed international cooperation on the new lunar programme. Where possible this latest development in EVA planning has been incorporated in this book, with the full development of a new EVA programme awaiting more detailed plans to be released.

Indications of pending amendments to ISS operations from the Shuttle were also forthcoming in January 2004, as there was news that the number of EVAs per Shuttle assembly flight was to be reduced, which in turn would increase the number of Shuttle flights to complete assembly. Apparently there was concern that medical and operational requirements for completing a heavy EVA programme on each flight were pushing the limits too frequently. The requirements to rest between EVAs, rather than support another EVA or prepare equipment, were pushing against safety limits over several days of intense EVA operations.

Monitoring the next few decades of EVA operations should be equally as interesting and varied as it has been throughout the past four decades.

David J. Shayler
West Midlands, England
www.astroinfoservice.co.uk

February 2004

Acknowledgements

Many individuals and organisations have provided me with information on EVA hardware, operations and training over a period of more than thirty years, and I would like to thank the following for their friendship, help, encouragement and support, and for providing me with access to archives and resources.

The staff of the Public Affairs and History Offices at the Johnson Spaceflight Center in Texas, Marshall Space Flight Center in Alabama, Kennedy Spaceflight Center in Florida, and NASA Headquarters in Washington; the staff of the Public Affairs departments at Boeing, Hamilton Standard, United Space Alliance and Rockwell; the staff of Rice University and University of Clear Lake, both in Houston, and NARA at Fort Worth; at NASA, John Charles, Joe Kosmo, Jim McBarron, John Uri, Glenda Laws, Chuck Shaw, Garry Kitmacher and Glen Swanson; at ESA, Andrew Wilson; and the staff at the Cosmonaut Training Centre, Moscow, and Zvezda, also in Moscow.

On the personal side, my interest in EVA operations was fired by the excellent television programmes generated by the BBC and ITV during the Apollo era – notably the reporting by Sir Patrick Moore, James Burke and the late Peter Fairley. The staff members of the British Interplanetary Society, London, continue to support my work with interest and friendship; and I must, as always, thank my friends and colleagues Rex Hall, Bert Vis, Phil Clark, Andy Salmon, Michael Cassutt, Bart Hendrickx, Colin Burgess, Anders Hanson, Brian Harvey, Neville Kidger, David Portree and Keith Wilson, who over many years have continually offered help, information and suggestions.

I must once again thank my brother Mike for his experience, skills and talent in crafting the original draft into the submitted text; Project Editor Bob Marriott, for his continued long hours spent in editing and preparing the text, monitoring consistency, authenticity and accuracy, and scanning and processing the illustrations; Jim Wilkie, for his cover designs; Arthur and Tina Foulser, at BookEns; the support teams at Praxis and Springer, for the production and promotion of the book; and, of course, Clive Horwood, Chairman of Praxis, for his understanding and patience during a difficult two years.

I must especially thank my mother Jean Shayler, who continues to provide support and encouragement for her family after a time of change for all of us. This book is also for my sister Karen and her family, who missed out in all my other books.

I could not have written this book without the help of a certain small and unique group of people, and I therefore extend my warmest appreciation to those who have been able to include the skill of EVA on their list of life's accomplishments, and have been willing to share their experiences.

List of illustrations

Space complexes

Next steps

Prologue

You are cocooned inside your spacesuit ... hearing only the crackle of the radio in your earphones ... breathing deeply from the oxygen supply ... your nose just millimetres from the faceplate of your helmet. You are floating inside the pressure garment, about to open the hatch of your spacecraft to the void of space. You thrilled at the rollercoaster ride of launch, grinned as you experienced microgravity minutes later, and stared fascinated at the view out of the windows of your spacecraft. But now, dressed in your modern-day suit of armour, you are poised for a new challenge: to enter space as a satellite{?} of the Cosmos.

As the hatch is cracked, blinding rays of sunlight flood in, reminding you of the piercing blue light from electric arc welders. As the hatch is swung open, small items of debris mysteriously appear all around you. All you have to do is push off from the hatch, and you too can be called a spacewalker. You poke your head out of the hatch, and for a few fleeting seconds you stare at the Earth passing by a couple of hundred kilometres away. But you are tethered to the spacecraft, and know that all you have to do is let go and trust to the orbital mechanics of spaceflight that you will not fall to Earth.

In the early part of the twentieth century, Konstantin Tsiolkovsky – the Soviet 'Father of Cosmonautics' – produced drawings illustrating a spacesuited human using an air-lock to exit a spacecraft and then 'dancing' around it. These drawings were published in his book *Album of Space Travels*. But it was not until some four decades later, on 18 March 1965, that such an activity was undertaken. Soviet cosmonaut Alexei Leonov – the first man to walk in space, and ever the artist – later recalled his first steps into the void: 'The boundless expanses of outer space unfolded before me in their indescribable beauty. I took my first look at the Earth. It sailed majestically before my eyes and seemed flat. It was only the curvilinear shapes of the edges that reminded me that it was a globe. In spite of a pretty dense light filter, I saw bright clouds, the azure of the Black Sea, the fringe of the coastline, the Caucasian range. The moment came to leave the ship and step into space, the moment for which we had prepared for so long, of which we had thought so much. Unhurried, I climbed out of the hatch and pushed myself away from it gently, moving further and further away. I saw the Universe in all its grandeur. The view of untwinkling stars on a dark violet background changing to the velvet black of the abysmal sky was followed by views of the Earth. I recognised the Volga, the mountain range of the Urals ... then I saw the Ob and Yenisei rivers, as though I was swimming over a vast colourful map.' (From *Man on the Moon*, by Peter Fairley, 1969.) Leonov was outside his spacecraft for only a few minutes, but being first he 'opened the door' to all who followed. In the same way that all space explorers follow in the wake of Yuri Gagarin, so all who conduct EVA follow the pioneering achievement of Alexei Leonov.

Tsiolkovsky realised that to explore other planets, a suitable spacesuit would have to be developed; one that would allow the wearer to move and walk across a planetary surface. Forty-nine years later, on 20 July 1969, two American astronauts conducted the first excursion on another celestial body. Neil Armstrong described, to a spellbound TV and radio audience on Earth, his impressions of that first human step on another world: 'I'm at the foot of the ladder. The LM footpads are only depressed in the surface about one or two inches ... The surface appears to be very, very fine-grained as you get close to it. It's almost like a powder ... Now and then it's very fine. I'm going to step off the LM now. That's one small step for man, one giant leap for mankind. The surface is fine and powdery ... I can pick it up loosely with my toe. It does adhere in fine layers like powered charcoal to the sole and sides of my boot. I only go in maybe an eighth of an inch, but I can see the footprints of my boots and the tread. There seems to be no difficulty in moving around.'

From the Earth, through the vacuum of space, to the Moon, and on to Mars. We have been impressed by Leonov's wonderful description of the Cosmos, and have listened with awe to Armstrong's words as he stepped onto the Moon ... but at the moment we can only guess the words that might be spoken by the first human to step onto the dust of Mars. Undoubtedly, the new wonder of seeing humans walk across the plains of Mars will inspire a new generation of explorers to continue the voyage ever deeper into space.

Theory put to the test

The idea that space explorers might work outside their spacecraft has long been a feature of science fiction and adventure stories, but its practicality was not seriously considered until the first half of the twentieth century. During the Mercury and Vostok programmes, all crew activity was restricted to the inside of the space vehicle – intravehicular activity (IVA). In these programmes the intention was to prove that humans could not only survive a flight into space, but could return home alive. Survival outside an orbiting spacecraft would be a challenge for future programmes.

It became clear from the early flights that the period of time that crews spent in space could be increased from a few hours or days to weeks and eventually months or years. For future crews to obtain maximum benefit from these longer flights, some of their work might involve activity outside their spacecraft – extravehicular activity (EVA). This became more commonly known as 'spacewalking', although walking in space can more accurately be described as 'floating at 8 km per second'. By 1965, new spacecraft types stood poised on the launch pads, ready to take the first crews into orbit to investigate if this so-called 'walking in space' was possible, and whether it would be valuable for future operations. It had taken a long time to include EVA in the mission objectives, but from then on the activity would in some way become an integral element of most manned spaceflights. During March and June 1965, Soviet cosmonaut Alexei Leonov and American astronaut Ed White put years of theory to the test in two uncomplicated demonstrations of walking in space.

DREAMS AND DESIRES

It is generally thought that EVA is a modern 'space age' concept, but the idea of humans exploring worlds other than our own is as old as our study of the stars. The human desire to personally explore the heavens is also centuries old, and is intertwined in religion, myth, fear and superstition, as well as science and discovery. The Moon became the earliest target of man's dreams of visiting the heavens, and once the nature and scale of the Universe began to be revealed, so, naturally, dreams of going there began to emerge. During the time of the ancient Greeks around 500

BC, Xenophanes of Colophon explained lunar eclipses, describing the Moon 'stepping into the void' between the heavens and Earth. This is also an apt description of what we now call EVA.

Less than a century later, Lucian of Samasota wrote one of the earliest stories that would now be termed 'science fiction'. Describing a voyage towards the Western Ocean (the Atlantic), he wrote of a whirlwind that carried the open-decked sailing galley high into the sky until caught by a strong wind that filled the sails with enough strength to carry the ship and her fifty-strong crew through the air for seven days and seven nights until they came upon the Moon. Lucian recorded the ensuing adventures of encounters with strange lunar beings in his inappropriately titled work *True Stories*. Lucian and his colleagues apparently remained on the Moon for a week, and then returned home via Venus, where they visited a new colony!

The idea of 'flying to the Moon' was taken literary in Lucian's other lunar story, *Icaromenippus*, in which his hero decides to wear artificial wings to fly into space, much to the displeasure of the gods residing on Olympus, who strike him down 'in flight'. Other recorders of myth and legend continued to explore the idea of the personal exploration of space. Perhaps the most famous of these early attempts is the story of Icarus, who flew too close to the Sun and melted his wax wings, and so fell to his death. This, at least, was an early demonstration of the need for adequate materials testing and thermal protection.

Dreams of discovery

Between the fourteenth and sixteenth centuries, humans gained a greater under-standing of the workings of the Solar System, although ideas about who or what inhabited the Moon remained a mixture of science and pure fantasy. The German theoretical astronomer Johannes Kepler wrote a scientific story called *Somnium* (*Dream*), in which his Icelandic spacefarer Duracotus used the skills of his witch mother Fiolxhilda to reach the Moon during a lunar eclipse, when the shadow of Earth falling on the Moon provided a 'bridge' that demons used to travel between the worlds. The work was published in 1634, four years after Kepler's death. In what might be described as early selection criteria for astronauts and scenarios for spaceflight and EVA, he wrote: 'The whole trip remains a grave risk to life [where] no inactive persons are accepted [nor] no pleasure-loving ones. We choose only those who have spent their lives on horse back, or have shipped often to the Indies and are accustomed to subsisting on hard tack, garlic, dried fish and such unpalatable fares.' Aware that the atmosphere did not reach the Moon, Kepler recorded one of the earliest descriptions of 'EVA suit components' in this tale: 'There will be intense cold, and the traveller will find it hard to breathe, so that he will have to apply wet sponges to his nostrils' – an apparent welcome relief and aid in the thin air! Once landed on the lunar surface, Kepler's hero continued on an expedition to describe his interpretation of features and inhabitants.

In 1638, John Wilkins described a voyage to the Moon in his book *Discovery of a New World*, including a description of exploring the surface in low gravity: 'He might stand there as firmly in the open air as he can now upon the ground; and not only so, but he may also move with a far greater swiftness than any living creature here below.'

Over the next two centuries, such stories of human exploration of the heavens continued, mostly retelling earlier myths and many restating the belief in inhabited worlds and man's capability of reaching such worlds. At the same time, between 1600 and 1800, there were significant advances in astronomy and the understanding of the Universe. In the early part of the seventeenth century, Galileo discovered four bodies (moons) that orbited Jupiter; and half a century or so later, Isaac Newton applied precise mathematics to describe orbital mechanics and the gravitational forces that affected the planets. Up to 1781 only six planets were known (out to Saturn), but in that year William Herschel discovered Uranus. In 1846, J.C. Adams and U.J.J. Leverrier used Newton's laws to reveal an eighth planet (Neptune); and 84 years later, in 1930, Clyde Tombaugh, studying photographs taken at Lowell Observatory, noted that one speck had changed position on photographs taken several nights apart – and thus discovered the ninth planet, Pluto.

Science enters science fiction
By the end of the eighteenth century a new awareness of science led many writers to incorporate scientific facts in their novels, largely due to the demands of their readers, who were becoming increasingly aware of the wonders of science around them. Comfort was also now becoming important to these new explorers of the cosmos. In 1827, Joseph Atterlay's *Voyage to the Moon* included a 'spacecraft' consisting of a copper vessel 1.8 m in diameter, constructed of double-sided panels for retaining an inner atmosphere, and lined with quilted cloth for comfort. Significantly, there was also 'an opening large enough to receive our bodies, built into the vessel to gain entrance and exit' – an 'EVA hatch'.

In 1835, Edgar Allen Poe wrote of his hero Hans Pfaall making a trip to the Moon in a home-made balloon that included apparatus for providing air for breathing in the rarefied atmosphere all the way there – a life support system. The flight was unexpectedly jolted by the explosion of kegs of gunpowder below the ascending balloon, the force of which sent the machine out of control, so that it spilled out its occupant for an unexpected 'tethered EVA', hurling him over the rim of the car (gondola), and leaving him dangling by a piece of slender cord about three feet in length.

As science fiction changed with the discoveries of the day, so did the understanding of travelling through the air. Balloons were being used to explore the region above the clouds, and were beginning to gather scientific data on the structure of the upper atmosphere. At the same time, 'air breathing diving suits' began to be used to explore the depths of the oceans. These two branches of science would come together in the early years of the twentieth century, as a significant step on the road to space. The first model hot-air balloons were constructed by Gusmão early in the eighteenth century, but it was not until the Montgolfier brothers made a trouble-free flight in 1783 that ballooning literally 'took off'. For the next century, balloons were limited to relatively low altitudes, but in 1898 the Frenchman Teisserenc de Bort began to use unpiloted balloons to carry a range of scientific instruments up to heights of about 15 km, to gather significant data on the structure and composition of the atmosphere at such heights. By this time, balloonists had reached the physical limits of altitude and survivability without artificial protection and supplies of oxygen. At 8.5 km only a very few intrepid balloonists could survive, and new developments in science and technology would be required to go higher – except, of course, in science fiction.

By the nineteenth century, the idea of the Moon being inhabited remained only in the realms of science fiction, and Jules Verne's *From the Earth to the Moon*, published in 1865, became a credible, if flawed, account of human flight to the Moon in the nineteenth century. Travel through space in projectiles, as described by Jules Verne and H.G. Wells, became the genesis of today's pressurised spacecraft, though their methods of operation remained firmly in science fiction.

In *From the Earth to the Moon* (1865) and *A Trip Around the Moon* (1870), Verne used a huge space cannon to launch his projectile towards the Moon, while in *The First Men on the Moon* (1901), Wells suggested the use of a gravity-shielding substance called Cavorite (named after its inventor, Cavor) to defy gravity and 'fly to the Moon'. The apparel of these Victorian explorers was one of top hats and long coats, relaxing in the 'spacecraft' in velvet smoking jackets, but Verne at least wrote that his space explorers would require a pressure suit for protection. Another aspect of future EVAs that Jules Verne identified was the minimalisation of air loss from the main spacecraft by using a 'trap door' (an air-lock) for depressurisation and repressurisation. He also indicated his awareness that anything leaving the vehicle without an additional boost would travel alongside in the same direction and velocity as the vehicle it had left. After one of the canine companions died during the flight, it was 'buried' in space by being ejected though the 'trap door', and afterwards kept pace with the vessel as it travelled to the Moon.

The theorists take over

At the turn of the twentieth century there emerged more and more theories about the exploration of space, and slowly the realisation dawned that rockets – in development for more than 1,000 years – could at last be the key to taking humans into space and in exploring the Moon and planets. One leading exponent of this theory was a Russian teacher, Konstantin Tsiolkovsky, who not only established credible studies for using rockets in space and creating space stations, but also devised methods of leaving such vehicles (air-locks) in protective suits (pressure spacesuits) and working outside in vacuum (EVA).

Tsiolkovsky's 'games on a tether'

In the early decades of the twentieth century, Tsiolkovsky produced many studies of his visions of the future of space exploration, and in Russia he is now regarded as the 'Father of Cosmonautics'. In his book, *Album of Space Travel*, he produced simple line-drawings of a spacesuited 'cosmonaut' conducting an EVA by using an air-lock system. Developed from this, the Voskhod 2 air-lock device design allowed Leonov to perform the world's first EVA. Tsiolkovsky's drawings were part of his assignment as a film consultant, and he depicted his figure performing somersaults while tethered to the spacecraft, in what he described as 'games on a tether around a rocket'.[1]

In his book *The Call of the Cosmos*,[2] Tsiolkovsky described his design for a spacesuit that would cover the whole body; a flexible, light, 'gas-tight' garment that still allowed for full freedom of movement. This design would have to be strong enough to withstand the inner pressure of gases, and also incorporate both reserves for urine and special cylinders to supply the wearer with oxygen for up to 8 hours. Special (Sun) visors would be attached to the helmet to protect the eyes.

Tsiolkovsky expanded his thoughts on suits for exploration in his novel *Beyond Planet Earth*, published in 1920. In this, the suit was used during excursions from a space station, and was also a requirement for the exploration of new worlds. 'The time will come when we shall have to land on a planet in an atmosphere unsuitable for us to breath, either because of its special composition or because it is extremely rarefied. In order to keep us alive in a vacuum we need special suits. The same kind will do for either contingency.' This suit covered the wearer from head to foot, was

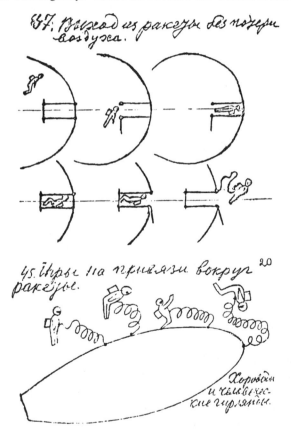

Some of the earliest and most prophetic illustrations of EVA techniques appear in the works of Russian theorist Konstantin Tsiolkovsky, who at the beginning of the twentieth century forecast the use of pressure suits, airlocks and tethers for working outside a vehicle travelling through the vacuum of space. These items have became standard EVA equipment.

impervious to gases, and was strong enough to withstand its internal pressure. The headpiece was heated, and had partially transparent plates to allow the occupant to see out. Urine was collected in a reservoir, and oxygen was constantly supplied from a 'box'. Other 'boxes' dealt with expelled carbon dioxide, water vapours and 'other bodily wastes', and automatic pumps constantly circulated the internal gases and vapours. Wearing this suit, a man would consume 1 kg of oxygen per hour, and would have supplies for up to eight hours to complete his work.[3] On Earth, such a suit would weight up to 10 kg, and on the Moon it would weigh about 1.66 kg. On Mars it would weigh about 3.33 kg, but in free space it would weigh almost nothing.

Tsiolkovsky's vision of what is now termed EVA remains inspiring: 'To place one's feet on the soil of asteroids; to lift a stone from the Moon with your hand; construct moving stations in ether space ... to observe Mars [or] descend to its surface ... a great new era [for a] more intensive study of the heavens.'

Outside in empty space
In July 1929 – forty years before the Apollo 11 astronauts walked on the Moon – the first issue of *Science Wonder Stories*[4] contained an article (translated from German) by Hermann Noordung (real name Herman Potocnik), in which he described how a 'spaceman' could wear a special protective suit for exit into space and use a manoeuvring gadget for traversing across the void.

The same year, Noordung's book *Das Problem der Perfahrung des Waltraums (The Problem of Space Travel)* focused on the creation of a manned space station in a 35,000-km geostationary orbit. Initially criticised in several areas, this concept has since been cited as of important historical significance in the theoretical road to the creation of a space station. In this work there are several references to activities outside of the station that are worth recording here.[5]

In his book, Noordung explained that by using enclosed capsules supplied with air consisting of a mix of gases in the correct proportions, and at the correct pressure, the people inside such craft would survive in space. Here he echoed the writings and theories of several other pioneers in the field. He also indicated that to exit such vehicles into 'empty space', air-tight suits would have to be worn, which would have to be pressurised and supplied automatically with air. Such supplies could come from 'attached devices', and he likened the operation to submarine technology and the equipment used by deep-sea divers. With the experience already gained in the submariner field, Noordung reasoned that the problem of a stay in 'empty space' was solveable. He called his garments 'spacesuits'.

Noordung's spacesuit was similar to the design of diving suits and gas-protective suits of the day, but it also had to be air-tight and resistant to external influences, and yet flexible enough to allow movement and as unrestricting as possible. It also had to be strong enough to retain the internal air pressure of 1 atmosphere, and be able to protect the wearer from the extremes of high and low temperatures that were already known to exist in space. Yet it could not be come brittle as it cooled. All these parameters would pose extreme design considerations on the chosen materials. Noordung considered that at least parts of the suit could be constructed of highly polished metal (and thus be rigid), but this restricted movement, so he also suggested an 'appropriately prepared flexible material insensitive to very low temperatures ... coloured bright white on the outside and as smooth as possible.'

He recognised that such a suit, when pressurised, would not be flexible and would restrict freedom of movement, which would require special construction of the joints, just as with armour. A special lining would be required to thermally regulate the suit, the highly polished surface would reflect unwanted heat, and mirrors would attract heat to parts of the suit in shadow. The air would be supplied in the same way as for modern deep-sea divers, with the necessary oxygen bottles and air purification cartridges carried in a metal back-pack. Communication by 'telephones' and connection by wire would be impractical, so each spacesuit would be equipped with radio communication gear, with sending and receiving equipment powered by storage batteries located in the back-pack. A small wire aerial, or even the metallic surface of the suit, would serve as an antenna. These radio systems would allow the 'spacefarers' to talk with each other and with the rest of the crew back on the station. These ideas showed remarkable foresight into real EVA and lunar suits.

Also addressed in Noordung's book was safety. This included the ever-present danger of floating away into outer space. Noordung proposed tethers for restraint, and special alarm devices located across the station capable of receiving even the weakest call for help. Each suit's radio would be tuned to different frequencies for emergency, inter-suit communication or telemetry, and small hand-held thrusters could be used for motion control, supplied by propellant tanks located in the wearer's back-pack.

In his description of the space station, Noordung proposed building the structure from components assembled in Earth orbit by spacefarers using spacesuits. In order to transfer from the inside of the completed station to the outside into open space, he stated: 'Since all rooms are connected with one another and are filled with air, movement is easily possible throughout the inside of the space station. Space travellers can only reach the outside into empty space by means of so-called air-locks. This equipment [is] familiar from underwater construction [and] consists primarily of a small chamber that has two doors sealed air-tight, one of which leads to the inside of the station and the other to the outside.'

To operate this air-lock, the crew-men would lie in restraint hammocks while the air-lock is counterspun to the rest of the station to induce zero g. They would then don their spacesuits and pump out the air. The outer hatch would then be opened to exit into space. To return to the station, the process would be reversed. The location of this air-lock would be in the hub of the circular station, and be designed to rotate with the rest of the construction to create a centrifugal gravity force around the rim of the station. When used as an air-lock it would operate independently of the station, turning to induce microgravity for easier access to and from open space. Noordung stated that with training, a crew-member might be able to dispense with counter-rotating the air-lock due to its relatively slow speed. (He did not, however, discuss any method of dealing with the disorientation produced by entering and leaving the hatch. Indeed, a sophisticated tether system would also have to have been devised to prevent tangling the lifelines).

The 1950s: men from Mars and the influence of science fiction
During the 1930s and 1940s several developments in exploring the stratosphere for scientific and military purposes were also milestones in the development of turning theory into reality for spaceflight. These included development of pressurised vessels for carrying an air crew, pressure garments, life support systems, and studies into the ability of working at altitude. The advent of the Second World War created the environment for rapid advances in technology and research, and following the end of hostilities in 1945, those who had worked on creating weapons of war and the methods of transporting them became influential in moulding the emerging space programmes in the East and in the West. It was also the period during which science fiction moved closer to science fact.

In 1953, Hugo Gernsbeck collaborated with Professor Donald H. Menzel, the Director of Harvard College Observatory, to design a spacesuit for use during the repair of spacecraft *en route* to the planets. The suit (which only appeared in print, was multilayered, and included padding for comfort and 'flexible metal joints' for mobility. The wearer would don a loose-fitting inner garment, and then climb into the suit, which would be made from a flexible metal inner layer and an outer 'special' plastic layer coloured white to reflect the solar rays. Vision would be enabled by means of a double-walled window faceplate, and an adjustable rear-view mirror so that the wearer would be aware of activities and obstructions behind. There was also a thermostat and a set of deployable metal fans on the chest, back and leg area to dissipate body heat and provide an air-conditioning system for the man inside. Mounted on the back at shoulder level was an oxygen tank and an air purifier to circulate the oxygen around the suit, and a walkie-talkie radio set and antenna for communications. Features of this suit were incorporated into the EVA suit designs of the 1960s.

This suit was intended for use only in a zero gravity environment, and not for surface exploration. For this, Gernsback proposed a dedicated surface exploration vehicle that was both lightweight and manoeuvrable, and offered better protection for the surface explorer. The so-called 'homobile' (*homo*, man; *mobile*, moveable) featured a pressurised crew compartment capable of tilting forward or backward, and a pair of robotic grab arms for sample collection. It traversed on a pair of tank-like caterpillar tracks, and carried additional thermal protection in the form of a stowable parasol. While not a true spacesuit, it was an early design for a lunar or planetary roving vehicle to support manned surface explorations (a technique discussed in subsequent chapters).

Destination Moon and *Collier's*
In 1952 the American *Collier's* magazine published a series of articles that featured a selection of papers by authorities in the field of space research, illustrated by some of the leading artists of the day. This collection reflected the Western predication for expansion above the atmosphere and the beginnings of the exploration of space over

the next few decades. In addition, many of the engineers and technicians assigned to the development of pressure garments were also working on technical support of some of the science fiction feature films being made during this time – *Destination Moon* being typical. In this motion picture, the storyline features a flight to the Moon that includes trans-Earth EVAs and surface activities to take scientific measurements and deploy experiments. In the trans-Earth EVA the crew use a large oxygen bottle to expel gas in a rudimentary manned manoeuvring unit, whilst wearing pressure suits with small oxygen tanks on their backs. A tether was supplied to prevent drifting too far from the space vehicle.

The work of the British Interplanetary Society
In 1937, four years after the British Interplanetary Society was formed, its Technical Committee initiated a two-year study into a manned lunar lander. The results, published in 1939, were a focus for other subsequent studies from leading members of the Society over the next two decades.

In the days of Buck Rogers and Flash Gordon, in comic books and emerging in feature films, the image of an heroic space explorer dressed in fishbowl-like helmets and bright costumes, fighting evil in the Universe, paid little heed to the scientific realities of walking in space or on other planets. In the BIS study, space pressure garments were recognised as essential for exploring the surface of the Moon: 'Constructed of thin but tough rubber or leather and provided with a roomy headpiece, they will contain heating arrangements and a supply of oxygen.'[6] It was realised that when pressurised, such suits would form a rigid and immobile balloon, so the BIS suit would be fitted with constant-volume ball-and-socket joints at the elbows and knees to provide flexibility. Hand dexterity would be especially difficult in such suits, and so one suggestion was that the wearer should be provided with very thin surgical operating gloves – and a puncture repair kit. For mobility on what was thought to be deep layers of lunar dust, each lunar explorer would be provided with a pair of snow-shoes.

The task of exploring the Moon would be the investigation and gathering of samples of lunar material, and for this a geological hammer, spring balance (for mass measurements), microscope and gravity pendulum would be included in the explorer's tool-kit, together with cine-cameras and still cameras to record activities on the surface. It was suggested that if a landing were to be achieved in the shadow, then lights could be carried to illuminate the surface for working. The importance of such an expedition might even attract the BBC into providing a running commentary (over the radio) of the progress of activities! Unwanted equipment would be left on the Moon at the end of the foray (a foretaste of the Apollo missions).

The work of H.E. Ross and R.A. Smith
The above report was published in 1939, and over the next decade H.E. Ross studied the design of the lunar suit in more depth. In 1949 he presented a paper entitled 'Lunar Spacesuit' at the BIS-sponsored Symposium of Medical Problems Associated with Spaceflight.[7] This was one of the first serious studies of lunar surface EVA equipment and procedures.

In the late 1940s, a series of dramatic and technically accurate artistic impressions

A 1950s painting by the English artist R.A. Smith, clearly showing EVA activities during the assembly of a space station. (Courtesy British Interplanetary Society.)

was produced by BIS member Ralph Andrew Smith, and these featured in several publications during 1950–54. Smith worked extensively with Arthur C. Clarke in three of his books, as well as for the BIS studies of the lunar spaceship.[8]

His work was also featured in *Space Travel*, published in 1953.[9] The authors explained the complexities of the prospects of space exploration to a wider audience, and explained to a new generation that pressure suits would allow personal access to open space: 'The crew would even find it possible to venture outside the spaceship, leaving the cabin through an air-lock before walking out on the surface of the vehicle (using magnetic boots). These procedures would actually be less hazardous than it appears at first consideration, and the essential point to realise is that, despite the vehicle's great speed, a man would be able to stand erect on the outer surface; there will be no rush of air to force him off! He must remember, however, that he possesses no weight, and unless anchored to the metal skin of the spaceship by magnetic boots, any slight movement will be sufficient to set him adrift in space. However, if guilty of such carelessness, he need have no fear of being left behind. Having the same speed as the vehicle he had recently left, he would travel alongside, though slowly drifting further and further away in the direction he had left the vehicle.' The book also described ideas for orbital rendezvous and assembly of large structures in space. Such spacecraft would be assembled by a team of 'three or four construction crew-members'. Artwork for the books, provided by J.W Wood and R.A. Smith, provided a visual impression of what had yet to be achieved. This included spacesuited crew-members tethered to the main spacecraft by light nylon safety lines, and wearing a version of Smith's lunar suit design, using hand-held rocket pistols for manoeuvrability.

In the July 1958 issue of the BIS magazine *Spaceflight*, J.C. Guignard reviewed some of the physiological aspects of crews exiting into space and performing space construction work, and updated some of the earlier thinking of the decade regarding EVA and exploration suits. The article emphasised that schemes for construction of large objects in space at that time focused on manual assembly of prefabricated sections in space – in much the same way that current ISS construction is supported by teams of EVA astronauts working with the use of the remote manipulators on the Shuttle and space station.

In the 1950s it was obvious that increasing the size of orbital structures would prevent them from being launched in one piece. Contemporary literature had already illustrated spacesuited astronauts manhandling huge items of hardware. It all looked easy on paper, in the art and in the film studies, but Guignard pointed out several potential pitfalls that awaited the construction teams of the Space Age. 'Even the more erudite accounts of satellite building are often illustrated by artists' impressions of confident, helmeted artisans heaving, hammering, welding and working against an awe-inspiring background [of *Earth*]. But just how easy will such work prove in practise?' Warning about taking too much for granted, he reviewed probable difficulties in working in open space.[10] These included:

Orbital dynamics Although artists could convey varying dimensions and shapes of hardware, it was not easy to determine the huge masses and various velocities involved. Being caught between unplanned collisions of these objects would certainly give the astronaut a far shorter career in astronautics than originally planned. Extremely high degrees of manual dexterity, judgement and teamwork would be required, although this was before the idea of the Remote Manipulator System (RMS) to aid the astronauts in their work. In examining EVA training for such operations, Guignard suggested that the astronaut trainee spend part of his apprenticeship as a rigger in a big engineering yard, where he could really learn the 'feel' of large and potentially unstable masses of metal.

In developing propulsion 'guns' for manoeuvring around the spacecraft, Guignard suggested that the most useful would be strategically positioned, rather than hand-held, 'pistol-like guns or fire extinguisher devices', thus freeing up the hands of the operator. Orientation, centre of mass and spinning would all be factors in determining propulsion devices. Guignard stated that a 1.84-m (6-foot) astronaut spinning at 80 rpm would probably experience 6 g negative at his head and 6 g positive at his feet, which he would be able to tolerate for only a few seconds. Even if the spin were to be stopped before the effects became too serious, the residual mental confusion and disorientation created by such a spin could last for several minutes, and be potentially dangerous. He also suggested that the most effective 'spacesuits' would in fact be rigid capsules, with operators inside using manipulator hands. Such a device, in itself, would entail a massive development programme.

Control of such capsules would require accurate redundant systems, a 'black box' type of control system and adequate safety and escape systems, further complicating the design. At the time of this paper, the most common impression of spacewalkers – tramping across space hardware wearing magnetic boots – was dismissed as a viable option, as it did not simulate body weight. The astronaut merely floated inside the boots, but in reality the force required simply to move them would also require significant expenditure of energy. Additionally, if the spacecraft were made from plastics and non-ferrous metals, magnetic boots would be useless.

Survival To provide an adequate environment for the wearer, the spacesuit or space capsule life support system would have to provide support for prolonged periods of work. This included oxygen, air purification, pressurisation to at least 4.5 psi absolute (the minimum level determined to prevent decompression sickness – the bends), and an efficient system of regulating internal and external temperature. If semi-automatic systems were incorporated, then drive and stabilising units, communication, mechanical handling equipment, power, and a caution, warning and monitoring system, would add to the overall mass and complexity.

The smaller the cabin, the greater the mass of the equipment required to ventilate and maintain its environment in relation to that of a larger enclosure. Over longer sorties, air circulation and venting systems would have to be more reliable than those in a larger capsule, to remove hazardous fumes from hydraulic coolant fluids and body gases before they reached toxic levels.

Science fiction meets science fact as a lunar transporter crawls across the lunar landscape. During the late 1950s and early 1960s, this was standard material for artists in indicating where EVA operations might progress after the first lunar landings.

There would also need to be adequate protection against micrometeoroids. The risk – though estimated to be small in comparison to a larger vehicle with a bigger surface area – was still potentially fatal in a one-person unit such as a spacesuit or sortie vehicle. A faster rate of decompression would be experienced in a suit or small capsule compared to a large vehicle with a similar small puncture (as evident during the Mir collision in 1997). Guignard stated: 'The time of useful consciousness following total decompression is only 5–7 seconds, and rupture of a spacesuit or pressure capsule is unlikely to be survivable. The belief is held by some enthusiasts that because man can draw and hold a deep breath for a minute or more, an experienced astronaut might be able to scramble to safety in such an emergency, or even make brief excursions into space without protection, living on the oxygen held in his lungs and blood stream. This is of course fallacious, as it is quite impossible for a man to maintain atmospheric pressure within his chest against a hard vacuum outside. All air would burst from his lungs, and since the oxygen gradient between his alveolar air and the blood would be reversed, the time of useful consciousness would again approximate to the time which it takes the blood to travel from the lungs to the brain.'

Radiation At the time of publication of the article, the biological effects of cosmic radiation had yet to be established, and studies of its intensity beyond the atmosphere would have to await new satellites. Guignard noted that unless these rays are demonstrated to be life-threatening, 'it is unlikely that any form of shielding would be used in space vehicles. If this is so then the flux through a man in a spacesuit outside a ship will be no greater than he receives on board.' In the 1950s it was understood that the absorption of solar radiation on the body in space depended upon the overall area presented to the solar disc, the heating of a smaller object being rotationally greater than that of a larger object. During EVA, direct exposure to the Sun (or shadow) would place demanding loads upon heat regulation systems in the support equipment.

Vision Filters for the protection of the eyes would be required for all visors of helmets worn in open space. 'Some ingenuity will be needed to devise one [transparent visor] which stops all harmful rays while preserving moderate and neutral density cut-off in the visual range necessary if planetary surfaces are to be inspected and coloured light investigated.' Drawing upon experiences from the stillness and silence of high altitude and the monotony of straight and level flight or night flying, Guignard observed: 'When cut off from the frame of reference of normal sensation, [a pilot] tends to become abstracted or withdrawn from his surroundings, and may report a loss of his sense of time and space. This condition might become troublesome in isolated astronauts floating in the void. With nothing to look at but an unmoving panorama of starry black sky [he] would be reduced to a very low ebb of sensory awareness. Under these circumstances, any visual or vestibular stimulus will have a diminished and possibly confusing effect. For a space traveller, this may mean that if the Moon appears abruptly somewhere beyond his feet, he may – although weightless – experience an alarming sensation of hanging upside down, because life on Earth has conditioned him to associate celestial bodies with regions above his head. This disturbing illusion has already been reported by pilots at high altitudes, where stars appear below the aircraft.' Working in cooperation with fellow astronauts and colleagues on the ground via radio communication would, it was expected, keep the EVA astronaut's mind occupied to help prevent this eventuality. A 'radar' screen to determine the location of objects in the absence of light would also be a practical device for future work in space. Guignard stated: 'If today's [1958] television-addicted children are to be the builders of tomorrow's satellites, it may be that their education is more practical than we think.'

SUITED FOR SPACE

In the book *Spaceflight and Satellite Vehicles*, published in 1957, authors R.B. Beard and A.C. Rotherham discussed current thinking on methods of achieving spaceflight and exploration, including a flight by a crew from Earth to Mars. In the authors' opinion, spacesuits worn by crew-members would not only have to provide maximum protection and comfort for their wearer by means of insulation, heating and cooling, but would also have to supply oxygen, incorporate a two-way radio for communications, and possibly supply of food and water. Reference was made to the development of high-altitude stratospheric pressure suits in development around the world, from which 'the first true spacesuits will be developed.'[11]

First spacesuits
When the first cosmonauts and astronauts ventured into space onboard Vostok and Mercury they wore pressure garments that were designed only as a back-up to the spacecraft life support system. In the event of cabin decompression these suits would provide independent life support and pressurisation for the crew-member until the earliest landing opportunity. These suits were not ideal for leaving the vehicle in orbit and walking in space, because each lacked the additional layers for protection against the huge variations in the intense solar heating and radiation and the freezing temperatures in the shadows. For ventures outside of the protective cocoon of the spacecraft, a crew-man would have to have a suit that could retain the inner pressure and include a life support system, an adequate heating and cooling system, a shielded helmet, facilities in the design for mobility, and a method of communication and monitoring the parameters of the suit during operation. The space explorer's pressure garment needed to be a true suit for space.

Science fiction in the Space Age
In 1958, Wernher von Braun published the first edition of his book *First Men on the Moon*,[12] in which – only a few months after the real Space Age opened with the launch of the first Sputniks – he described a future flight to the Moon. For the ascent, descent and in-flight phases of their mission, the two-man crew wore partial pressure garments with removable helmets and oxygen masks, but for exploring the Moon a full-pressure garment would have to be worn on top of the partial garment.

Wernher von Braun discusses a one-man EVA servicing vehicle for the making of a series of Walt Disney films during the early years of space exploration.

Like the earlier BIS lunar suit, it featured additional equipment and facilities to explore the lunar surface.

During the 1960s and 1970s, as the technology of space programme evolved, it transferred to feature films of the era. In their story-lines, films such as *Marooned*, *2001: A Space Odyssey*, *Alien* and others featured 'EVA' as a natural element of space exploration in which strange new worlds were discovered. These cinematic EVAs were carried out with an ease and frequency that was not always reflected in the early EVAs that the 'real' space programme was demonstrating as astronauts and cosmonauts began to leave their spacecraft. From the late 1950s, science fiction became forever entwined in the 'real' space programme. (These mixed impressions of science fiction and space fact created a very personal and lasting impression on this author, and stimulated an interest in EVA that led to the writing of this book.)

REFERENCES

1 Fedorov, A.S., 'Tsiolkovsky as a film consultant', *USSR Academy of Sciences*, No. 5, May 1970, 117–120; NASA Technical Translation (NASA TTF-14, 890, May 1973, Figs. 37 and 45), NASA JSC History Archive, Houston, Texas.
2 Tsiolkovsky, K.E., *The Call of the Cosmos, ed.* V. Dutt, Foreign Languages Printing House, 1960, p. 222.
3 Abramov, Isaak P. and Skoog, Å. Ingemaar, *Russian Spacesuits*, Springer–Praxis, 2003, p. 100.

4 Noordung, Herman, 'The Problems of Space Flying', *Science Wonder Stories*, **1**, July 1929, 170–180; August 1929, 264–272; and September 1929, 361–368.
5 NASA History Online: http://www.hq.nasa.gov/office/pao/History/SP-4026.
6 Parkinson, B., *High Road to The Moon: From Imagination to Reality*, British Interplanetary Society, 1979, p. 22.
7 Ross, H.E., 'Lunar Spacesuit', *Journal of the British Interplanetary Society*, **9**, No. 1, January 1950, 23–37.
8 Clarke, A.C., *Interplanetary Flight*, 1950; *The Exploration of Space*, 1951; *Exploration of the Moon*, 1954; and *High Road to the Moon*, British Interplanetary Society, 1979.
9 Gatland, Kenneth W. and Kunesch, Anthony M., *Space Travel: an Illustrated Survey of its Problems and Projects*, Wingate, London, 1953, p. 88.
10 Guignard, J.C., 'Spaceman Overboard', *Spaceflight*, **1**, No. 8, July 1958, 282–287.
11 Beard, R.B. and Rotherham A.C., *Space Flight and Satellite Vehicles*, George Newnes, London, 1957, p. 59.
12 Braun, Wernher von, *First Men on the Moon*, Frederick Muller Ltd., 1958.

Pioneering the techniques

After centuries of writing about and planning man's excursions into space, the first forays were missions into Earth orbit to qualify the integrity of spacecraft design, verify the use of rockets as launch vehicles, and demonstrate that a properly trained human could survive the rigours of launch, orbital flight, entry and landing. The Soviet Vostok and the American Mercury were designed initially for one person on relatively short missions lasting from only a few hours to a few days.

On 12 April 1961, Soviet cosmonaut Yuri Gagarin became the first person to leave Earth by rocket and orbit the planet. Two American sub-orbital flights followed, and the year was completed by a Russian spending a full day in space. On 20 February 1962 astronaut John H. Glenn became the first American in orbit, and in May 1963 he was followed by Gordon Cooper, who set a US endurance record of 32 hours. A month later, Russian cosmonaut Valeri Bykovsky set a new solo endurance record of 116 hours, which is still the record some forty years later. For part of his flight he was joined in orbit by the first female cosmonaut, Valentina Tereshkova, on a three-day mission. Then, in October 1964 the Soviets launched a modified Vostok capsule, called Voskhod, which could carry three cosmonauts, but with consumables sufficient for only a little over 24 hours. The pace of advance meant that new spacecraft, with greater capabilities for longer flight and a variety of mission profiles, would be required to expand understanding of human spaceflight.

Further modifications to Vostok were already in the pipeline to accomplish some of these goals on Voskhod missions, while the Americans were examining ways of improving the Mercury design to accommodate two astronauts, extend the mission, and perform rendezvous and docking. This was originally called Mercury Mark II, but in 1961 became known as Gemini. While these new spacecraft were under development, the Russians were also evolving their Soyuz design to carry up to three cosmonauts for a variety of mission roles, including flights to the Moon – a role that the American three-man Apollo spacecraft would also undertake.

STEPPING STONES

In both the Mercury and Vostok programmes (including Voskhod), there was neither the capability nor the need for the single occupant to exit the spacecraft in orbit. The role of these missions was one of survival, to clearly demonstrate that a human could live and operate in orbit for a few days, operate equipment, and make basic scientific observations. But with both countries planning to send crews to the Moon, new techniques, including rendezvous and docking, would have to be learned. To really explore the Moon, the astronauts and cosmonauts would have to leave the spacecraft and walk around. While the techniques and skills needed to walk on the Moon were being evolved in both countries, plans were also in the pipeline to prove the theory that a spacesuited crew-man could open the hatch and work outside.

The media interpretation of opening the door and 'stepping outside' into the void at 28,100 kph helped popularise the term 'spacewalking' in a way that no author or feature film had managed to convey. In the Cold War era of the 1960s, the generally open nature of the US space programme was well documented and the intention of Gemini missions to include spacewalks (EVAs) had been known for some time. In contrast, most of the early Soviet programme was shrouded in secrecy unless one of the numerous 'space spectaculars' grabbed the headlines. It was assumed, however, that the Soviets were also working on equipment and procedures to achieve their own EVA capability, probably before the Americans. It is now clear that such activities had been in the minds of Soviet planners for some time.

Cosmonauts on Mars

The constant target for Soviet space scientists for most of the twentieth century was the planet Mars. As far back as Tsiolkovsky's time, plans were laid to explore the Red Planet, and they formed part of the major decree signed by the Central Committee and the USSR Council of Ministers on 23 June 1960. This became the blueprint for the Soviet programme of the 1960s and the long-range planning at that time. It had the grand title 'On the Creation of Powerful Carrier-Rockets, Satellites, Space Ships and the Mastery of Cosmic space in 1960–67.' Korolyov had developed a personal interest in sending Soviet probes (and eventually cosmonauts) to Mars, probably more so than to the Moon, and a group of engineers headed by Gleb Maksimov had been working on large interplanetary spacecraft in Department 9 at the OKB-1 facility since 1959. In Korolyov's 1960 plans, Object KMV was to have sent cosmonauts on a martian orbital mission (without a landing) and back to Earth, and on 30 April 1960 one of the leading engineers at OKB-1, Konstantin Feoktistov (later a cosmonaut who flew on the first Voskhod mission), put forward a plan called TMK (Heavy Interplanetary Ship), which envisaged a ten-person crew on a three-year trip to Mars. Its landing craft would be a mobile vehicle that could survey the planet for a year, using the onboard drilling device and probably multiple EVAs across the surface before heading home. Although it never proceeded beyond the early planning stage, this focus for Mars also affected preparations for other aspects of the Soviet programme, such as unmanned planetary craft and larger booster rockets.

After Vostok
With the Soviets achieving success after success in space between 1957 and 1961, it seemed inevitable that by the 1970s Soviet cosmonauts could be walking on the surface of the Red Planet. The Moon, though much closer, was a valid target, certainly for robotic exploration, and would be explored as a stepping-stone to Mars. The development of a large heavy lift booster in the Soviet Union was therefore aimed to support manned martian missions, not a manned lunar programme. When President Kennedy committed the US and NASA to the Moon in 1961, it was some time before the Soviets realised that the Americans could indeed reach the Moon long before cosmonauts landed on Mars, and so the focus therefore shifted to a race, against the Americans, to the Moon.[1]

Plans for a small spacecraft to follow Vostok had existed at OKB-1 since 1958. It was designed to master two techniques that were thought crucial to long-term space exploration including piloted flights to Mars – circumlunar flights to the Moon, and rendezvous and docking. The achievement of these objectives would provide experience in a wide variety of technical areas valuable to the more ambitious plans aimed at Mars. In the early 1960s, therefore, work began on a multi-role spacecraft capable of achieving these two objectives, and although these designs would be revised between 1959 and 1963, the basic plan emerged for what essentially became the Soyuz programme.[2]

NASA's advanced manned space programme
During a four-day NASA Staff Conference in Williamsburg Virginia (2–5 April 1959), the advanced follow-up to Mercury was discussed. Its purpose was to create a stepped programme towards 'manned interplanetary spaceflight,' extending the duration of orbital flight and supporting military space objectives.[3] Most of these objectives also indicated the need to bring one or more spacecraft together in orbit (orbital rendezvous), and in both the Soviet and American plans, orbital rendezvous was therefore seen as key to future spacecraft programmes and, as displayed in the artwork of the 1950s, in some scenarios spacesuited crew-members would be required to go outside and construct some of the new vehicles required.

By July 1960 the new space programme had been named Apollo, and was designed to conduct manned spaceflights in Earth orbit and on circumlunar missions by 1970. This would lead to a follow-on programme with a manned lunar landing and the creation of a space station. Over the next year, continuing studies also indicated that Apollo might be part of an early lunar landing programme, and this was confirmed in President Kennedy's speech in 1961. By November 1961 the original Apollo (Command and Service Module) statement of work indicated that pressure suits would be carried onboard for use during unspecified ventures outside the spacecraft.

Mercury Mark II becomes Gemini
Studies had also continued into ways of improving the Mercury design. Between 1959 and 1961 these were mainly focused under the Mercury Mark II programme, and featured a two-man spacecraft with a variety of objectives to increase skills and

techniques for later manned orbital and deep-space flight. In March 1961, as designs for the new spacecraft continued, it was suggested that if a two-man version was being developed, then one of the crew should be allowed to open the hatch while in orbit, and step outside. From then on, EVA became an integral part of Mercury Mark II planning. Now that two men would be in orbit at the same time, an element of safety was built into the design, allowing planning for the first American EVA to proceed with more confidence. This was helped partly by the design of the spacecraft, in that the Mercury Mark II would employ ejection seats for astronaut escape in the event of problems during launch or landing. Despite concerns about pressure integrity and structural strength, as well as heat protection of the spacecraft during entry, the ejection seat design represented a convenient method of exiting the spacecraft directly from the crew station. Simulations of a spacesuited test subject, suspended in a harness to simulate zero g, demonstrated how an astronaut could open the hatch and exit the spacecraft.

In December 1961, Mercury Mark II became Gemini. It would be used to evaluate techniques for long-duration spaceflight, rendezvous and docking, pinpoint landing and EVA. For Apollo these were also seen as significant milestones in the road to the Moon. Early studies concerning a suitable EVA programme from Gemini were established to determine the requirements for a 15-minute EVA capability from a Gemini spacecraft. These included studies into astronaut protection and life support, the engineering requirements and capabilities for opening and closing the hatch in orbit by one spacesuited astronaut, and suggestions for manoeuvring systems close to the spacecraft. These studies defined the EVA process for Gemini, with only one EVA astronaut (the Pilot) making a foray outside, and only one hatch at a time being open. This was determined by the number of times the hatches could be opened, the limitations of the spacecraft's life support system in replenishing the atmosphere, and the design of the crew station layout. However, several early artists' impressions of EVA from Gemini, including the cover artwork of *National Geographic* in March 1964, featured EVA by the Command Pilot.

By 3 February 1963, guidelines had been established by Crew Systems Division at the NASA Manned Spacecraft Center, which would form the basis of EVA planning across the ten manned Gemini missions manifested. Initially, the plan was for the Gemini 3 crew to test the desired procedure of an astronaut standing up in a spacecraft seat in the altitude chamber, prior to the Gemini 4 astronauts performing a stand-up EVA in space, with a full exit (spacewalk) planned for Gemini 5.[4]

Soviet Vykhod (Exit) plans
In the early planning for the follow-up to the one-person Vostok, Korolyov had sketched out a series of missions, some of which included experiments on EVA techniques. One of the earliest featured the placing of a spacesuited dog inside a container in a Vostok that could be depressurised, and then exposing the canine passenger to the vacuum of space (whether this meant inside the capsule or by opening the outer hatch of the capsule is not clear). Thus, valuable experience of a living creature enduring EVA, plus the suit materials and systems could be gathered

prior to attempting a human EVA. This was similar to the way in which canine flights in Vostok capsules pioneered the first cosmonaut missions.

Although this mission was cancelled long before it was authorised, the idea of EVA from a Vostok capsule remained on the drawing boards. These plans moved a stage further in the spring of 1963, when further Vostok missions, including EVA tests by cosmonauts, were authorised for flights in 1964 and 1965. The problem arose in determining how the cosmonaut should exit the pressurised spacecraft and enter it again safely. As with Gemini, the crew escape system on Vostok consisted of an ejection seat, although unlike Gemini the hatch was not hinged, but instead was blown away as the seat system was initiated. The brunt of Gemini re-entry heating was in the blunt aft end of the spacecraft, away from the hatch seals, so that a less rigorous thermal protection system was feasible around the crew area to allow Gemini astronauts to crack the hatch, reseal it, and still maintain sufficient thermal and pressure integrity in the spacecraft. On Vostok the whole spacecraft was sealed for thermal protection, and the provision of an opening hatch and repressurisation system was a major design difficulty. Although not officially authorised, plans were made at OKB-1 to allow these extra missions to carry more than one cosmonaut, extend the flight, and perform EVA (duplicating Gemini plans). To achieve this, the single ejector seat had to be taken out, and for three cosmonauts there would not be enough room for them to wear pressure suits. For EVA this was of course essential, and so only one or two cosmonauts would fly EVA Voskhod missions.

When NASA announced plans to allow the Pilot to open the hatch and stand up on his seat during Gemini 4 in February 1965, Korolyov was spurred to ensure that it would be a Soviet cosmonaut, and not an American, who would make the first exit into space. By March 1964 the EVA from the redesigned spacecraft (now called Voskhod) was restored, and with pressure from the Kremlin it was planned to fly the mission by November 1964, three months prior to the American attempt from Gemini 4. Air Force General Staff Deputy Nikolai Kamanin was in charge of manned spaceflight in the Air Force, including the selection and training of the Air Force cosmonaut team (he would later become Director of Cosmonaut Training). Kamanin had put forward a proposal to orbit a lone cosmonaut in a test of EVA equipment and procedures, but Korolyov rejected this idea almost immediately – not, as might be expected, on the grounds of technical difficulty or even safety, but on the lack of political effect of such a flight. A solo one-day test flight without EVA would not been seen as a major advance in manned spaceflight, and Korolyov insisted that each new flight should be seen as an advance over the previous mission.[5] Delays in preparing for the three-man Voskhod mission pushed back the EVA mission on the second Voskhod. Korolyov pushed his team for an end-of-year launch, and Kamanin wrote: 'As always, Korolyov is in a hurry. He prefers a cavalry charge to well conceived and methodically prepared offences on the 'space fortress' [America].'

Official approval for the Vykhod (Exit) project from Voskhod came from a governmental resolution dated 13 April 1964, and later that month, in a meeting between Korolyov and specialists from OKB-1 and Zvezda, it was decided to approve the air-lock concept for the mission. Following two months of concept studies and evaluations of the pressure suit, the back-pack, exit hatches and air-lock, the specification document was signed on 9 June 1964, with governmental approval a month later on 8 July. Zvezda would be prime contactor.[6]

Korolyov's initial plan may have been linked to a promise to the Kremlin to fly the second mission in November as part of the celebration for the anniversary of the October Revolution, but political changes in the Soviet Union would see Khrushchev overthrown in October 1964, at the time of the first Voskhod mission. The pressure to meet the deadline diminished, and the second Voskhod and its EVA experiment were then set to fly in the early months of 1965, just before the Americans attempted EVA from Gemini. The Soviets were still determined to ensure that a Soviet cosmonaut would be the first to leave his spacecraft in orbit.

EARLY HARDWARE

In 1961, human spaceflight was achieved in spacecraft that were originally conceived without provision for EVA. In their original design, both the American Mercury and the Russian Vostok were incapable of flying a mission to include the simple opening of a hatch.

At launch, the sole Vostok Pilot was sealed into the spacecraft on an ejector seat that was used during the final phase of landing to allow the cosmonaut to descend by parachute. Because of the position of the cosmonaut in the spacecraft, he could not reach the hatch from the seat, as it was above his head and mostly behind him. More importantly, it was an explosive hatch that was to be used only in ejection exit, and with the ejection seat included, there was very little room for the cosmonaut to move around. Indeed, he merely floated a little above his seat for the duration of the mission. Finally, the pressure garment was an evolution from the pressure garments developed for high-altitude air-crews, and was not suitable for prolonged EVA activities in a vacuum. There were also the safety considerations for a lone crew-member to perform EVA and leave spacecraft monitoring to onboard computers or to remote monitoring on the ground. Korolyov and his team did not have time to wait for their new Soyuz spacecraft, and so went about adapting the solo Vostok into a multi-crewed vehicle with the additional but limited capability to perform EVA.

In the American Mercury too, it was impracticable to develop an EVA capability, as the spacecraft was simply too small. (It was said that the astronaut did not so much 'get into' the spacecraft; rather, he 'wore it'!) The hatch was again not designed to be opened in orbit, and the pressure garment was an American development from the military pressure garments used by air-crews. With Apollo in mind, it was decided to end the Mercury programme and move to the new Gemini spacecraft that could include EVA capability, extended flight, and rendezvous and docking

Apart from the hatches and pressure garments, another problem was to ensure that once the spacecraft had been evacuated of its breathable atmosphere and internal pressure, the life support system could supply both breathable air and safe pressure to allow the crew to work in the suits and cope with the increased thermal and humidity levels that such exercise would produce. The life support system would also have to be able to replenish the cabin atmosphere after the EVA, and, if possible, do so more than once to allow multiple EVAs.

Methods of exit: air-lock or hatch
The additional mass and volume required to fly three cosmonauts in Vostok created its own problems for the designers as they evolved the Voskhod variant. It was determined that three cosmonauts could fly a short 24-hour mission if the seats were turned 90° to the entry/exit hatch and if the centre one was moved slightly forward, provided that both the ejector seats and the pressure garments were also removed. This, of course, also eliminated all methods of crew escape or protection in the event of launch or landing difficulties or decompression during flight. Those who argued against this radical move were overruled in light of the success of the Vostok missions, during which no such incident had occurred. However, for Voskhod 2 and the EVA experiment, because of the addition of a method of exit and the need for pressure suits it would carry only a two-man crew on a short 24-hour mission.

For Gemini, the two-man design already included ejector seats for escape during launch and entry, and these would be modified to allow for opening and closing the hatch for EVA operations. There was just one crew area inside the spacecraft, with both astronauts sitting side by side, but there were two hatches for entry and exit from the small spacecraft. The atmosphere of the entire crew area therefore had to be vacated, thus exposing both astronauts to the vacuum of space. (Because of the spacecraft's dimensions, a flight was often likened to spending up to two weeks in the front seats of a Volkswagen Beetle.) History books record that of the ten manned Gemini missions flown between March 1965 and November 1966, six were planned to include EVA by one of the crew. In the event, however, only five missions completed EVA, and those five astronauts are listed as the first Americans to perform a spacewalk. What is often overlooked is that each of the other members of those crews were exposed to the vacuum, even though they remained inside the capsule with the Commander's hatch firmly closed.

For the Russians, due to the less than fully reliable operation of the life support systems on Vostok it appears that the spacecraft was insufficiently robust to allow full depressurisation and repressurisation. The Vostok design did not accommodate lengthy exposure of cabin equipment to a vacuum, and, indeed, some instrumentation could not operate in a vacuum. A set of requirements was therefore established for the EVA at OKB-1. These included the stipulation that the crew cabin could not be depressurised, which necessitated the design of an alternative 'compartment' with inner and outer hatches so that the cosmonaut performing EVA could pass from the crew cabin into the compartment via an air-lock. Wherever possible, existing equipment had to be used, and there had to be a limit in the changes to the spacecraft. The smallest items (in both mass and size) had to be incorporated in the already tight spacecraft. This all had an effect on the planned duration of the EVA and its activities. The cosmonaut could be outside for only a few minutes, and would therefore be restricted to proving the capability of the system and the suit, taking photographs, and verbally describing his activities and condition. More adventurous activities outside the spacecraft would have to wait for the larger Soyuz.

The Volga air-lock
To achieve the first Soviet EVA it was decided to create an air-lock system, called Volga, to allow the spacewalking cosmonaut to leave the spacecraft without depressurising the main compartment. The air-lock had to be more than 2 m in length to allow the suited cosmonaut to enter, close the hatch and change the air-lock pressure levels. Attaching this air-lock to the side was impractical, as the aerodynamic forces at launch would simply rip it off as the vehicle ascended, and so a collapsible air-lock was chosen. This attached to the spacecraft hatch, and could be deployed in orbit and then jettisoned after the EVA to clear the hatch area for entry.

The air-lock consisted of a rigid outer ring, into which was fitted an inner opening EVA hatch. The main structure of the 'air-lock' shell consisted of a series of forty soft rubber inflatable cylinders along the length of the structure. Inside was a pressure bladder, and covering the 'air beams' was a soft enclosure of a high-stretch fabric. At the base of the unit was a lower assembly ring that affixed to the spacecraft entry/exit hatch and opened inwards. Inside the support ring were the deployment system, the air pressure operation system in four spherical canisters, control panels and back-up systems, and umbilical connections. The whole unit was stowed folded against the side of the spacecraft for launch, under a special cover assembly on the Voskhod launch shroud. Shortly after entering orbit and separating from the upper stage of the launch vehicle, the cosmonauts deployed the air-lock to its full extent.[7]

When fully collapsed, the inflatable cylinders measured just 73.9 cm, but when fully extended the air-lock measured 159.7 cm. When combined with the hatches, the deployed structure measured 2.49 m long, with an internal diameter of 0.97 m and an external diameter of 1.18 m. Inflation, which took seven minutes, produced an internal volume of 2.5 m^3. In total, seven air-lock units were fabricated: five for the ground test programme, and two as back-ups to the two flight units. The first flight unit was tested on Cosmos 57 in February 1965, during which the automated deployment of the Volga hatch was monitored by an onboard TV camera during the first orbit.[8] The second flight unit was prepared for Voskhod 2.

The Gemini EVA hatch
The development of Gemini had for some time indicated two spacecraft hatches – one for each astronaut – as the first choice for entry and exit. These were primarily designed to support emergency ejection from the launch pad, or during powered ascent up to 60,000 feet (18,275 m) or descent below the same altitude. At 50,000 feet (15,230 m) the parachute deployment system initiated, and had the parachutes failed then the crew could have ejected. For launch, performance monitoring of the Titan II focused on the two Malfunction Detection Systems (with redundancy capability) to allow time for the astronauts to initiate the opening of the hatch and ejection of the seats in the event of a malfunctioning booster. It was determined that the non-explosive propellants of Titan would burn but not violently explode, allowing the quick reflexes of the astronauts to initiate the escape system – hopefully before the

destruction of the vehicle. These hatches had to be capable of opening under maximum dynamic conditions of launch, and remain locked open for clearance of the ejector seat. In one unmanned ground test, witnessed by astronaut John Young, the system worked perfectly apart from the opening of the hatch. The seat ploughed straight through the hatch door, which would give any astronaut, according to Young, 'one of the biggest but shortest headaches they would experience.'

During EVA this same hatch would have to be manually opened and closed, isolated from the ejection system until after re-entry. Following Gemini 3, a security pin was added to the ejection control handle to prevent inadvertent activation.[9] In early 1964, simulations of opening and closing the hatch (during zero-g flights on the KC-135 indicated that the hatch would have to be manually closed for the final 10–15 cm. As a result, a lanyard system was added, designed for either crew-member to operate while wearing a full pressure suit. The size of the D-ring housing on the ejector seat was reduced to make it flush with the seat to avoid causing a hindrance during EVA operations. The size of the control lever for the hatch operating mechanism was also reduced, as tests revealed that it could damage the EVA visor during ingress.

The hatch also had to be designed to ensure correct sealing after closure at the end of each EVA, to ensure pressure integration and thermal safety during the rest of the mission and re-entry and landing. The sill of the hatch on both sides was a 1.9-cm wide and 1.27-cm deep groove, which was manufactured under intense scrutiny to enable perfect sealing and to ensure that it was able to endure several opening and closing cycles.

The first EVA suits
One of the criteria in selecting suits for the first excursions into space was that, due to the restricted volume inside the crew compartment of both Voskhod and Gemini, the EVA Pilot would need to wear the EVA pressure garment throughout the mission, not just for the EVA. It was also a requirement that the suit could support the astronaut in the event of cabin depressurisation. The detailed development of every pressure suit for EVA is not covered in this volume, but the description of the first two types of suit for the Russian and American programmes in 1965 are presented here due to their historical significance.

Berkut
The EVA suit for the initial Soviet excursion into open space was designated Berkut (Golden Eagle). It was a development of the SK-1 full pressure garment worn by aviation crews at high altitude and the first cosmonauts onboard Vostok, and elements of the S-10 experimental full pressure garment developed at Zvezda during the late 1950s.[10] Features developed from the SK-1 garment included a body enclosure with a restraint layer, and the arm and glove assemblies. The enclosures for the legs and the internal ventilation system were developed from the S-10. The suit was a four-layer design, with the innermost restraint layer fabricated from a nylon-type fabric with high-strength capability. A primary and back-up pressure bladder surrounded the restraint layer, and featured a sheet of rubber with an inner nylon liner that also incorporated an internal ventilation system. Due to the requirements for thermal

control outside the vehicle, a new feature was added: the external layer was, in fact, a multilayered covering that insulated the suit from thermal variations in the vacuum of space. Separate pressure gloves and over-boots were added to the garment, together with a non-turning metal helmet that featured a hinged visor and light filter/sunshade, which was a development of the basic GSH-8 air-crew pressure helmet.

In Voskhod, the oxygen–nitrogen mix was at a relatively high pressure of 400 hPa in the primary mode and 270 hPa in back-up mode, so that the decrease in pressure would render the suit more easily flexible. Due to the limiting of the changes to the life support system used in the basic Vostok design, minimal changes were required to convert to the two-seater spacecraft, with LSS components located in a pair of modules next to each crew couch. Together with the use of the air-lock design, this restricted the use of umbilical connections for EVA without major modifications for looping the supply of oxygen through the hatches to the suit. A back-pack was therefore designed for the cosmonaut to wear on EVA. Designated KP-55, this was manufactured at design bureau SKB-KZ in the city of Orekhovo-Zuyevo. Donned by the EVA cosmonaut inside the crew compartment before entering the air-lock, it was secured to the back by a harness system. It featured a 2-litre bottle of oxygen at 22 mPa pressure supplied directly to the helmet, then into the suit, and finally dumped into the vacuum of space by means of a pressure control valve. The oxygen flow also pressurised the suit, provided breathing oxygen for 45 minutes, and removed carbon dioxide. The vented oxygen also carried away excess heat and moisture. The back-pack had three operating modes. Nominally it provided a 16–20-litre per minute flow rate of oxygen, and during periods inside the air-lock this was increased to 25–30 litres per minute (550 hPa). The emergency mode would be initiated automatically when internal pressure dropped below 240 hPa, and supplied oxygen at 30 litres per minute. As a back-up procedure there was also a supply of oxygen from a small bottle located in the air-lock support ring. This supply was attached to the 7-m safety tether, which also featured a shock-absorbing device and electrical cables to transmit medical data, technical parameters and voice communications to the crew compartment.

For Voskhod 2, the Commander remained inside the crew compartment and monitored the systems of the spacecraft and the progress of the EVA. He also wore the same type of suit, and in theory could, in the event of difficulty, render assistance to his colleague on EVA.

G4C

The Gemini G4C pressure garments used for the EVA operations were a development of the intravehicular G3C suit fabricated at the David Clark Company in Worcester, Massachusetts. It was constructed in several layers. Innermost was the constant wear (underwear) garment and comfort layer, and on top of this was a pressure bladder and link-net restraint layer. Over this was two layers of high-temperature nylon for micrometeoroid shock absorption, and then seven layers of aluminised Mylar and woven Dacron super-insulation material. The outermost layers consisted of HT-1 (High Temperature) nylon felt for protection against micrometeoroids. The EVA cover layer was fabricated in two parts: an upper torso coverall and a jacket that covered the arms and shoulders and was removable for the rest of the flight following the EVA.

The helmet featured a detachable EVA visor that also had two over-visors. The first was the Sun visor, constructed of grey-tinted Plexiglas with a thin gold coating on its outside. This was flake-resistant, and helped reduce the temperature of direct sunlight into the helmet by 88%. The second visor was a polycarbonate material providing thermal control and with an ultraviolet inhibitor and impact protection for the inner pressure shell of the helmet. A pair of over-gloves were worn for Gemini 4, with palm protection that allowed direct palm contact with objects ranging from +250°F to –150°F for up to two minutes.

To control ventilation inside the suit and maintain a pressure of 202 mm Hg, a Ventilation Control Module (VCM) was mounted on the chest with restraint straps attached to the parachute harness and held by Velcro to the front of the suit. There was an emergency supply of oxygen for nine minutes via the VCM to the feed port in the helmet. The Berkut design required a back-pack, but because of direct exit from the crew compartment on Gemini it was possible to incorporate an umbilical connection to directly supply oxygen from the spacecraft to the astronaut on EVA. The 7.62-m umbilical used on Gemini 4 featured an 8.22-m oxygen line and a 7.22-m safety tether. This design prevented undue stress on the oxygen line, while the tether was designed to absorb a 44.99-kg load. Like the Russian tether, it also prevented the astronaut from floating away from the spacecraft, and allowed the astronaut to fully test the hand-held 'zip-gun', exhaust the limited fuel supply, and still be able to return to the spacecraft. To protect the umbilical from the solar thermal heat and distribute its energy, it was covered with a thin layer of gold film.

The experiences and results from both EVAs would have an influence on the design of future equipment and procedures for both pioneering EVA programmes.

FIRST STEPS INTO THE VOID

The first EVA took place when Alexei Leonov exited the Soviet spacecraft Voskhod 2 on 18 March 1965, one week before the first manned Gemini flight. Less than three months later, on 3 June 1965, he was followed by Ed White on the first American EVA, from Gemini 4. These first steps into the void of space turned theory into practice, and the door to space was literally open. But although the technique was proven on these two missions, there was still a long way to go hone these new skills. The years 1965 and 1966 marked the beginning of the learning curve.

Vykhod: the first exit

Preparations for Leonov's exit began almost as soon as the spacecraft attained orbit, with the extension of the air-lock from the side of the capsule. The Vostok spacecraft, designed for only one cosmonaut, was now modified into Voskhod for two, and did not have room for them to remove pressure suits. The Berkut EVA pressure suit worn by Leonov was put on before launch, and would not be taken off until after landing. As much as he could while wearing his own pressure suit and in

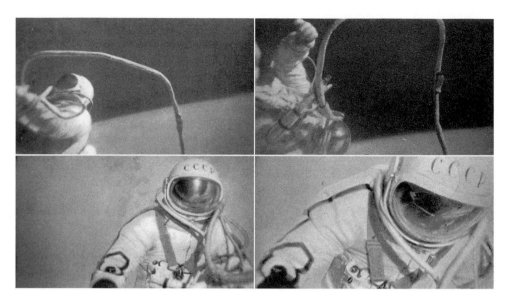

Alexei Leonov becomes the first man to walk in space, 18 March 1965.

the confines of the spacecraft, commander Pavel Belyayev helped his colleague put on the life support system back-pack and tried to curb his friend's enthusiasm.[11]

Leonov eagerly floated head-first through the hatch into the even smaller air-lock, with Belyayev sealing him inside. After a systems check and the reduction of air-lock pressure, the time came for Leonov to open the hatch and step outside. It had been only 90 minutes since the spacecraft had left the launch pad, and now, nearing the end of the first orbit, Leonov was ready to step into history.

As the air was taken out of the air-lock, his suit became rigid and springy. Then it became very quiet before he heard Belyayev inform him that there was a vacuum in the air-lock and he could open the hatch.

Leonov later wrote about his experiences: 'First the lock opened, then the opening mechanism started to turn and the hatch shuddered and gently opened. There before me was the sky of outer space. I had never seen it like this, with its myriad of stars, through the light filter of my visor instead of a view port of the spacecraft. The hatch moved slowly upwards, opening the window to outer space wider and wider. Holding my breath, I watched the star pattern change rapidly against the black sky of outer space. The hatch came to a standstill. Now everything was ready for my spacewalk. I floated towards the hatch and poked my head out and saw a limitless black sky and stars ... There were many more than can be seen from Earth ... They were brighter and did not twinkle. 'If you want to be a spacewalker, get out into space,' I said to myself.'[12]

Pushing off, Leonov grabbed the hand-rail as he floated outside. 'It was so quiet I could hear my heartbeat and my breathing,' he recalled, looking back at the Voskhod light in the bright sunlight, which looked like something from science fiction. His view of Earth was spectacular as he passed over the Black Sea and the Volga River, eagerly looking for places of his childhood, even covered with the ice and snow of a Russian winter. Earth was 'spread out like a velvet carpet. What a picture! I was amazed at its vast size.' Belyayev's voice in Leonov's earphone brought him back to the job in hand, and he concentrated on the task of setting up a cine-camera and photographing the spacecraft. He took the lens cover off the camera and, releasing he had nowhere to store it, cast it off into space, 'sparking like a star' as it disappeared.

Leonov recalled that he moved about 7 m from the spacecraft, twisting and turning on the end of the umbilical. His suit had ballooned, which restricted his movements, including the activation of the still camera shutter switch attached to the leg of his spacesuit. This ballooning of his suit made it increasingly difficult to manoeuvre his arms and legs without great effort, and his actions imparted movement on the spacecraft, which was being controlled by Belyayev. In his later accounts of the EVA, Leonov wrote that through his earphones he heard that Radio Moscow had announced that a man was floating free in space and wondered who this could be, until he realised that it was him they were talking about! His realisation of the enormity of his achievement, and the significance of what only he could see and feel, must have been similar to that experience by Gagarin on the first trip into orbit just four years earlier.

The Soviet propaganda machine then went into full swing once more, claiming a great advancement in Soviet technology and space exploration. In the post-flight press conference, Leonov reported: 'In space you can float about as you like. For example, I stretched out my arms and legs and soared ... it's more convenient. There's more room to move in. You breathe easily, even better than on Earth. It is true that my pressure suit resisted [any] changes in the form of my body, [the] flexing of arms and legs, and so it required an effort to work.' The impression given was that EVA was pretty easy and comfortable to accomplish, and the world's media picked up on the ease with which Leonov had been able to complete simple tasks, make intelligent observations, and perform in the hostile vacuum of space. But would walking on the Moon be just as easy? It was some years before the details of Leonov's few minutes outside became clearer.

Hurdles to overcome?
Shortly after the mission, a report in *Life* (14 May 1965) quoted Leonov as stating that he had become tired from working in a spacesuit over a long period of time, with each movement requiring more effort than on Earth. Working against the internal pressure of the suit was an effort and caused fatigue, and after just ten minutes in open space, Leonov was tired.

Following the flight, NASA's Crew Systems Division collected all possible data concerning Soviet pressure suits, and compared it with those developed or under development for the American programme.[13] The report stated: 'It becomes increasingly apparent that precise hardware descriptions, such as those available in open literature for American spacesuits, is not in existence for their Soviet counterparts.' For the analysis they called upon the expertise of Major C.L. Wilson, USAF, who was a recognised authority on Soviet pressure garments. Wilson viewed an 8–10-minute colour motion picture and two TV documentary films of Leonov's EVA, which appeared to be a blend of both training and in-flight views, further confusing the analysis of his actions in the suit. Trying to determine the components and systems of the suit was difficult, but in general Wilson concluded the following:

- Apparent rapid arm movement made by Leonov involving the shoulder and elbow did not reveal visible rebound (by means of any restriction by internal pressure or poor joint mobility) back to its original position, which would not be desirable.
- Views of the rigid glove hand, with palm and wrist straps for apparent ease of hand grasp and finger flexing to approximately 40%, indicated that it would be very tiring to hold tools in this glove for a prolonged time.
- The use of a long umbilical connected to the spacecraft supported literature stating that Belyayev could rescue Leonov in an emergency, perhaps 'by automatically pulling Leonov back into the hatch'.
- The impression of the authors of the report was that 'the mobility of the cosmonaut in the pressure suit is only moderately good'. But given all the relevant information available, 'it is speculated that the effort in doing this [arm and leg movement] was not excessively great, and it is possible that a cosmonaut could work on the EVA assembly of space vehicles or other work for perhaps an hour or two in the suit under these circumstances, if all other life support requirements were met.' It was suggested that the Soviet state-of-the-art in joint mobility and all other aspects of pressure suit design was on a par with those of the US.

This recent Soviet 'space spectacular', the lack of clear information, and the suggestion that the suit technology of the USSR was as good, if not better, than the Americans, did nothing to change NASA's overall EVA planning for Gemini, and in fact contributed to a change of GT-4 planning from a 'simple' stand-up EVA in the hatch to a full exit, and use of the zip-gun for manoeuvring.

Later, of course, more details of the difficulties that Leonov encountered were revealed. Leonov himself wrote that at one point he was rotating around several axes at once, and was unable to control his movements as he had no manoeuvring device available. This also caused the umbilical to begin to entangle him 'like an octopus', and he struggled to free himself, realising that sudden movements in open space were not a good idea. He soon learned that it takes effort to maintain a position if there is nothing to brace against, and nothing with which to orientate. 'It became hotter inside the spacesuit. I felt sweat on by back, my hands became wet, and my pulse quickened.' The sweat was beginning to float into his eyes as he struggled to return to the hatch and as he tried to push the camera into the air-lock with one hand while steadying himself with the other. His suit had ballooned to such a degree that it was almost impossible for him to enter the hatch, and he therefore had to reduce the internal pressure to allow himself enough movement and reduction of volume to squeeze inside the air-lock. He now had to close the outer hatch, but having entered head first he struggled to turn around to do so. He finally sealed himself inside, having been outside for only 12 min 9 sec during an EVA of 23 min 41 sec. He was a national hero and a space pioneer, but he was also soaked in sweat, close to passing out due to heat stroke, and exhausted. He had completed the very first exit into space, and survived – just.

At the time of Voskhod 2 there were other missions planned in the series, some of which would include EVA. These daring plans evolved mainly on paper prior to Leonov's exit, and included having a Voskhod crew complete two or three EVAs on one mission up to 100 m from the spacecraft. For a while, planning for a Voskhod 5 mission included having a female cosmonaut perform EVA, while a Voskhod 6 EVA would demonstrate the UPMK manoeuvring unit. In the event, due to pressure to end the Voskhod flights and move on to the new Soyuz programme, Voskhod 2 was the final flight of the series. Voskhod's objectives, however, were not cancelled in their entirety, and some would be reassigned to Soyuz (including the EVA plans for a while); but Soyuz would not be ready until 1967, by which time the Americans would have completed Gemini and several EVAs. So, although the USSR had taken the lead in EVA with Voskhod 2, it would be four more years before another Soviet cosmonaut 'stepped outside'.

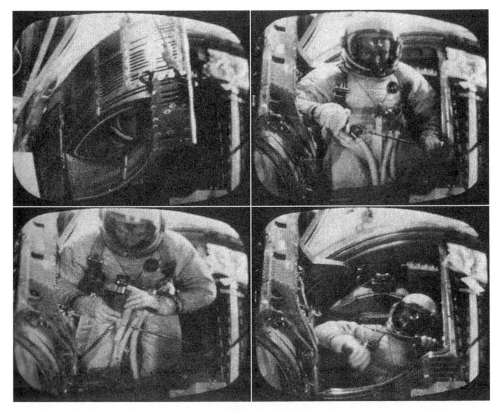

Hatch opening/closing and stand-up EVA demonstrations conducted in an altitude chamber prior to the first Gemini EVA.

Zip guns

Leonov's EVA clearly demonstrated a need for a system to maintain a cosmonaut's orientation while outside on the end of an uncontrollable umbilical. The Gemini astronauts quickly referred to Leonov's 'octopus' as a 'bag of snakes'. Astronaut Ed White was assigned to lead the way for the Americans, with a hand-held manoeuvring device called the Hand Held Manoeuvring Unit, or more commonly, the 'zip-gun'. During its development, MSC engineers used the air-bearing floor to evaluate a number of experimental hand-held devices using gas expulsion systems for thrust, and from these evaluations the final design of the HHMU for GT-4 emerged. In the tests, it was found that during translation the pulling mode of the gun was by far the most stable and easy to control, and it was also learned that placement of the exhaust nozzles parallel to each other at the ends of a long arm produced less loss of thrust from the impingement of the gas emitted from the opposite nozzle, than if the two nozzles were placed side by side and canted outwards. Because of the lack of finger dexterity in the pressurised EVA glove, control of the push/pull trigger was best achieved by solid pressure from the hand and by placing the handle of the HHMU on the top of the device, helped by limiting the necessary arm and hand movements. To ensure that the thrust was aligned to the centre of gravity, each thruster would be orientated at specific angles for accurate aiming by the operator. Thrust levels for attitude control would be improved by using proportional thrust systems rather than an off/on switch.

GEMINI EVA PLANNING

The Gemini EVA plan, published in January 1964, set out the objectives of the Gemini EVA programme:

1. To evaluate man's capability to perform useful tasks in a space environment.
2. To employ EVA operations to augment the basic capability of the Gemini spacecraft.
3. To provide the capability to evaluate advanced EVA equipment in support of manned spaceflight and other national space programmes.

To meet those objectives, the Flight Crew Operations Directorate evolved the following schedule for flight-crew activities:

GT-4   Depressurise the cabin, open the hatch, and stand up.
GT-5   Perform complete egress and ingress manoeuvre.
GT-6   Egress and proceed to the interior of the equipment adapter and retrieve data packages.
GT-7/GT-8   Evaluate manoeuvring capabilities along the spacecraft exterior by using tether and hand-holds.
GT-9   Evaluate the Astronaut Manoeuvring Unit (Department of Defense experiment 14C).
GT-10–GT-12   Evaluate other advanced EVA equipment and procedures.

Leonov's exit into space, followed by the success of Gemini 3 the following week, prompted MSC Director Robert Gilruth to discuss the possibility of allowing Ed White to complete a full exit into space. Various NASA departments argued for and against such a move, but in the end approval was given on 25 May 1965, and just nine days later the first American to walk in space achieved a 21-minute excursion outside Gemini 4.

Go for EVA

The first Gemini EVA was delayed by one orbit because of difficulties in completing all the pre-exit checklist requirements, which caused the Pilot to become hot and sweaty in the confines of the crew compartment; but the extra orbit allowed him time to relax and cool down a little. Exiting the hatch by using the hand-held manoeuvring gun, White found that the gun was not difficult to use, and he was able to maintain his orientation, turn to face the spacecraft, and move away or toward the spacecraft as required. He was in far better control of his actions than Leonov had been. Then, evaluating movement only by tugging on the umbilical, White found that this was far more difficult, and the umbilical had a tendency to place him over the rear adapter section of the spacecraft, out of line of sight of Command Pilot Jim McDivitt, who was inside.[14]

When White was in view, McDivitt could orientate the spacecraft using particular thrusters to ensure that his colleague was nowhere near the thruster vents. White also found no difficulty in orientating himself by using the spacecraft as a reference point, but once the fuel of the HHMU had been used up his control or orientation was lost, and once again he found it difficult to orientate himself using only the umbilical. After taking photographs of the spacecraft, White returned to the hatch before loss of signal with the ground station. He then encountered difficulty in moving sufficiently low into the seat area to close the hatch easily, and McDivitt had to help to bring him in low enough to close and seal the hatch. However, the first exit into space by an American had been as successful as the first Russian EVA.

A LEARNING CURVE

With the success of Gemini 4, mission planners wanted to achieve the other important mission objectives of long-duration flight and rendezvous and docking with a target spacecraft, and so EVA was removed from Gemini 5, 6 and 7. In-flight difficulties with the thruster system on Gemini 8 aborted any attempt for EVA on that mission, during which there was an emergency re-entry and landing just ten hours after launch. EVAs were accomplished – with varying degrees of success – on Gemini 9, 10, 11 and 12. (The Gemini EVAs are detailed in the author's *Gemini: Steps to the Moon*, and are summarised here.)

Gemini 9 An attempt to use the Astronaut Manoeuvring Unit was abandoned when the efforts of the Pilot (Cernan) led to overheating and fogging of the visor. He achieved hook-up to the AMU, but it was not released from the back of the Gemini Adapter Module, where it was stored. Although declared a success, the EVA did not accomplish its objective of a tethered evaluation of the AMU (which was

Mike Collins checks his EVA equipment prior to the Gemini 10 flight.

subsequently removed from Gemini 12), but it revealed the problems of simply maintaining position in space to complete a prescribed task – an expensive but useful lesson.

Gemini 10 This mission was more successful, although a mysterious fogging of the visors of both astronauts (Collins and Young) caused eye irritation and blurred vision (a potentially dangerous situation) for a period during a stand-up EVA. During the second umbilical EVA, Collins again found that the lack of body restraints and hand-holds caused difficulty when trying to retrieve experiment packages from an Agena target. He also experienced difficulty with pre- and post-EVA preparations and stowage, and a checklist that numbered more than 130 items. This took time and effort before the hatch was opened.

Gemini 11 The task here was for the Pilot (Gordon) to attach a tether to a docked Agena for subsequent tethered spinning exercises between the two spacecraft to create a small degree of artificial gravity. The attempt was successful, but Gordon found real difficulty in trying to stay close to the Agena long enough to firmly attach the tether. He repeatedly floated away, and only by jamming his legs into the docking adapter and grasping the nose of the Gemini could he finally achieve his goal. In preparations for the EVA, he and the Command Pilot (Conrad) once again found it extremely difficult to complete all of the pre-exit tasks, which created a rise in suit temperature even before the EVA began. Already tired, Gordon exerted even more effort at the Agena, and once again it became clear that to perform tasks in one location, suitable foot-holds and hand-holds had to be provided to prevent over-exertion by the EVA astronaut.

Gemini 12 The most successful EVAs of the programme resulted from the efforts of the other Gemini EVA astronauts and the lessons learned from their difficulties. By the time EVA preparations were finalised for GT-12 (Lovell and Aldrin), underwater EVA training had come to be used as an excellent simulation of weightlessness, although too late for the crews of GT-9–GT-11. Aldrin trained in the water tank for his mission, and Cernan conducted a post-flight repeat of his Gemini 9 EVA in the pool, as well as additional training as GT-12 back-up Pilot. The AMU was removed, and instead a programme of testing tethers, foot and hand restraints and simple manual tasks was included to evaluate improvements to EVA equipment and support procedures prior to moving on to Apollo or the Apollo Applications space station programme.[15]

RESULTS OF GEMINI EVAS

Gemini ended with the splash-down of the tenth manned mission (GT-12) on 15 November 1966. In early February 1967, a programme summary conference was held at the Manned Spacecraft Center in Houston, at which were presented twenty-one technical papers highlighting the ten manned missions, with a major section evaluating the results of Gemini EVAs.[16]

The report praised the achievements of the Gemini EVAs, which 'met or exceeded the original objectives of EVA,' but was critical of *some* of the limitations. In eleven hatch openings and more than twelve hours of EVA (including eight periods of night-time operations), the programme clearly established the feasibility of EVA. Additional observations were also included:

- The ability to control EVA work-load within the limits of the life support system and the capabilities of the Pilot were demonstrated, but the placing of the EVA chest-pack was an encumbrance, and the use of gaseous oxygen as a coolant was a limiting factor in heat rejection. Work levels and metabolic rates could not be recorded during the missions, although experiences during EVAs prior to Gemini 12 clearly revealed the limits of suit design in coping with the added excursions of the astronauts.
- The need for hand-holds for transit was shown, and the use of several types of hand-holds and hand-rails satisfactorily demonstrated the ability to perform complex EVAs. [At least on Gemini 12. It was the lack of such hand-holds and restraints that plagued both Gemini 10 and 11. Only on Gemini 12 was it demonstrated that with adequate foot-holds and hand-holds, it was possible to pace the EVA task and prevent overheating of *that* particular suit design.]
- The ability to perform tasks of varying complexity was demonstrated and the limitations of certain tasks were identified. Transfer between vehicles was demonstrated in a variety of ways: by surface transit while docked; free-floating between two undocked vehicles in close proximity; self-propulsion

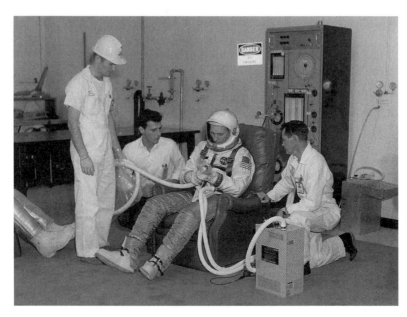

A break in Gemini EVA equipment checks and suiting up prior to entering the altitude chamber.

between vehicles; and tether or umbilical pull-in from one docked vehicle to another. Each of these was conducted within 5 m maximum separation between the vehicles. Retrieval of objects from outside the spacecraft was also demonstrated on four of the missions, including on an unstable passive target [Agena 8 during the Gemini 10 umbilical EVA].

- Hand Held Manoeuvring Units were only briefly evaluated on two missions, and the AMU could not be analysed at all. When the HHMU was used, the astronauts were easily able to orientate and control their movements as required. Dynamics using the short tethers were evaluated on two missions in a demonstration of a distance-limiting device. The requirement for body restraints was firmly established, with foot and waist restraints evaluated and improved as a result of previous flight experience in time for Gemini 12 to flight-prove a number of solutions.

- Training in 1 g and on zero-g aircraft flights was limiting, as revealed by the experiences on Gemini 11, but the use of underwater simulations was identified, used in training, and proven in the execution of flight work-loads, and post-flight evaluations vindicated its use for accurate simulation of EVA techniques on Gemini 12.

- Careful coordination of close-proximity flying was demonstrated on Gemini 10 to allow the Pilot to reach an unstable target safely and easily, while avoiding hazardous thruster-plume impingements. Working as a coordinated team was essential for Gemini, and spacecraft control by the Command Pilot was also important for EVA photography on an experiment performed by the Pilot.

- Most Gemini activities progressed smoothly as far as the limitations of suit design would allow, but despite the excellent physical condition of the Pilots, more arduous tasks both exhausted the astronaut and put considerable strain on the suit and LSS. The design of the suit had been established *before* any EVA had been accomplished, and it was not possible to include major design changes with a flight programme progressing at such a rapid rate [ten manned missions completed in twenty months, and four EVA missions completed in five months].
- The sequencing of EVA scheduling was also a factor. Where a Pilot conducted a stand-up EVA first, they appeared more acclimatised on subsequent umbilical EVA, indicating that a period of acclimatisation was preferable, rather than attempting complex and involved activities on the first exit. It was also made clear during Gemini that retaining equipment in the cabin with an open hatch [a spare glove on Gemini 4; a camera on Gemini 10] or within easy reach was a serious problem if it was not fastened down or restrained. The extensive use of lanyards on Gemini 11 and 12 helped reduce the loss of equipment.

CONCLUSIONS FROM THE PIONEERS

The results of Gemini (which also applied in part to the Voskhod EVA) led to conclusions applicable to future planning for operations in orbit and, to some degree, for planning activities on the Moon. The post-flight summary recorded:

- EVA in free space was feasible and useful for productive tasks, but only if there was enough attention given to supplying body restraints, sequencing each task carefully, controlling the work-load for each crew-member, and providing a proper and effective training programme with realistic simulations. *Future planning*: EVA should be considered for future missions 'where a specific need exists and where the activity will provide a significant contribution to science or manned spaceflight'.
- Restrictions in the mobility of the pressure suit were seriously limiting factors in tasks applied to Gemini EVA objectives and for the quick and safe return of Leonov on the first EVA. *Future planning*: Priority should be given to the improvement of suit mobility, in particular to the arm and glove area and, for future lunar surface applications, to the legs.
- The chest-mounted Gemini LSS performed satisfactorily, but caused encumbrance to the Pilots. Umbilicals provided an alternative but restricting option. *Future planning*: The use of gaseous cooling was undesirable for increased work-load, and alternative [liquid cooling] devices should be investigated for future EVA. In addition, back-pack supply of LSS would extend the range of the EVA astronaut, with necessary tether devices as required.
- Underwater simulation provides a high-fidelity duplication of the EVA environment, is effective for procedures development and crew training, and provides evidence that tasks performed underwater can also be easily

accomplished in orbit. *Future planning*: Underwater simulation should be used for procedures development and crew training for future [orbital] EVA objectives.

- Loose equipment *must* be restrained at all times to avoid loss, and body restraints are a requirement for extended EVA tasks and work-stations. *Future planning*: A range of tool and equipment restraints and aids should be produced to assist in a specific EVA task, and foot and waist retention devices should be incorporated in future EVA planning.
- A hand-held manoeuvring system appeared to be a promising mode of personal transportation, but evaluations were far too brief to define both the full capabilities and any limiting factors of this type of device. Umbilical orientation [on both Gemini and Voskhod] is not a suitable mode of crew-man orientation. *Future planning*: Further evaluation of orbital manoeuvring and mobility devices should be conducted in orbit.
- Gemini [and to a degree Voskhod] provided a foundation of technical and operational experience on which to base future EVA operations in future programmes.

Clearly, the EVAs completed during Voskhod and Gemini proved that EVA was possible. What was also revealed was that performing EVA was not as simple as the science fiction stories implied. More work was required in Earth orbit to master the technique, but in both the American and Russian programmes the next phase of EVA would not to be in open space, but on the Moon. The 1/6 gravity would help alleviate some of the difficulties encountered on Voskhod and Gemini, but would reveal other difficulties. The learning curve was becoming steeper.

REFERENCES

1 Siddiqi, Asif, *Challenge to Apollo*, NASA SP-4408 (2000), p. 333–337.
2 Shayler, David J. and Hall, Rex D., *Soyuz: A Universal Spacecraft*, Springer–Praxis, 2003.
3 Ertel, Ivan, and Morse, Mary, *The Apollo Spacecraft: A Chronology*, Vol. 1, NASA SP-4009, 1969, p. 19.
4 Shayler, David J., *Gemini: Steps to the Moon*, Springer–Praxis, 2001, pp. 269–273.
5 Siddiqi, Asif, *Challenge to Apollo*, NASA SP-4408 (2000), p. 446–447.
6 Abramov, Isaak P. and Skoog, Å. Ingemaar, *Russian Spacesuits*, Springer–Praxis, 2003, pp. 59–74. A detailed evolution of the development of Volga and its systems.
7 Shayler, David J. and Hall, Rex D., *The Rocket Men*, Springer–Praxis, 2001, pp. 236–251.
8 *Ibid.*, p. 243.
9 Gemini Program Mission Report, Gemini IV (U) MSC-G-R-65-3, June 1965, pp. 3–6.
10 Abramov, Isaak P. and Skoog, Å. Ingemaar, *Russian Spacesuits*, Springer–Praxis, 2003, pp. 67–69.

11 Shayler, David J. and Hall, Rex D., *The Rocket Men*, Springer–Praxis, 2001, pp. 236–251. A detailed description of Voskhod 2 EVA activities.
12 Leonov, Alexei, *I Walk in Space*, Malysh Publishers, Moscow, 1980.
13 Spacesuit Technological Developments, NASA General Working Paper No. 10-058, (Copy No. 11), 31 January 1966, prepared by Crew Systems Division, MSC, Houston, Texas.
14 Gemini IV Mission Report, MSC-G-R-65-3, June 1965, 7.1 Flight Crew Activities, EVA, pp 7-7–7-8.
15 Several Gemini astronauts wrote their own accounts of EVA. These accounts were used as reference sources in *Gemini: Steps to the Moon*, and are here listed in the Bibliography.
16 Gemini Summary Conference, 1–2 February 1967, NASA SP-138, MSC, Houston, Texas; also, Summary of Gemini EVA, NASA SP-149, MSC, 1967.

Tools of the trade

Throughout the history of EVA, astronauts and cosmonauts have used a variety of tools and specialised equipment. This tool-kit has modified and expanded as programme requirements have changed. In the early days, the basic tools were the spacesuit and the life support system, but there are numerous additional items of hardware designed specifically to assist the crew in their tasks outside their vehicles.

During the development of Gemini and Apollo in the early 1960s, a wide range of tools was designed for crew-members to make repairs to their spacecraft in orbit, including during a spacewalk. As with many aspects of the exciting new world of spaceflight, NASA devised a new language to describe the equipment and procedures – some with acronyms, and others with strange new words. So, a space hammer became the 'spammer', a combination of pliers and wrench became the 'plench', and a zero-reaction tool became a 'zert'. In order to prevent an astronaut spinning around as force is exerted on a wrench, a special tool was developed: a 'nab wrench' (nut and bolt), which allowed the bolt to be twisted without moving the astronaut.[1] Many of these tools developed into devices that are still used today (fortunately with more appropriate names), and much of the technology devised to help astronauts in orbit has found its way into the DIY stores and tool-kits found in homes around the world.

Towards the end of the Gemini programme, restraint and manoeuvring devices were devised to assist the EVA crew-member, and these have continued to be improved over the years. The camera is also a useful tool for EVA, and missions have included still cameras, movie cameras, television, and specialised photographic equipment. Since the 1980s a variety of manipulators and cranes have also been used to move the EVA crew and the equipment around the place of work. All of this equipment has become an integral part of a launch manifest, and as the space station era has progressed, an ever-expanding supply of EVA tools and equipment has been left by successive crews for the use of those to follow.

The previous chapters trace the development of EVA from the earliest theories of flight into space to the first EVAs, and recall the triumphs and challenges that those pioneering EVAs revealed to the crews, trainers, mission planners and hardware designers. The formative years of the mid-1960s saw early demonstrations of the ability to work in open space, but also the difficulty of accomplishing such work without adequate aids in restraint, mobility, training, support equipment and planning. This chapter reviews the options available for later EVAs, for mastering the task of spacewalking as the space programmes gathered experience and changed their requirements. This is by no means a comprehensive review, but is representative of the depth and range of often overlooked implements that assist the EVA crew. It also reveals the amount of development and ingenuity required to provide the EVA crew with the 'right tools for the right job', in order to ease the work-load.

EXPANDING THE CAPABILITIES

Although some equipment was planned for evaluation on Gemini (and to some extent Voskhod), it was the Apollo programme that really required EVA tools for exploring the surface of the Moon. Early tools were limited by the restrictions of the pressure suits, the length of the EVA, and the lack of surface mobility. Later, Apollo 14 was supported by the Modularised Equipment Transporter (MET), and from Apollo 15 the Lunar Roving Vehicle (LRV) provided greater range and capability in mission planning and surface activity.

The next phase of EVA work came with Skylab in 1973, and this initiated a long programme of space station support EVA that continued through the Russian Salyut and Mir programmes to the current International Space Station. These EVAs were designed to expand the capabilities of the station (space construction) and to prolong the life of the orbital base with repair, maintenance or expansion EVAs.

The advent of the Space Shuttle introduced a third element of spacewalking activity: space vehicle servicing, or repair on orbit. The Shuttle's ability to capture, support and redeploy payloads added to the range of EVA tasks, and teams of astronauts have retrieved and repaired, redeployed and serviced large space payloads in orbit, as well as returning some to Earth.

These three specialist areas of EVA work have required the inventiveness of countless teams of engineers, designers and testers to evaluate the best method of supporting the crew from the point of view of productivity, mobility and, of course, safety, and though many of these items have proven very useful, the designs occasionally do not work as advertised, and human ingenuity is called upon to solve the problem. Perhaps the best tools any EVA crew-member possesses are eyes, brain, a good pair of hands, and the ability to record experiences post-flight.

METHODS OF EXIT

There have been two methods of exiting the spacecraft to begin a spacewalk. Direct exit via a hatch system (Gemini and Apollo) during which the whole spacecraft crew compartment is decompressed for EVA, and via an air-lock system (Voskhod, Soyuz, Skylab, Salyut, Mir and the ISS) during which the rest of the spacecraft retains its pressurised environment whilst the EVA hatch is located in a smaller area that is depressurised for exit into space.

Hatches

The first Americans to leave their spacecraft in orbit did so through the right-hand Gemini crew hatch. This exposed both astronauts to the vacuum of space, and though only Gemini Pilots performed the actual EVAs, each Commander was technically performing an internal EVA on spacecraft LSS.

The next direct exit into space took place via the Apollo spacecraft. The Command Module (CM) hatch was first used for EVA during Apollo 9, when the CM Pilot stood up to place his upper body through the hatch to photograph the Lunar Module Pilot exiting from the Lunar Module (LM) forward hatch. The Commander did not exit the spacecraft, but was again exposed to the vacuum of space. The hatch on the front of the LM was also the only way to and from the lunar surface, down the ladder on the front leg of the LM. The upper hatch on the LM was used only once for an EVA. During Apollo 15 the Commander stood up to survey the landing site panorama shortly after landing, while the LM Pilot remained inside the LM crew compartment (but was again performing an internal EVA). During the final three Apollo missions, the CM Pilot performed the first deep-space EVAs to retrieve film canisters from the SIM-bay experiment area of the Service Module (SM). On these excursions, the LM Pilot remained in the hatch area on a stand-up EVA, while the Commander monitored the spacecraft systems from inside the spacecraft, again exposed to the vacuum of space and performing an internal EVA.

The first EVA at a space station (and the latest direct-exit EVA) was performed during Skylab 2, when the Pilot stood up in the CM hatch to try to release the station's stuck solar wing. During the attempt, the Science Pilot helped restrain his legs inside the CM while the Commander controlled the attitude and positioning of the Command and Service Module (CSM).

All subsequent EVAs were conducted via an EVA air-lock facility or module, but even here, opening the door to space necessitated the use of an outer hatch. On Skylab this was a left-over from the Gemini programme, and for the Russian Soyuz the side hatch of the Orbital Module (OM) (used to enter the spacecraft on the launch pad) doubles as an EVA hatch, although this has been used only once for that purpose, in 1969.

With the advent of the Salyut series of space stations an EVA hatch was incorporated into the design of the station, but was not available until the Salyut 6 programme from 1977. However, the initial EVA, for damage inspection, was actually carried out through the forward docking port, and not through the specially built EVA hatch. With the launch of Mir in 1986, early EVAs were carried out via the forward nodes, but the Kvant 2 module, added in 1989, included a specially designed air-lock facility with an enlarged outer EVA hatch for use with the Soviet MMU.

Before operating the hatch on an air-lock it is important to ensure that the internal atmosphere is below the safe operating level to allow the hatch to be opened safely and not be sprung open by evacuating air. This happened during EVAs from Mir in 1990, and damaged the hatch. On a later EVA, excessive pressure applied in attempts to open the hatch resulted in a broken latch tool, and the EVA was therefore cancelled.

The Shuttle has an air-lock facility that has proven very reliable, but on one mission (STS-80) a foreign object caused the EVA hatch to stick. It could not be opened, and so the EVA had to be cancelled. To date, there have been no reported difficulties with the hatch systems of the Pirs or Quest air-locks on the ISS.

Air-locks

In order to retain the atmosphere inside the main habitable section of the spacecraft, the air-lock separates the EVA team from the rest of the crew, who can remain in flight overalls inside the pressurised environment and monitor operations by TV, or visually through the spacecraft windows. Cosmonaut Alexei Leonov followed this procedure, using the inflatable air-lock facility on Voskhod 2 to perform the world's first EVA.

During the American Skylab space station programme, EVAs were accomplished via an air-lock module that could be isolated from the rest of the spacecraft, thus avoiding complete depressurisation and repressuration. This system was also incorporated in the Russian Salyut and Mir space stations, and has been incorporated into the ISS design with the attachment of the Quest air-lock and Pirs docking and air-lock facilities in 2001.

The Shuttle's air-lock facility can be located either inside the mid-deck area adjacent to the exit facility into the payload bay, outside in the payload bay to allow more room on the mid-deck, or as part of the Spacelab tunnel or Shuttle docking facility (see p. {?}).

SPACESUITS AND LIFE SUPPORT

In order to complete any period of activity outside the spacecraft, the spacewalker requires a pressure garment and a method of providing life support throughout the duration of the EVA, with a margin of contingency in case something should delay the entry back into the spacecraft.

Spacesuits

The story of the development of pressure garments for spaceflight, and in particular for EVA, is long and complicated. It has been reviewed in many publications, and is only briefly summarised here. For details of American suit development, Lillian D. Kozloski's *US Space Gear: Outfitting the Astronaut*[2] is recommended, and for the Russian suits the excellent *Russian Spacesuits*,[3] by I.P. Abramov and Å.I. Skoog, is an invaluable source of reference.

The suits that followed the two pioneering EVA suits were as follows:

Gemini Following the Gemini 4 EVA, the next EVA was planned for Gemini 8, for which the G4C suit was modified to include two neoprene-coated nylon layers in the EVA cover layer instead of the nylon felt and HT-1 nylon micrometeoroid layers, and to include integral thermal gloves to help prevent conductive heat transfer by external surfaces ($250°$ to $-150°$ F). Unfortunately, the EVA was cancelled due to early termination of the mission (see the author's *Gemini: Steps to the Moon.*). For Gemini 9 and the planned first flight of the Astronaut Manoeuvring Unit (AMU), extensive modifications to the cover layer of the suit were required. A stainless steel fabric was incorporated into the suit legs as a protection against AMU thruster firings. Eleven layers of aluminised H-film, sandwiched with fibreglass cloth, were also incorporated. The suit was not put to the test on Gemini 9 due to early termination of the EVA prior to detachment of the AMU from its storage location. The suits for Gemini 10 and Gemini 11 were very similar to those used on Gemini 8, but with the arms and legs of the underwear removed at the torso seams. A helmet visor anti-fogging kit of saturated wet wipes proved troublesome when, in flight, the crews' eyes became irritated. Refined locking tabs were added to the wrist disconnects, neck ring and pressure sealing zipper, whilst locking tab guards were added to the suit gas connectors to prevent inadvertent disconnection. For Gemini 12 the suit resembled that of Gemini 9, but was modified to remove the super-insulation layers when the AMU was removed from the flight.

Manned Orbiting Laboratory Development of USAF suits for the MOL evolved from high-altitude research in the 1950s, and plans for an IVA suit for the cancelled X-20 Dyna Soar. When the MOL was authorised, using Gemini spacecraft, the USAF opted not to purchase Gemini suits (although these were used for training), but instead developed a new suit design that featured components made from hard material instead of completely soft materials, but remained lightweight. Experience from fatigue on Gemini EVAs indicated the requirement for more studies on working in suits during EVA. The MOL could have provided that opportunity, but the programme was terminated in June 1969 before any firm manned launch dates were scheduled.

Apollo Each Apollo mission required fifteen pressure suits, each individually tailored for each of the six members of the prime and back-up crews. Each flight crew-member required a training suit and two flight-qualified suits, whilst the back-ups each had one flight suit and a training suit. With the attached Portable Life Support System, the suit weighed 81 kg on Earth and 13.6 kg on the Moon. The suit's prime contactor, International Latex Corporation (ILC), was chosen in November 1965, and developed a series of seven suit designs resulting in the suit designated A7L (Apollo suit design version 7 from ILC), which was used on missions from Apollo 7 and up to and including Apollo 13. It consisted of a total of twenty-one layers, and had twenty-two layers in the improved A7LB used for Apollo 14–17, with a modified version, without additional thermal protective layers, for non-EVA operations. The suit included a faecal waste-collection garment under a liquid-cooling garment, while the urine transfer assembly was worn over the coolant garment, which provided cooling by direct conduction, and reduced perspiration. It also included an integrated thermal and micrometeoroid garment, a Portable Life Support System in the back-pack, which contained coolant liquid, the communication radio systems, and an emergency oxygen supply. The polycarbonate helmet was bubble-shaped to provide all-round visibility, and a lunar EVA Sun-visor assembly

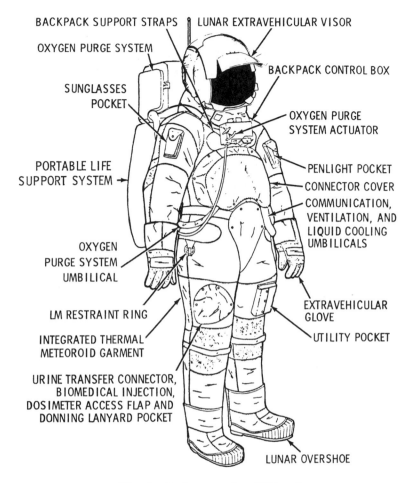

BACKPACK SUPPORT STRAPS   LUNAR EXTRAVEHICULAR VISOR

OXYGEN PURGE SYSTEM

BACKPACK CONTROL BOX

SUNGLASSES
POCKET

OXYGEN PURGE
SYSTEM ACTUATOR

PORTABLE LIFE
SUPPORT SYSTEM →

PENLIGHT POCKET

CONNECTOR COVER

COMMUNICATION,
VENTILATION, AND
LIQUID COOLING
UMBILICALS

OXYGEN
PURGE SYSTEM
UMBILICAL

LM RESTRAINT RING

EXTRAVEHICULAR
GLOVE

UTILITY POCKET

INTEGRATED THERMAL
METEOROID GARMENT

URINE TRANSFER CONNECTOR,
BIOMEDICAL INJECTION,
DOSIMETER ACCESS FLAP AND
DONNING LANYARD POCKET

LUNAR OVERSHOE

The Apollo lunar surface EVA suit.

was attached for use on EVA. Individually moulded IVA gloves were provided for each astronaut, and these were worn under the EVA gloves, attached by wrist connection rings. The unit was completed with lunar over-boots with moulded silicon rubber soles and additional thermal-layered soles that attached over the integrated pressure garment boot. Modifications to the suit for later Apollo missions included improved joints to ease mobility, and an improved back-pack to increase the amount of consumables stored inside. The suit was pressurised at 3.7 psi.

Skylab For Skylab, the astronauts wore a modified version of the Apollo A7LB design, but instead of a back-pack, umbilicals were used for connection between the spacecraft environmental control system and the suit's life support system. On relatively short EVAs in Earth orbit, astronaut's life support assembly was worn on the lower chest area, and was regulated at 3.7psi. This system supplied water, oxygen and electrical power as required. Additional mobility was incorporated into the

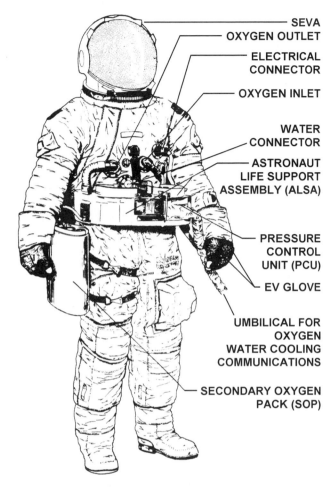

SEVA
OXYGEN OUTLET
ELECTRICAL
CONNECTOR
OXYGEN INLET

WATER
CONNECTOR
ASTRONAUT
LIFE SUPPORT
ASSEMBLY (ALSA)

PRESSURE
CONTROL
UNIT (PCU)
EV GLOVE

UMBILICAL FOR
OXYGEN
WATER COOLING
COMMUNICATIONS

SECONDARY OXYGEN
PACK (SOP)

The Skylab EMU, derived from the Apollo suit.

joints, including the neck and waist area to reduce fatigue, a modified thermal and micrometeoroid garment, and a Skylab EVA visor (SEVA) instead of the lunar version.

Early Russian EVA suits Since the 1950s the Russian company Zvezda has been the primary contractor for all Russian pressure garments, including IVA and EVA suits. In developing their pressure garment for the cancelled lunar programme, the Soviets followed a different route from the Americans by opting for a semi-rigid design over a fully soft suit that the Apollo design featured. The Berkut suit, used on Voskhod 2 for the first EVA, was thought for many years to be a development of a potential lunar suit, but it was a separate development during 1964 and 1965, designed to achieve the first EVA from Voskhod, and not associated with the lunar programme, although the experience was useful. The other type of soft suit was Yastreb, designed to test the EVA transfer method during Soyuz 4/5. This suit consisted of three layers, with a protective coverall and a self-contained life support system operating at 400 hPa, allowing for a 2.5-hour EVA. In 1961 the Soviets began the development of an 85-kg semi-rigid suit (glass-fibre-reinforced plastic) designated SKV (EVA space-suit), designed to be used in the Orbiting Heavy Satellite Space Station (OTSST). Because of the adoption of an integrated helmet and rear-entry door, the suit could be donned in five minutes, and the crew wore a cooling garment and communication cap before entering the hard upper torso and softer arm and leg components. All life support systems would be incorporated in the hinged door/back-pack. Suit regulation was controlled via a chest panel, and a handle controlled the opening and closing of the rear hatch once the cosmonaut was inside. The SKV was designed for operation at Earth-orbital altitudes from 450 km up to 36,000 km, including within Earth's radiation belts. Operating positive pressure was to have been 400 hPa, with the capability of an 8-hour EVA using onboard systems or a 4-hour EVA using the self-contained systems. This design did not progress past a number of working mock-ups, and was retired in 1965. It was modified for both the lunar programme and the space station programme, and remains the basic design for the Orlan suits used on the ISS four decades later. Only a small number of suits were built for these early models. For Berkut only nine models were built in the development and crew training phases, and only four flight models were made. For Yastreb, eighteen training and test suits were supplemented with six flight-qualified garments, and five SKV mock-ups were fabricated. Between 1967 and 1969, the Yastreb suit was also evaluated for use on Almaz space stations, until it was decided that the more advanced Orlan suit was more flexible for space station EVA operations.

Russian lunar suit designs Following the SKV experience, in 1966–67 the Krechet semi-rigid suit was used in the development of a lunar surface EVA suit. This was superseded by the Krechet-94, which was planned to be worn by cosmonauts on the Moon during N1/L3 missions. Its development lasted from 1967 to 1972, after which the lunar programme was suspended, and was finally cancelled in 1974. The 106-kg Krechet operated at a positive pressure of 400 hPa, and could support EVAs of up to ten hours. Its design and operation was based on SKV. In the development programme, three Krechet suits were manufactured by Zvezda, and to support the lunar surface programme a total of twenty-two test and training suits were supplied, with another nine in production at the time of the termination of the lunar programme in 1974. During the period 1966–70, a second lunar suit, Oriol, was evaluated. This featured a soft-suit design, with a back-pack similar to that used on Baikal during Soyuz 4/5. This multilayered suit had a mass of 56 kg (suit, 20 kg; back-pack, 36 kg), had 400-hPa positive pressure, and had an EVA operational limit of four hours. The design was evaluated with three test suits as a unit that could support both in-space and surface EVAs, but it was decided not to continue with its development in the lunar programme, although its technology was used in the development of pressure garments for rescue suits on Soyuz and Buran. The third Soviet lunar programme EVA suit, Orlan, was designed to be used in the N1/L3 programme. Orlan featured an integrated life support system and the semi-rigid design that was pioneered on SKV and used on Krechet, but was designed for use in

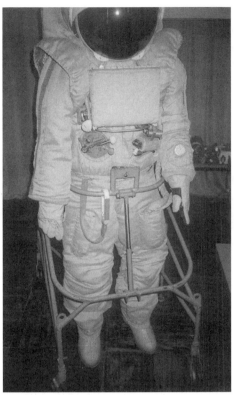

The Soviet Krechet lunar EVA suit. (Astro Info Service collection.)

lunar orbit and not on the surface. This one-size unit had adjustable features in the soft elements of the design to fit cosmonauts with a chest measurement of 96–108 cm and standing 164–178 cm, had a mass of 59 kg, and operated at 400 hPa. Development continued between 1967 and 1971, until its use was identified for the developing space station programme, but it was never used in its original lunar orbital role. A total of eleven development and training suits were manufactured, but no flight models were fabricated.

Salyut and Mir In 1969 it was decided to use the Orlan on the Salyut and Almaz stations, and there then followed eight years of development in upgrading the design from the original Orlan developed for lunar orbital EVA, to the new Orlan D for the (DOS) space station programme. It featured a regenerative closed-loop life support system with expendable consumables that could be exchanged and reserviced. The height was increased to 180 cm, but it was basically the same as the Orlan, with modifications to suit a longer useful lifetime of up to 3.5 years (in contrast with 2.5 years for Orlan), and a minimum number of EVAs increased from two of 2.5 hours to six of five hours each. The improvements increased the mass to 73.5 kg, and included provision for a 20-m electrical/communication/telemetric umbilical that doubled as a safety tether. This suit design was used on both Salyut 6 (1977–79) and

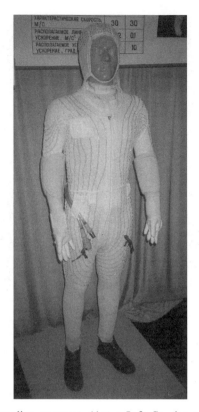

The Orlan cooling garment. (Astro Info Service collection.)

Salyut 7 (1982–84), and for the Salyut 7 units, upgrades allowed an increase of each unit to ten EVAs with a maximum of seven hours each, and a four-year operational lifetime. In total, twenty-seven test and training units were supplemented by seven flight models. For the final EVAs from Salyut 7 in 1985 and 1986, and for the first five EVAs from Mir between 1986 and 1988, an upgraded Orlan DM was introduced. This featured modified components, and a rearranged internal arrangement was used with increased capabilities and a mass of 88 kg, allowing up to ten EVAs per unit, each of eight hours. Only ten models were made, five of which were for test and training and five for flight. In 1988 another upgraded version, the Orlan DMA, was introduced, but unlike the earlier versions this unit did not need to rely on umbilical connections, which allowed EVAs over greater distances across the Mir station. The sixteen development and training models supported a flight production of twelve models that completed fifty-six two-man EVAs between 1988 and 1997. The 105-kg unit could support nine-hour EVAs, and the height adjustments allowed for cosmonauts with a 96–110-cm chest measurement and a height of 164–185 cm. The latest development of this series of suits is the Orland M, which incorporated lessons learned on Mir and the requirements for increased EVAs at the ISS. This unit was used in the final EVAs of

The Orlan EVA suit. (Astro Info Service collection.)

Mir between 1997 and 2000, and provided a useful evaluation of the suit before it was incorporated on Russian-based EVAs from the ISS. To allow larger astronauts to wear the suits, the size of this unit was increased to 96–112 cm chest measurement and a height of 164–190 cm. There was also provision to fit SAFER units to the Orlan M, and the installation of an overhead viewing port in the helmet to aid vision.

The Shuttle and the ISS The development of the Shuttle Extravehicular Mobility Unit (EMU), which features a hard upper torso and soft arm and leg units, commenced in the early 1970s, when the design of the Shuttle was finalised. The contract to Hamilton Standard was awarded in 1976. Instead of the individually tailored units, the Shuttle EMU reflected the multi-mission role and the increased number of astronauts who could perform an EVA. Because of NASA's decision to use the suit only if EVA was required, the crews had to pre-breath prior to donning the suit, and the training of the assigned Shuttle astronauts reflected specialist EVA training for only part of the crew. The suit is carried on every Shuttle mission, whether or not EVA is planned, thus allowing for contingency EVA. At least two EMU units, and sometimes three or four, are flown, depending on mission requirements. The mass of the original EMU was 117 kg, and it features the basic

Astronaut Mike Foale prepares to enter the rear hatch of the Russian Orlan EVA suit during training for Mir EVA operations.

pressure garment in the form of a hard upper torso with a separate leg assembly attached by a locking seal ring at the waist. The Portable Life Support System allows for 6–7-hour nominal EVAs, and despite the early planning documents suggestion that the nominal six-hour EVA would probably not be exceeded, improvements and upgrades to the system and procedures have resulted in several EVAs of 6–7 hours. Since 1985, all gloves have been custom-made for each astronaut. The mass of the eleven-layer suit is approximately 47 kg, with the LSS at 67 kg, and the operating pressure is 29.6 kPa. The suits have been used since 1981, with the first EVA being completed in 1983. Over the years the design has been improved and upgraded to encompass new technology and experience of flight operations, but for current use on Shuttle EVAs and American EVAs at the ISS it essentially remains unchanged.

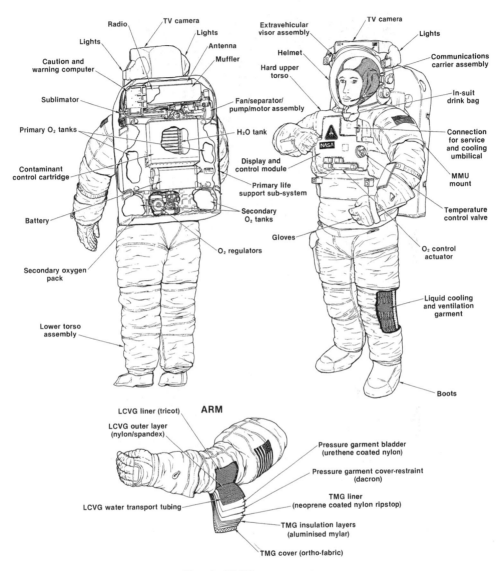

Radio — TV camera — Lights — Antenna — Muffler — Lights — Caution and warning computer — Sublimator — Primary O₂ tanks — Contaminant control cartridge — Battery — Secondary oxygen pack — Lower torso assembly — Fan/separator/pump/motor assembly — H₂O tank — Display and control module — Primary life support sub-system — Secondary O₂ tanks — O₂ regulators — Extravehicular visor assembly — Helmet — Hard upper torso — TV camera — Lights — Communications carrier assembly — In-suit drink bag — NASA — Connection for service and cooling umbilical — MMU mount — Gloves — Temperature control valve — O₂ control actuator — Liquid cooling and ventilation garment — Boots

LCVG liner (tricot) — **ARM** — LCVG outer layer (nylon/spandex) — Pressure garment bladder (urethene coated nylon) — Pressure garment cover-restraint (dacron) — TMG liner (neoprene coated nylon ripstop) — LCVG water transport tubing — TMG insulation layers (aluminised mylar) — TMG cover (ortho-fabric)

Shuttle EMU components.

Umbilicals

Umbilicals are used to provide life support supplied by the spacecraft consumables and/or for the supply of electrical, telemetric and communication connections. They can also double as safety tethers, but seriously restrict the EVA astronaut's traverse distance.

Back-packs

The back-pack provides an EVA astronaut with greater mobility, but is limited by the amount of consumables available inside the unit. The back-pack has occasionally

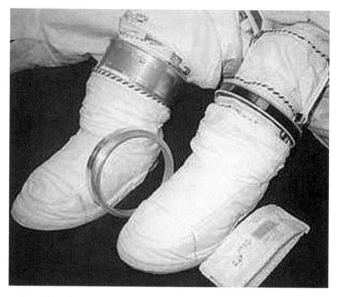

Sizing rings and small adjustments in the Shuttle EMU allowed more astronauts to use the same suits rather than the earlier customised suits.

been supported by umbilicals. The use of back-packs has allowed surface exploration of the Moon, and untethered EVAs using manoeuvring units. This self-contained design has allowed EVAs to be carried out at Mir and the ISS, where the EVA path can be cluttered by spacecraft appendages. The lack of umbilicals has also assisted in the Shuttle servicing of satellites and the Hubble Space Telescope (HST).

MANOEUVRING DEVICES

The notion that a space explorer, while outside the spacecraft, might require a propulsion system in addition to tethers and magnetic boots, was recognised in the early twentieth-century writings of spaceflight pioneers. Plans for so-called 'jet-packs' were first assigned to the Gemini programme in the mid-1960s.

Gemini
In the original mission planning for Gemini, the astronauts were to evaluate a Hand Held Manoeuvring Unit (HHMU) on Gemini 4, 8, 10 and 11, and the USAF Astronaut Manoeuvring Unit (AMU) on Gemini 9 and 12. In the event, only the HHMU was demonstrated in space on Gemini 4 and 10. Due to problems with other systems, evaluation of the manoeuvring devices was not completed on Gemini 8, 10 and 11, and because of difficulties during the Gemini 9 EVA the AMU was not fully evaluated. It was removed from Gemini 12 due to a renewed emphasis on the evaluation of body restraints before the end of the programme.

Science fiction stories of the 1920s often featured the notion of manoeuvring devices for crews in open space. Such devices attracted the attention of the USAF, which was evaluating potential military uses of space, including the inspection, resupply, servicing and repair of large space objects and satellites during EVA. By December 1960 a report entitled *Self Manoeuvring for the Orbital Worker* was published by the USAF Aerospace Medical Division.[4] In this paper, authors John C. Simmons and Melvin S. Gardner summarised the efforts of the USAF in three areas of study: a discussion of the expected tasks and requirements of such devices; the mathematics of EVA dynamics when moving through space; and discussion of current (1960) developments.

These studies also included the results of the first rudimentary tests of early linear propulsive devices, including the Air Jet Propulsion Unit Mark I. This was a hand-held single-nozzle air-gun, supplied by six oxygen bottles mounted on a frame and worn on the back. It was tested on an air-bearing table and in a C-131B aircraft flying parabolic curves. Both tests demonstrated that such a unit was useable, but its thrust – only 1.36 kg – was inadequate for manoeuvring. However, it led to the 6.8–7.71-kg-thrust Mark II, and these studies, together with those completed at NASA, evolved into the HHMU planned for use on Gemini.

Self-Contained Hand Held Manoeuvring Gun (Gemini 4)
This was a 3.4-kg device that had evolved from earlier concepts and the availability of already qualified components helped speed the development of the device. Engineers used the emergency oxygen bottles from early Gemini ejector seat designs to provide the 4,000-psi twin oxygen storage tanks, while the pressure regulator came from the Mercury environmental control system. The design had to conform to Gemini mission constraints, in that it had to be stowed inside the spacecraft crew compartment, which presented a safety issue. The specifications stated that if the system leaked it should not create a hazardous environment for the crew. Therefore, gaseous oxygen (0.31 kg) was selected. To ensure the that unit fitted in the already cramped cabin, it was produced in two sections: the hand assembly section, and the high-pressure section. The astronaut joined the two sections with a connection coupling at the regulator, and by inserting a pin adjacent to the pusher nozzle. The tractor arms were also folded for ease of storage, and were deployed prior to use. From the storage tanks, the gas flowed through a manifold to a shut-off and fill valve. As the valve opened, the pressure of the gas was reduced to 120 psi. The low-pressure oxygen then passed through the handle of the device and through a filter and then to two valves. The rearmost valve directed the gas through the trigger guard to the pusher nozzle, while the forward valve passed the gas through a swivel joint, out the two extension arms, and to the tractor nozzles. Both the pusher and tractor nozzles were activated the hand trigger. To use the device, the astronaut squeezed on the trigger while pointing the nozzles in the required direction of firing. The gas flow to the nozzles required an initial force of 6.8 kg to the poppet valve, and increased to the maximum 9 kg. The thrust level increased from 0 to 0.9 kg. The storage bottles held only 0.31 kg of gaseous oxygen, which allowed a total impulse of 18.1 kg/s, or an increase in velocity of 1.82 m/s.

Back-pack supplied HHMU (Gemini 8)

The propellant used in this system was changed to Freon-14, stored at 5,000 psi in a [439 cu. inch] 7,199.6-cm^3 storage tank in the EVA support package mounted on the astronaut's back. This also housed a second 3.17-kg tank of life support oxygen. This system gave an increase of total impulse to 272.16 kg/s. Despite its lower specific impulse (33.4 sec as opposed to 59 sec for oxygen and 63 sec for nitrogen), the density of Freon-14 was almost three times as great, which produced a significant increase in total impulse from only a 4.98-kg weight penalty. The calculation was that Freon-14 provided a 45% increase in total impulse over oxygen at the same tank pressure. The HHMU was designed to operate in a similar way to that on Gemini 4, but the expansion from 500 psi to a regulated pressure of 110 psi resulted in temperatures of 150 F in the handle assembly. Initial tests revealed that the poppet valve stuck in the open position when the Freon-14 was activated. To solve this, Teflon cryogenic seals replaced the elastomer seals. After this, the valves operated satisfactorily in subsequent testing, but as a precaution, two shut-off valves were incorporated into the system. In the upstream of the coupling, one valve was designed to shut off the gas supply in the event of poppet-valve closure failure, while the second was located in the flexible feed line of the back-pack upstream, which shut off the gas flow in the event of a leak in the hose. The handle of the HHMU had a better grip than the previous model. These modifications were incorporated to prevent the possibility of uncontrolled thrust from the device, which could result in severe tumbling and loss of control by the astronaut during EVA. Unfortunately, this system was not operated on the flight due to other in-flight system failures that led to early termination of the mission.

Umbilical-supplied HHMU (Gemini 10 and Gemini 11)

On Gemini 10, the propellant for the HHMU was stored in two 11,150.6-cm^3 tanks located in the adapter section of the spacecraft. The twin hose arrangement in the 17.4-m dual umbilical provided life support oxygen to the astronaut and nitrogen to the HHMU. Nitrogen was used in this system because further development of components would have been required to use Freon-14, whereas the system was already qualified for oxygen and nitrogen. Aluminium tubing was used to route the nitrogen from the storage tank to a recessed panel behind the hatch area. To provide heat in order to warm the gas, these tubes were clamped to the spacecraft at several points to absorb heat from internal components. The recessed panel also featured a quick disconnect and shut-off valve assembly that connected the umbilical nitrogen line to the nitrogen supply. On this mission the HHMU was improved to allow easier movement of the Pilot's hand when moving between pusher and tractor activation. This was achieved by sloping the handle to accommodate the palm contours of the EVA glove. Instead of a rocker switch, Gemini 10 was fitted with two short trigger switches, pivoted at the end. This also reduced acceleration forces of 6.8–9.07 kg on the earlier design to 2.26–3.62 kg on Gemini 10 and Gemini 11. It also resulted in less hand movement between the pusher and tractor positions.

During Gemini 11, the HHMU was stowed in the spacecraft adapter section instead of in the cabin. A quick disconnect coupling replaced the former screw-on coupling, and simplified the connection of the hand device to the umbilical, since this was planned to be accomplished one-handed by the Pilot, wearing a pressure suit and in a limited area. The propellant stage tank supply was identical to that on Gemini 10, except for the use of a 9.14-m umbilical instead of the 17.4-m umbilical. Unfortunately, this device was not evaluated during Gemini 11 due to other EVA difficulties.

USAF Astronaut Manoeuvring Unit (Gemini 9, and removed from Gemini 12)
While the USAF evaluated the HHMU system for linear movement, it also evaluated a separate device – Skyhook – for rotational movement. This device had a mass of 7.25 kg, and featured a weighted aluminium wheel inside a housing, and two operator handles. By torquing the spinning wheel, the subject rotated around a perpendicular axis. The airborne zero-g flights had revealed difficulties in separating the movements created by using the Skyhook, from those actually created by flying the aircraft in the parabolic loops. To alleviate this, higher thrust levels were introduced for the Air Jet Propulsion Unit Mark II, and this led to higher thrust levels in the later USAF AMU under development for Gemini. Moreover, the AMU recorded the highest thrust levels of any subsequent MMU.

The Gemini AMU allowed an astronaut to manoeuvre in space, independent of the Gemini spacecraft systems. The hardware was developed under a USAF experiment designated D012.[5] The prime contractor to the Air Force was Ling-Temco-Vought, which had been evaluating in-house EVA propulsion studies since 1959. The same year, confident of its work in that area, the company approached the USAF with an unsolicited proposal for an AMU, and two years later received a contract for an AMU feasibility study from the Research and Technology Division of the USAF Propulsion Laboratory. There then followed development of a nitrogen-powered engineering test bed, which was flight-evaluated onboard a KC-135 during a parabolic flight programme in the summer of 1962. These tests proved successful, and in May 1964, LTV won the Gemini AMU contract and authority to build three flight units. Two of these would fly on Gemini 9 and Gemini 12, and the third would be used as a back-up unit. The project officer was Ed Givens, who was later selected by NASA as a Group 5 astronaut in April 1966. Tragically, in June 1967 he was killed in an automobile accident while off duty, and he never flew in space.

The unit required a significant amount of effort and training to prepare it for evaluation, but due to in-flight difficulties on Gemini 9 and removal from Gemini 12 it was never flight-evaluated in orbit. The AMU featured an aluminium back-pack shell structure, with a form-fitting cradle in which the astronaut positioned his back to use the unit, plus two folding side arm controllers and folding nozzle extensions. Its mass was 75.29 kg fully laden, and it measured 81.2 × 55.8 × 48.2 cm. The supply tanks determined the size of the structure, and the thrusters were located around the corners of the unit to provide adequate controlling forces and maintain the centre of gravity when in use. There were six major sub-systems included in the design:

Propulsion Provided by 10.88 kg of hydrogen peroxide at 1,360.8–1,587.6 kg/s, and fed by a nitrogen supply, with pressure against a bladder in the propellant tank separating the two gases. Supply to the thrusters was controlled by manual valves that also controlled electrical power to each thruster's valve. Two manual control valves were fitted – one for the primary and one for the alternative system – and there were twelve thrust chambers of a nominal 1.04 kg thrust, and sixteen solenoid-actuated control valves. In each control system there were eight thrusters – two each for forward, aft, up and down transition. Forward and aft thrusters operated in pairs for moving forward and aft and control in pitch and yaw, while up and down thrusters were available for vertical movement and roll control. The alternative system used different thrusters for forward and aft firing, but the same thrusters for up and down movement and separate control valves for roll, pitch and yaw. Outboard vents and relief valves were provided for safety in the nitrogen and hydrogen peroxide lines.

Flight control Two arm controls held the flight control system. This provided manual or automatic three-axis attitude control and stabilisation, and manual-only translation in two axes. Two redundant systems were also provided. The left-hand control was for translation commands, while the right-hand control was used for attitude control. Also installed on the left controller was a voice communications volume control, a vox disable switch that prevented keying of voice-operated switches, and the selection switch to choose either the automatic or manual system. Immediately prior to Gemini 9, thermal shields were added on each arm for protection against plume heat, because the addition of extra thermal covers to the gloves was unworkable in an already restricting pressurised EVA glove. During manual control mode, acceleration could be achieved, but direct translation was not possible, due in part to the offset centre of mass. In automatic mode, however, it was possible, because the level of acceleration was approximately halved due to control limitations on the thrusters. Built into the jet select system was a priority firing system in both the forward and aft thrusters. This gave first priority to yaw (left or right), second priority to pitch (up or down), and third priority to translation (forward or backwards). Manual rotational mode inputs produced gave 11°/s roll, 13°/s pitch, and 25°/s yaw. Again, pure manual rotation was not possible due to the same limitations on the centre of mass. In automatic mode the astronaut could command acceleration to the stated levels, but when the rate of about 18°/s in pitch and yaw or 26°/s in roll was achieved, the acceleration would have stopped, and further input from the astronaut would be required. When released, the controller was designed to return to the neutral position, and the rate would slow down. When rotation stopped, the system moved to the attitude control hold mode, and was capable of maintaining attitude to within $\pm 2.4°$ in each of the three axes.

Oxygen supply A total of 3.31 kg of oxygen at a pressure of 7,500 psi was stored in a supply tank. The oxygen supply system was routed via a heat exchanger for initial heating, then via a pressure regulator, and then to a thermostatically controlled heater. These temperatures and pressures were closely regulated. The minimum level of oxygen that could be delivered to the ECLSS was 2.31 kg at 97 ± 10 psi at temperatures of $65 \pm 10°$ F. The flow rate was designed for 2.26 ± 0.09 kg/hr, up to a peak of 3.81 kg/hr. In planning for Gemini 12, the AMU would have had the thermostatic heater switch replaced by a manual switch.

Power supply Enclosed in a sealed unit, two silver–zinc batteries (one ± 28.5 V and one ± 16.5 V) provided electrical power to the AMU systems. Two such units were installed for redundancy, with independent distribution sub-systems. The 16.5-V batteries supplied the control system electronics, telemetry signal conditioners, thruster valves and gyros, while the 28.5-V batteries supplied the voice and telemetry transmitters, telemetry multiplexer encoder, telemetry signal conditioner, warning lights, tone generator, oxygen heater and position light. The system also included provision to prevent a short circuit, in one battery draining the other, and isolation of the batteries from the AMU systems, with a main power switch activated by the Pilot during pre-donning procedures. Installation of the batteries in the AMU was among the last operations prior to mating to the second stage of the launch vehicle, so that later access would be impossible without demating the spacecraft.

Alarm To provide the Pilot and the Command Pilot with warnings of certain critical out-of-tolerance parameters, an audible warning system was linked to both crew-members' headsets, with a warning light on the ECLSS chest-pack display panel. The four warning lights indicated out-of-tolerance readings for the oxygen pressure and temperatures, the hydrogen peroxide pressure, the fuel pressures, RCS excessive usage in the automatic mode, and decreases in voltage indicating imminent loss of control.

Communications This included a telemetry and voice system. The telemetry reported certain back-pack and biomedical data to the spacecraft, for storage in the onboard recorders and post-flight evaluation. UHF voice transmission utilised the standard Gemini microphone and earpiece in the helmet of the spacesuit. While the AMU was stored in the aft of the Gemini spacecraft, a limited number of parameters were automatically transmitted to the ground through the spacecraft transmission system. In addition, the hydrogen peroxide pressure and temperature were displayed inside the spacecraft cabin.

Interfaces and provisions To attach the AMU to the spacecraft, four claw assembly structures were attached to the back-pack, and tension bolts pressed the claws down onto a sheet-metal structure on the blast-shield door on the adapter module. An electronically operated pyrotechnic guillotine would have severed the connections after donning and power up. Other connections consisted of simple pull-away electrical connectors. To assist in donning the unit, a foot rail, two handlebars, an umbilical guide and two floodlights for night-side operations were installed on Gemini 9. These were deployed when the thermal cover was released.

Unfortunately, although the Gemini 9 Pilot reached the AMU and achieved most of the steps prior to release, due to persistent fogging of his suit faceplate he could not separate, and so the exercise was cancelled. The unit was destroyed upon re-entry at the end of the mission, and because of its location on the adapter module it could not be recovered for reuse. The Gemini 12 unit was removed from the mission – partly as a result of the difficulties which arose during the Gemini 9–11 EVAs. It did not fly again.

Soviet AMU (Voskhod/Almaz)

With designs for heavy orbital stations on the drawing boards in the early 1960s, Soviet designers also evaluated manoeuvring and transportation systems for cosmonauts outside of these space stations. From 1961 the design of a suit and manoeuvring unit was integral to these station EVA studies. In defining the specifications of a cosmonaut manoeuvring system, the equipment – the Cosmonaut Transference and Manoeuvring Unit (*Russian*, UPMK) – was to provide a suited cosmonaut with the capability of undertaking an EVA and detaching himself from the station's surface to repair a payload, for examination and inspection of the exterior surfaces of the station, to transport loads, and to perform rescue operations.[6]

The engineers at Zvezda decided that in order to be compatible with the spacesuit and its limited field of vision, the unit should incorporate controls similar to those with which the cosmonauts were accustomed on the spacecraft. Like the Americans, they recognised the problems of mass, dimensions, fuel supply, and rescue capabilities, but unlike the Gemini unit it was planned that this unit should be stored inside the spacecraft. It therefore had to be small enough to move through the spacecraft hatches, and safe enough to be stored inside the habitable volume of the station. Work on the theoretical studies was completed by Zvezda and Scientific

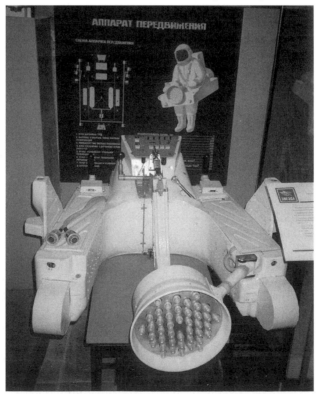

The early Soviet MMU for Voskhod and Almaz. (Astro Info Service collection.)

Research Institute 2 (NII-2) during 1962 and 1963, and for the purpose of flight dynamic studies, an air-jet mock-up was developed and test-flown on a Tu-104 flying laboratory during 1964. Following a review in 1965, work continued to develop the unit at a faster pace.

The early design of the unit suggested a theoretical translation from the spacecraft of 50–100 m. The twelve engines would use asymmetric dimethyl hydrazine, with a nitric acid oxidiser and iodic additive. Compressed nitrogen was used to pressurise the tanks. Wearing the SKV hard suit, the cosmonaut had two hand-operated controllers for attitude control and translation. In front of the cosmonaut, four jets were positioned for aft thrusting, and at the rear there were four for forward thrusting. At each side, two canted thrusters provided up, down or roll capability. Combination firings would enable pitch or yaw movement. The total mass of this unit was 65 kg, with a fuel mass of 6 kg. It provided a total impulse of 15,000 N/s for an experimental velocity of 50 m/s. There was also a redundant back-up system with a 2-kg fuel reserve and an impulse of 5,000 N/s.

A directive issued on 27 July 1965 ordered two additional Voskhod spacecraft with soft air-locks and use of the soft Yastreb pressure suit. Some reports have indicated that such a mission on Voskhod (with an EVA and a demonstration of the UPMK) could have been tentatively assigned to 'Voskhod 6', but there is no evidence to indicate that any cosmonaut trained for this experiment.[7] By December 1966 the Voskhod series had been terminated, with only two missions flying with crews. However, in a directive issued on 28 December 1966, work continued on the UPMK – this time in support of the Almaz military space station programme, with plans for cosmonauts to wear the Yastreb pressure garment and RVR-1 back-pack.

In 1966 two ground test units were manufactured for evaluation at NII-2, and a 1-degree-of-freedom rotational test facility was fabricated at Zvezda. Work on developing the unit for use with the Yastreb suit for Almaz continued into 1968, when approval for an EVA demonstration was given. With the decision to use the unit on the space station, a TsKBM-developed control system and winch was designed and fit-and-function tests conducted in the orbital module of the Almaz spacecraft. By 1968, the design of the UPMK had evolved.

The new design featured two propulsion units in a horseshoe-shaped unit that surrounded the cosmonaut – again featuring two integrated control handles. Both the forward-facing and aft-facing thruster units consisted of forty-two individual solid propellant rocket micromotors, providing translation forwards and backwards. A system of fourteen air thrusters provided a six-degrees-of-freedom translation for angular and linear movement. Each of the thrust lines of the solid propellant motors was aligned with the centre of mass of the suited cosmonaut and UPMK unit. The unit could be initiated from the control panel, and could impart a forward or backward rate of 0.2 m/s. Impulse from the single solid motor was given as 45 N/s, while the thrust from the air-jet engines was 2.5 N for eight engines and 5 N for six engines. The total impulse for the unit was 4,000 N/s. The mass of this version had grown to 90 kg, while its speed was quoted as being 32 m/s.

Although tested in ground and airborne tests, this unit never reached space. In late 1969, the Yastreb suit was replaced by Orlan for all EVAs from Almaz stations (and later Salyut stations), and although work on the UPMK continued for a short time, the project had no specific objectives, and was terminated. It would be twenty years before cosmonauts had the opportunity to test the next Soviet MMU in orbit.

Manoeuvring Unit M509 (Skylab)

In the late 1950s, a young engineer named Ed Whitsett was looking for a topic for his Masters thesis at Auburn University. When Russia launched Sputnik he found his topic, but not his field of interest. However, working in the aeronautical medicine laboratories of the college, he became interested in how future human explorers of space might function. His studies included early computer models using geometric cones, cylinders and spheres to represent body parts. When propulsive systems were added, changes in the centre of gravity were displayed. When the centre for American human spaceflight was located at Houston in the early 1960s, Whitsett (by then in the USAF) began requesting that he be sent there. By June 1965 he was working in flight control for Gemini 4, and after the problems on later Gemini EVA missions his interest once more returned to his thesis studies and the engineering aspects of walking in space. This work led directly to his assignment as Principle Investigator of M509, the Skylab Manned Manoeuvring Unit (MMU).[8] As a development unit for a later model to be tested in space, the Skylab unit featured a few short cuts, as it would only be flown inside the Orbital Workshop (OWS) in an enclosed pressurised environment.

On 18 November 1966, during the monthly Apollo Applications Program meeting, a summary of NASA plans for using the Department of Defense AMU, as flown on Gemini 9, was discussed. George Mueller stated that the AMU could be incorporated into planning for the AAP missions without compromising the programme's objectives. Depending on USAF agreement, the unit could be carried on missions AAP 210, 211 and 212. The prime contractor for the USAF unit, Ling-Temco-Vought, was already working with Apollo CSM prime contactor North American and S-IVB prime contractor McDonnell Douglas, with a view to adapting the unit to fly on AAP missions. No formal request had been received from the USAF to assign the experiment – which would cost an estimated $2.5–3 million – to the AAP.[9]

Work continued on the development of an MMU for use in the AAP during 1967–69, when Martin Marietta became involved in developing the unit, with Whitsett as the Principle Investigator.

The objective of experiment M509 was to 'conduct an in-orbit verification of the utility of various manoeuvring techniques to assist astronauts in performing tasks which are representative of future extravehicular activity requirements.'[10] The M509 experiment consisted of several manoeuvring systems combined into a single test article. It included a back-pack with arm controls, a hand-held manoeuvring device, a spare rechargeable battery, a battery charger, a telemetry receiver, three rechargeable cold gas (nitrogen) propellant bottles, and support racks for stowing the unit and gas bottles. The back-pack was a welded aluminium frame covered with sheet aluminium, and included a nitrogen supply, battery and electrical power distribution, instrumentation and telemetry, a rate gyro-controlled automatic attitude control, a control moment gyro-controlled automatic attitude control system, and control and display sub-systems.

Structure The unit was 68.58 cm wide, 105.41 cm high, and 38.83 cm deep, and with the arms extended it was 121.92 cm deep). Its mass was 115.66 kg.

Propulsion This was provided by nitrogen gas stored at 3,000 psi in three spherical 24,600-cm^3 pressure tanks. The system also included a combination regulator/relief valve assembly, a solenoid emergency shut-off valve, a manifold, and a network of fourteen solenoid-operated fixed-position thrusters, any one of which could produce a thrust of 0.45–1.82 kg.

Instrumentation and telemetry This consisted of sensors, a transmit antenna, a data module that provided multiplexing, analogue-to-digital conversion, and radio transmission via a pair of antennae mounted on the interior workshop walls. The receiver collected twenty-nine analogue and fifty-four bi-level measurements from the data, and converted it for storage on a tape recorder for in-flight dumping to ground stations. Biomedical data was also recorded during pressurised suit runs, and was down-linked to the ground together with voice communications and post-flight evaluation. Additional documented experiment data was recorded through in-flight TV, post-flight still picture and motion picture media, and logbook entries for post-flight evaluation.

Stabilisation The rate gyro was a miniature three-axis integrating gyro, which sent rates of attitude to the control electronics and on to the thrusters. When data was input into the rotation controllers, the gyro system responded by firing the required thrusters in proportional rates to the input on the controller. If no command was entered, the gyros maintained the astronaut in a 'relatively' stable mode. The control moment gyro had a mass of 22.68 kg, and consisted of six gyros arranged in 'scissored pairs'. Each 7.62-cm gyro spun at 22,000 rpm, and provided 'rock solid attitude' to resist external disturbances. Again, input into the hand-operated controller resulted in torquing of the gyros to produce a rotational rate proportional to controller input.

Controls and displays Two hand-operated controller handles were available on arms, which could be lowered when not in use. There were also power switches, a mode selector, a voltage meter, a tank pressure meter, and control moment gyro status lights. On the left-hand arm was the translation controller – movement forwards, backwards, up, down, left and right – as well as switches, meters and indicator lights. The right arm housed the rotational controller for control in pitch, yaw and roll.

Operation There were four modes of movement available to the astronaut wearing the M509. *Direct.* The astronaut moved the left-hand controller in the direction he wished to travel, resulting in a normal velocity of 0.09 m/s up to a maximum velocity of 0.6 m/s (approximately the rate of a slow walk in 1 g). Rotation of the right-hand controller changed attitude – either pitch up or down, or roll left or right – at a maximum of 15°/sec. At all times, the astronaut needed to use visual cues to direct or stop his movement, as no systems were incorporated into the unit. *Control moment gyro*. The control moment gyros provided the torque to change the attitude rate in response to astronaut input on the rotational controller. In a neutral position, the gyros automatically reduced body attitude rates to zero. Attitude control was achieved by momentum exchange between the back-pack and the gyros rather than by firing thrusts as in the rate gyro mode (see below). The use of translation thrusters was exactly the same as in the direct mode. *Rate gyro*. Operation of the translation thrusters was very similar to that used for the direct mode. However, with the rotational hand-controller in neutral, the rate gyro system fired the proper thrusters automatically to stop movement. *HHMU*. This was similar to the unit used during Gemini. For Skylab it featured two 0.68-kg tractor (puller) thrusters and a single 1.36-kg pusher thruster. Nitrogen from the back-pack was fed to the unit through a flexible hose. The astronaut controlled the thrust with a throttle valve and a selector valve (pusher or puller thrusters). Control was achieved visually by estimating the relationship between the astronaut's own centre of mass and the line of thrust from the unit.

Foot Controlled Manoeuvring Unit (Skylab)
The idea for the Foot Controlled Manoeuvring Unit (experiment T020) to free the astronauts' hands for other tasks evolved from the Hillyer flying platform, which was investigated for exploration on the Moon during the AAP. In early 1965 it was developed by John D. Bird of NASA Langley Research Center, and in January 1967 it was evaluated for the AAP. Several astronauts evaluated an engineering model of 'jet shoes' on the air-bearing facility at the Manned Spacecraft Center, in

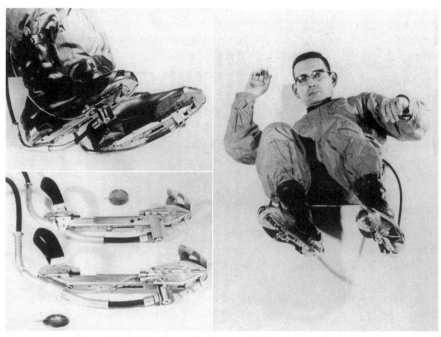

An early design for a foot-controlled manoeuvring unit.

cooperation with the Principle Investigator. Each of the astronauts tested the unit in shirtsleeves, but a test pilot from Langley evaluated the unit while wearing a full pressure suit. The tests proved unsatisfactory. This was a small unit with a thruster mounted under each foot, with the controls mounted under the toe area and orientated so that with the legs and feet in a comfortable position the thrust vector passed close to the wearer's centre of mass. On 6 April 1967, Deke Slayton, Director of Flight Crew Operations, requested these be removed from the experiment programme. Slayton suggested that their use while wearing a full life support unit would reduce visibility of the shoes, which posed a safety issue for the Astronaut Office.[11] The experimental nature of the unit caused some concern, even though it would be used only inside the OWS and not in open space.

Development continued, however, and it flew on Skylab as experiment T020. It consisted of a bicycle-like framework and saddle seat, with a back-pack that held the propellant and control equipment. Attached to the framework was a set of restraining straps, a camera for documentation of direction of flight, and a set of four nozzle thrusters (eight in total) assembled under each foot and controlled by foot-actuated switches. The framework was aluminium, with shoulder straps and a 38-cm diameter (1,500 inch^3) 24,600 cm^3 pressure tank of nitrogen gas at 3,000 psi attached to the back-pack, together with a battery supplying power. Both the tank and the battery were taken from experiment M509. Only 4° of freedom was possible. It could rotate around the three axes of roll, pitch and yaw, but directionally it could move in only one axis – either forward/backward or left/ right – but not forward and up or other combinations at the same time. The thruster assembly mounted outboard of each foot controlled the four nozzles, the top or bottom each producing 4.464 N of thrust, and the fore and aft each producing 1,339 N of thrust. Pedal combinations produced the following movements:

Foot movement	*Result*
Both feet up	Translation head first
Both feet down	Translation feet first
Right and left toes up	Pitch up
Right and left toes down	Pitch down
Right toes up, left toes down	Yaw left
Right toes down, left toes up	Yaw right
Right foot up, left foot down	Roll left
Right foot down, left foot up	Roll right

The Principle Investigator of this experiment was D.E. Hewes, of Langley Research Center. The unit was constructed by the Denver Division of Martin Marietta, and was assigned to the Skylab 3 and Skylab 4 missions. The thrusters were mounted on stirrup foot holders, and were attached to either the main frame or the saddle of the main unit. The back frame simulated the approximate mass of a full life support system.

MMU (STS-41B, 41C and 51A)

Following the use of the M509 unit inside Skylab its development continued at Martin Marietta, where it evolved into the MMU used on Shuttle missions in 1984. This device received a higher priority in 1980, when the MMU was evaluated as a possible aid in an EVA for tile repairs on early Shuttle flights. Although this soon subsided when tile repair was deemed unnecessary, the MMU engineers had quickly assembled the first flight units using equipment salvaged from other programmes, including spare parts left over from the Viking Mars-landing craft. Although not employed for tile repairs during the 1980s, the repair concept re-emerged following the loss of *Columbia* in 2003. The MMU was also evaluated for satellite inspection and repair (and was utilised in this mode in 1984), and as a space rescue system and an aid for space station construction and inspection (although none of these materialised).

The data produced by the M509 flights helped to establish the design criteria for the MMU for the Shuttle. The primary difference was that the new unit would be operated outside the habitable compartments in the vacuum of space. The MMU was modular in design, allowing attachment of the EMU to the unit. There were no telemetry requirements, as the EMU providing all voice communications during the EVA. The MMU was stored in flight support units in the forward end of the orbital payload bay, and could be put on or taken off by one pressure-suited astronaut during EVA. It had the capability of supporting a six-hour EVA.

Martin Marietta designed, built and tested three units at its plant at Denver and at JSC, and by 1984 the value of the contract was estimated to be in excess of $50 million. Flight unit 1 was the MMU/Flight Support Station qualification unit, completed in February 1981,[12] with thermal vacuum testing completed in March. The unit was then moved to JSC for a programme of manned thermal vacuum testing in May 1981. The reworking and upgrading of systems was completed during March 1982, and static load testing of the FSS was carried out during the following month. The electrical and mechanical connections were tested in May, and acoustic tests and random vibration and transition load tests were carried out in July. By December 1983 a re-test of the transient launch and landing loads had been completed, and during most of 1984 the first unit filled a mission support role for the three missions on which the flight units were used: STS-41B, 41C and 51A. It then became a baseline for future systems upgrades from 1985, and fulfilled the contract through the 1985 calendar year.

Meanwhile, work continued on flight units 2 and 3 and their support structure, intended for flight operations. In August 1983, both flight units completed pre-delivery system acceptance tests. JSC manned thermal vacuum testing for quality of workmanship was completed in October 1983, and by December, in-orbiter functional testing had been completed, thereby qualifying the units for integration as payload for their first mission. Flight unit 2 was installed in its FSS on the starboard side of *Challenger*, and unit 3 was installed on the port side.

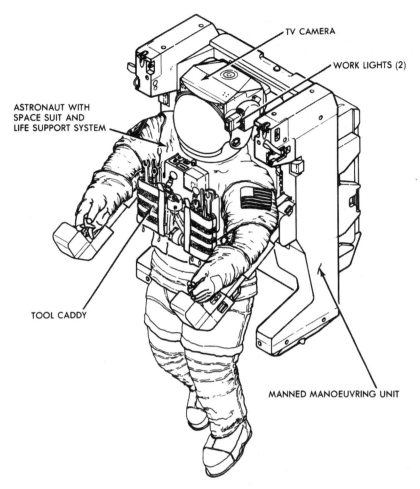

The American Manned Manoeuvring Unit.

Structure The significant feature of the MMU was that it was a self-contained back-pack propulsive device which allowed the astronaut to venture several hundred metres from the orbiter. Measuring 125.4 cm high, 82.7 cm wide and (with both controller arms extended) 120.9 cm deep, it was fabricated from aluminium, and featured two side towers, two deployable controller arms and a central structure. The towers were the support locations for some of the thrusters, crew-member displays and the MMU retention latches, while the central section housed two silver–zinc 16.8-V batteries, circuit breakers, the control electrical assembly and associated circuitry, two nitrogen tanks, and all the plumbing associated with propulsion lines and fittings. The total laden weight – including 11.8 kg of nitrogen propellant – was 153 kg, and its operational mass – loaded with propellant, plus the astronaut and the MMU – was 290–346 kg, depending on the mass of the astronaut. (For example, Joe Allen (STS-51A), who weighed 56.7 kg, was one of the lightest male astronauts, while James van Hoften (STS-41C) – also known as 'Ox' – weighed 90.72 kg, and was one of the largest.) The two nitrogen tanks, measuring 79 cm in length and 25.4 cm in diameter, were fabricated from aluminium with a Kevlar filament over-wrap. Each was filled with 5.9 kg of nitrogen pressurised 3,000 psi, providing each jet with 7.6 N of thrust. Attachments for ancillary equipment, such as cameras, were available on the two towers.

Propulsion The MMU featured twenty-four thrusters, affording a six-degrees-of-freedom control mode (+X, +Y, +Z, +roll, +pitch, +yaw). The thrusters – twelve on each tower – were powered by gaseous nitrogen from the two storage tanks in the rear central compartment of the unit. They were laid out with four thrusters on each face – top, bottom, forward, rear, left and right sides – and the hand-operated controllers could be used to fire the thrusters opposite to the required direction of movement, or in combination to pitch, yaw or roll. Under normal operation the two nitrogen tanks fed one set of twelve thrusters, but in the event of failure a cross-over valve would have ensured a constant supply of propellant from both tanks. In the event, operation of the MMUs in space was nominal. These same valves were also used to recharge the tanks between EVAs. Each MMU tower also featured a pressure gauge for Pilot monitoring of the propellant supply.

Electronics and electrical power Organisation of the electrical and propulsive sub-systems into redundant component sets created a fail-safe in the design, with two electrical power systems. In the normal mode of operation both sets were used continuously, but in the event of a failure the option was available to shut down the faulty set and either continue working or return to the orbiter on the remaining set. The Control Electronics Assembly (CEA) for the MMU consisted of three gyros (one for each of the rotational axes), control logic, thruster-set logic, and motor-driven thruster-valve drive amplifiers. Input from the hand-controllers, or the gyros during attitude hold, could be accepted by the CEA. The power system of the unit consisted of the two silver–zinc batteries combined with a power distribution system linking the circuit breaker switches and relays. The batteries fed connections in the CEA to provide power for the CEA and hand-controllers. Externally, there were three locator lights on the MMU to aid visual location of the MMU Pilot at a distance from the orbiter. Two further indicator lights on moveable stalks could be positioned in the view of the operating astronaut to indicate and confirm thruster operation.

Thermal control To assist in the reflection of heat, all surfaces were painted white, except those where surface wear and certain closely monitored thermal proprieties were required. On the rear panel and upper pressure vessel shroud, silvered Teflon was used to provide radiator surfaces for controlled electronic heat rejection. Interior surfaces were specially coated to prevent corrosion, while the CEA, batteries and other components requiring heat rejection had a high-emissive surface. Inboard surfaces of the rear panel and the battery box were also coated white. Many of the components of the propulsion system were fabricated from stainless steel or aluminium, with low-emissive surfaces. Heaters were used both for orbital storage and for EVA operations.

When exposed in space the surfaces of the MMU approached –118° C, but when in storage the MMU required a temperature no lower than about –23° C. During EVA operations, surface temperatures could reach as low as –84° C, although most of the components had an operating temperature of above –51° C. These heaters had to be thermally isolated from the structure wherever possible, by bonding the heaters directly to components. Only the battery storage heaters were mounted on the compartment itself. Each of the two batteries had one heater system for EVA free-flight phases, and one for orbital storage. When the MMU was in free flight the heaters were powered internally, but when in storage they were supplied by orbiter power routed through connections in the FSS.

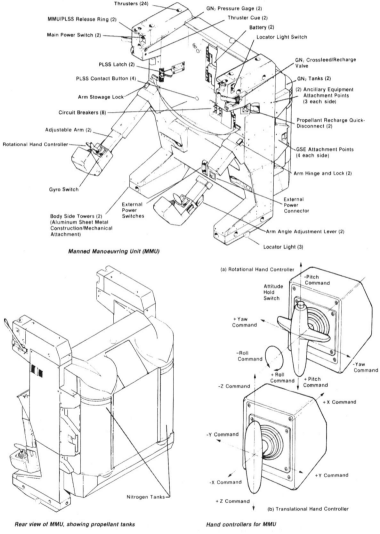

Features of the American MMU.

Operation A suited EVA crew-man would exit the air-lock and reverse into the MMU in its FSS. The inboard surfaces of the towers supported the EMU Portable Life Support System, with latches holding the unit firmly in place. These included two independent and manually activated latches with guide ramps and Portable Life Support System contact points. These latches attached to receptacles on the Portable Life Support System. To remove the MMU the astronauts pulled the manual release rings in the front of both towers, releasing the latches and freeing the MMU. Two levers attached the unit to the FSS and were released to free the unit once the astronaut had secured himself to it. To 'fly' the unit, the astronaut used the hand-

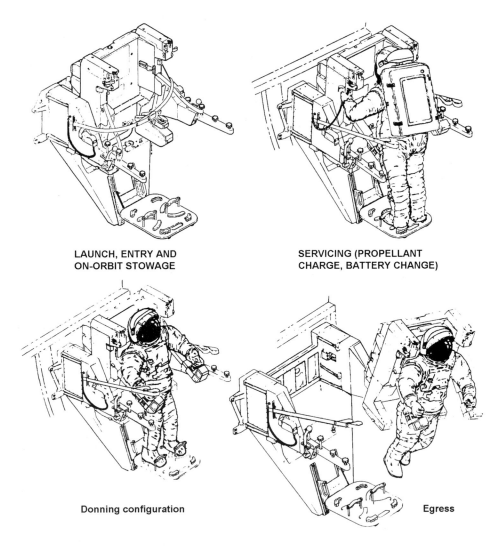

LAUNCH, ENTRY AND
ON-ORBIT STOWAGE

SERVICING (PROPELLANT
CHARGE, BATTERY CHANGE)

Donning configuration

Egress

The donning of the American MMU.

controllers to fire the MMU thrusters. The controller on the right governed the pitch, yaw and roll, while that on the left was used for forward and reverse, up and down, and left and right. On top of the right-hand controller was a button to activate attitude hold for the three rotational axes, allowing the astronaut to maintain angular position relative to another object.

From the Skylab M509 unit, designers determined that the most comfortable translation velocity was generally 1% of the initial separation distance in feet per second (1 fps for 100-foot initial separation; 0.3 m/s for 30-m separation). The astronauts were trained to accelerate to this velocity, hold the rate, and then gradually decelerate when approaching the desired target vehicle or object. Depending on the mass of the astronaut flying the vehicle, translation acceleration ranged between 0.09 and 0.11 m/s, with a total velocity difference of 20.1–25 m/s. During rotational acceleration the rate could vary between 7°/s for 'the heaviest system' (95 percentile male) to 9°.5/s for the 'lightest system' (5 percentile female – although no female ever flew the unit).

The whole unit was designed to conserve propellant by consuming negligible amounts in the attitude control mode and even in the most extreme cases of offset centre of mass of the astronaut. The minimal impulse from the thrusters and the select logic were designed to ensure attitude hold over a long period of time without seriously impinging on overall propellant consumption. Attitude hold using the automatic system allowed the astronaut to free one hand for other tasks while still fully controlling the attitude of the MMU. The orbital velocity of the MMU could not exceed that of the orbiter – an additional safety feature alongside the redundant system design throughout the unit and the decision to execute EVAs only in pairs, with a second unit available for use. It was also a flight rule that the MMU remain in visual contact range of the flight deck windows of the orbiter at all times, and venturing below the orbiter payload bay sill line and underneath the orbiter was not allowed. In the unlikely event of complete MMU failure, or inadvertent use of full thrust, the Shuttle orbiter could be rolled to 'rescue' the MMU astronaut by 'scooping' him into the payload bay, where, assisted by the second crew-man, he could be quickly disconnected from the faulty unit. Alternatively, the RMS could be used to ensure the retrieval of the stricken astronaut. Fortunately, none of the MMU operations required such a manoeuvre.

Flight Support Station Located in the forward end of the payload bay close to the forward bulkhead either side of the EVA hatch, the twin flight support stations housed the MMUs during periods of inactivity. It was also possible to recharge the batteries and propellant through these units. Adjustable foot-restraints (usually set prior to launch) could be set to the proper height for the astronauts intending to use the unit. The nitrogen recharging system was connected to the orbiter's nitrogen supply, with toggle valves on the FSS and on the flight deck used to control the supply. Electrical connections were used to supply orbiter heater power to the MMUs, and also to gather temperature-monitoring data from seven temperature sensors. The nitrogen recharge system was also used to sever gas-actuated nuts that secured the MMU to the FSS during ascent.

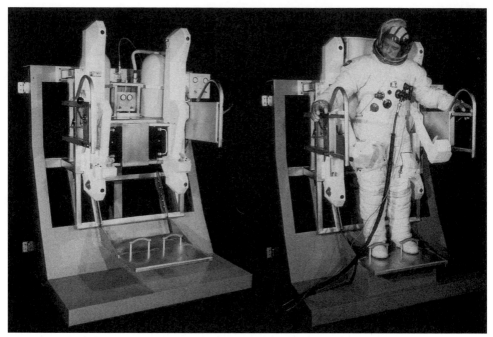

The Flight Support Station.

Soviet MMU (Mir and Soyuz TM-8)

It was not unit 1990 that the Soviets finally demonstrated their Cosmonaut Transference and Manoeuvring Unit (UPMK, or 21KS), during the fifth resident crew visit to Mir. The original idea was to use the unit for Mir and the Buran space shuttle, and on 22 March 1984 it was decided to proceed with studies into the device. Contractors were specified during 1985, and by 1986 development of the unit expanded when the demonstration device was tested and construction of several units began.

Structure The size and configuration of the unit was similar to the American MMU (although it appears that the Russians never entirely revealed all the specifications), and it was capable of an independent operating time of up to six hours. It had a total mass of up to 400 kg and a velocity of 30 m/s, with a $\pm 2°$/sec attitude stabilisation rate. The mass was quoted as being 180 kg maximum. The maximum safest separation distance from the spacecraft was given as 100 m from Buran (which presumably indicated a rescue scenario similar to that of the Shuttle/MMU), or 60 m from the Mir space complex, where any rescue would have to have been undertaken by a second crew-member. According to a recent source,[13] the 21KS/UPMK unit could be used for up to fifteen EVAs.

The method of attaching the suited cosmonaut to the unit was very different from that of the American system. On the front of the Orlan DMA EVA suit there was provision for the attachment of restraint devices to a rigid frame that was jointed and fixed to the UPMK. This frame split into two uneven 'rods', and rotated outwards to provide an opening for the crew-member to position himself in front of the unit. To

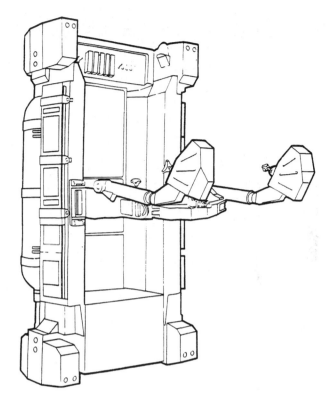

The Soviet Cosmonaut Transference and Manoeuvring Unit.

attach to the unit, the cosmonaut first fixed the locking device on the front of his suit to a central lock and a side rod, and then, moving back towards the UPMK, the other arm of the front rod came into position to lock him in place, with the back-pack firmly attached. This facility required no assistance from a second crew-member, and allowed for in-flight maintenance and donning and doffing of the UPMK inside the pressurised Kvant 2 air-lock. Access to and from the hatch at the back of the Orlan suit was made possible by swinging the suit attached to the front rod clear of the back-pack, but with it remaining mechanically linked to it.

As with the American MMU, the control arms could be moved to two positions – either lowered for storing inside the spacecraft or during close proximity work on EVA, or raised for piloting the unit. The left-hand arm carried the rotational controls, while the right-hand arm housed the translation controls.

Propulsion Two sets of sixteen thrusters, each with 28 litres of pressurised air at 32 mPa, and each of 0.5-kg thrust, arranged in quads of four at each corner of the unit, operating each set separately and offering redundancy. A total of six hours of operation was possible.

Electronic and electrical power Two silver–zinc batteries (a primary unit and a back-up unit). Approximately a hundred telemetry signals were transmitted to Earth on the operating performance of the unit, with a system of audio cues to the Pilot.

Thermal control Sub-systems were arranged inside a housing covered with vacuum thermal insulation.

Operation Similar to the American Pilot-controlled orientation and movement of the unit by means of the two control arms, with selection available in semi-automatic and automatic modes.

Flight Support Station A flight model of the UPMK was delivered to Mir in the Kvant 2 module with its onboard test and support equipment. Following the EVA demonstrations in February 1990, the unit was relocated to a support frame on the exterior of the hatch area of Kvant 2, where it remained unused until the deorbiting of Mir in 2001.

SAFER (Space Shuttle and ISS)

In order to address the need for a small, simple, low-cost EVA manoeuvring device for self-rescue during EVA, during the early 1990s the Automation and Robotics Division at NASA JSC developed the Simplified Aid For Extravehicular Activity Rescue (SAFER). This was much lighter and more basic than the MMU (which was designed for specific EVA tasks), and the idea was to attach the SAFER unit to the back-pack of each EVA astronaut involved in untethered activities during Shuttle and space station operations. It would be available in the case of an emergency during which the Shuttle or another crew-member would not be able to rescue them. The unit also has no in-built back-up systems, as it is intended only as a back-up rescue system and not as a primary propulsion device. Because of the guidelines to ensure that the SAFER would be as light and cost effective as possible, flight-proven

The SAFER unit mounted on the back of the astronaut's life support system.

'off-the-shelf' hardware would be employed in its design wherever possible. Modified spare MMU components, small commercial components, a modified Apollo translation hand controller and small alkaline batteries were therefore all integrated into the unit.

Initial tests were conducted under Detailed Test Objective (DTO) 661, assigned to STS-64 in September 1994, to help establish a common set of requirements for the SAFER in both the Shuttle and space station programmes, to validate the system on an early Shuttle flight and develop a flight demonstration test of SAFER, and to use the results to develop a production phase of SAFER units in time for the ISS. By 1994 and the first test flight, development costs had exceeded $2.18 million, with the majority of funding diverted from the Shuttle and space station budgets. The unit is currently used on EMU ISS missions, and has also been adapted to fit on Russian Orlan M units.

Structure SAFER is a small self-contained propulsive back-pack capable of attitude control and free-flying mobility for a single EVA astronaut in emergency situations. In order to minimise restrictions to mobility and to allow access to confined areas on the station or a crowded payload bay on the Shuttle (as in the case of the MMU), the SAFER is designed to fit around the EMU Portable Life Support System back-pack. Weighing just 29.48 kg (excluding its stowage interface), the unit is folded and stored in the crew air-lock stowage bag during launch and landing.

Propulsion Twenty-four fixed-position thrusters expel nitrogen gas at a thrust of 0.36 kg each. A nitrogen storage capacity of 1.36 kg (3,600 psi maximum), in four steel tanks in the unit, produces a change in velocity of 4.53 m/s for the operator. The nitrogen can be recharged from onboard supplies on the orbiter, and the nitrogen from the tanks to the thrusters is isolated by means of a manual switch rather than the motorised valve of the MMU.

Control and operation Control of the unit by the attitude control system includes an automatic attitude hold system (with a six-degrees-of-freedom control) via the single hand-controller mounted on the suit display and control module on the chest area of the Shuttle EMU. The force of thrust is 0.36 ± 0.01 kg, and acceleration is 0.06 ± 0.01 m/s, or $10 \pm 4°$/s. For the operator there is a sixteen-character LCD display and automatic attitude control and thruster LEDs.

Electrical power A set of twenty-four 9-V alkaline batteries is included in a battery pack (28 V), replaceable on orbit. The electronics are digital, with supporting software.

Thermal control There are no component heaters, and the units are white to reflect heat.

Recharge station The SAFER recharge station (SRS) is mounted in the forward part of the payload bay, using the attachments for the existing MMU recharging interface with the orbiter gaseous nitrogen system. For STS-64, which first demonstrated this facility, a cart storage assembly from STS-37 was modified with a saving of $130,000, compared with the design and fabrication of a new unit from scratch. It was capable (on STS-64) of performing seven recharges of the unit, and supplying up to 9.42 kg of nitrogen.

RESTRAINTS AND SUPPORT

Gemini

Various tethers and umblical combinations were employed during Gemini EVAs. These were used to provided linkages to the spacecraft and to support fluid and electrical lines, as well as to limit the distance which the EVA Pilot could venture from the hatch. The pure umbilical was again used for structural strength and for attachment to the spacecraft, and for carrying the oxygen supply line as well as electronics for voice communications and biomedical data telemetry. The 7.6-m umbilical flown on Gemini 4, 8, 9 and 12 was the original umbilical developed for Gemini, and based on the experiences on Gemini 4 the 9.1-m and 15.2-m umbilicals were developed. The 15.2-m line on Gemini 10 and the 9.1-m line on Gemini 11 also carried the nitrogen supply line for the HHMU. The 22.8-m line on Gemini 8 was to have supported only structural strength and electrical leads to the EVA support package.

Space Shuttle restraints and harnesses

Tethers During the early phase of the Shuttle programme, two 0.6-m waist tethers and two 36-cm wrist tethers were provided for each EMU, for safe tethering of the astronaut and equipment during EVA. These were designed for use when wearing full pressure gloves, and had a push-button locking device to prevent inadvertent opening.

Portable Foot-Restraint (PFR) This 13.8-kg device was originally designed for restraining the EVA astronaut in the payload bay of the Shuttle during contingency activities. The whole device consists of two telescoping booms (from 1.73 m to 2.44 m), a centre-line clamp, an extension arm, and a foot-restraint platform incorporating pitch and roll adjustment capability. The extension boom is locked into either the centre-line clamp or the fitting on either end of the telescoping booms, to provide restraint for work on any standard hand-rail. The foot-restraint platform is fitted to the extension arm, which in turn is fitted into the clamp and can be moved and locked along the telescoping booms. If required, a female fitting, designed to accept the extension arm. could be incorporated into the design of a payload to provide a payload work-site restraint using the Manipulator Foot-Restraint.

Manipulator Foot-Restraint (MFR) This is used to provide access to remote EVA work-sites using the remote manipulator systems. An RMS grapple fixture is used to attach the device to the robot arm, which is controlled from the aft flight deck of the Shuttle. A foot-restraint is provided on the platform, as well as hand-holds and locations for stowing additional equipment and tools to be taken to the required work-site.

Manipulators and booms

In EVAs performed around the Shuttle or space stations, the astronauts have been assisted in their activities by remote manipulators and booms which extend their reach and provide a stable working platform during long and complicated EVAs.

Remote Manipulator System (RMS)

One of the most successful and reliable systems of the Shuttle has been the Canadian-built Remote Manipulator System (Canadarm), flying for the first time on STS-2 in November 1981, and since then regularly used in support of EVA operations from the Shuttle. The arm is normally stowed on the port (left) longeron, and although it could be stowed on the right this has never been employed. It is also possible to carry two arms; but again this has never been utilised, as they can be operated only one at a time from the single set of software and using the single set of displays and controls. For RMS operations the fifth orbiter computer is used, as the system takes up 32% of the CPU, or 30% for manually augmented operations.

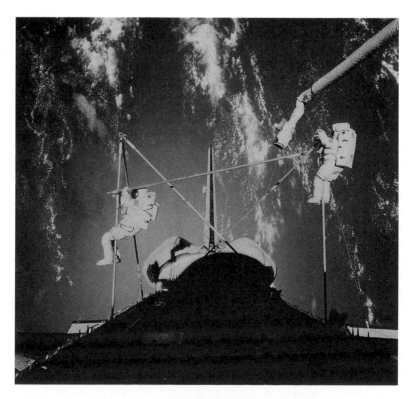

Using the Shuttle RMS for EVA support.

The RMS has a length of 15 m 7.62 cm and a diameter of 38.1 cm, and is capable of 6° of freedom. The system weighs 450 kg (410 kg being the mass of the arm), and due to the capacity of the arm it can lift and stow a 29,484-kg payload with a diameter of 4.5 m and a length of 18 m. When the payload bay doors are open, the arm rolls 31°.36 towards the doors for operational use, and then back 31°.36 towards the payload bay for stowage. The components are identified by their similarity to the human arm – shoulder (two joints), upper arm, elbow (one joint), lower arm, and wrist (three joints) – and include joint housings, electronic housing, the booms, and joints. The elbow is limited to a rotation of 160°.

The RMS can be fitted with the standard end effecter, which uses three wire snares to capture and release a target by rotating an inner cage assembly around the payload-mounted standard grapple fixture. This can also be used to support Manipulator Foot-Restraints, allowing the tethered astronauts on EVA to be 'carried' towards a payload to be serviced (such as the HST), or to translate the astronaut/payload (such as with ISS hardware transfer) to a target area, which provides a firm base platform upon which the astronaut can work while utilising both hands and speeding translation to the target area.

The RMS is controlled from the aft flight deck, by using the aft and overhead windows, and from the closed-circuit TV system, using both the arm and payload-bay TV cameras. These views are displayed on two TV monitors, each with split screen capability, on the aft flight deck. Passive thermal control of the arm consists of multilayered insulation blankets and thermal coatings, while the active system consists of twenty-six heaters on the arm, supplying 520 W of power (28 V dc). With redundant power buses, if there is a power failure in one heater then the other can supply the full heating requirement. The system automatically maintains the joint temperatures above –25° C, and the heaters are switched off as the temperature reaches 0° C. Temperature readings from twelve thermostats on the arm are displayed on the aft flight deck station.

Each arm joint is electronically driven, with a normal loaded movement rate of up to 0.06 m/s (0.6 m/s unloaded) and a controlled rate of movement to within 0.009 m/s and 0°.09/s. The arm can operate in any one of five modes: automatic, manual augmented, manual single jointed drive, direct drive, and manual back-up drive. Selection depends on the type of role about to be performed. Use of the RMS during EVA normally features manual modes, with the primary RMS operator (or back-up) at the controls, and other members of the flight crew conducting support (to the EVA coordinator), or visual roles such as planned photodocumentation.

The first flight of the RMS and an unladen demonstration took place during STS-2 in November 1982. This was followed by further laden and unladen evaluation during STS-3, 4 and 8, with deployment and retrieval of a small pallet satellite on STS-7. The RMS in EVA mode was first used during STS-41B in February 1984, and since then it has been utilised on most Shuttle-based EVAs. The use of the RMS during the retrieval, repair and redeployment of satellites, and during construction of the ISS, has demonstrated its valuable role as an EVA tool.

Part of the ISS RMS system, also known as Canadarm 2.

Space Station Mobile Servicing System (SSMSS)

Due to the size and scope of the ISS, a new robotic manipulator system had to be devised and installed on the station, for use in conjunction with the Shuttle RMS during construction and during periods of orbital operations when the Shuttle was not present. The experience gained by the Canadians from using their RMS on the Shuttle programme gave them the lead in the experience, operations and technology required to develop the station RMS, and it formed part of Canada's contribution to the international orbital construction project. The Canadian RMS (Canadarm 2) is but one element of a complex RMS system designed to support assembly and maintenance tasks, moving equipment and supplies around the station, supporting EVA crews, and servicing scientific instruments and other payloads attached to the station.

Canadarm 2 Termed a 'bigger, better and smarter' next-generation version of the Canadarm used on Shuttle, this manipulator was installed on the ISS during STS-100 in April 2001. It features seven motorised joints, and when fully extended it measures 17.6 m. It is used to assist in handling large payloads between the station and the Shuttle (and can also assist in Shuttle docking if necessary), and to support EVA work, and is controlled from a console inside the Destiny laboratory. Its mass is approximately 1,800 kg, and it can handle payloads of up to 116,000 kg. It requires 2,000 W of peak operational power, and a minimum 'keep alive' average power of 435 W. In motion it requires 0.6 m to stop when carrying a maximum load. This RMS is 'smart' in that it can self-relocate across the station by using a Latching End Effecter (LEE) to attach itself to complementary ports located across the exterior of the station. Under the control of an operator inside the station it can essentially 'walk' across the station from one location, and is limited only by the number and location of the Power Data Grapple Fixtures (PDGF) on the station, which accept one of the two LEEs and supply power, data and video links. Using the Mobile Base System (MBS), the RMS will be able to traverse the entire length of the ISS using an 'inch-worm' end-over-end movement. The design again resembles a human arm, with a shoulder (three joints), elbow (one joint) and wrist (three joints), but can change configuration without moving its 'hands'. With full joint rotation, the seven joints can rotate 540° – far beyond the capability of human arms.

Mobile Base System (MBS) This was delivered to ISS on STS-111 in June 2002, and features a work platform that moves along rails across the length of the station, supporting the Canadarm 2 as it traverses the main truss of the station. It measures 5.7 m × 4.5 m × 2.9 m, has a mass of 1,450 kg, and can transport or handle a payload of up to 20,900 kg. The operational peak power requirement is 825W, and an average power of 365 W is required to keep the system operating.

Special Purpose Dextorous Manipulator (SPDM) This is the final element of the MSS onboard the ISS. This element is manifested for delivery on an ISS Shuttle flight some time after 2005. The SPDM is also known as Canada Hand, and is a smaller twin-armed robot designed to take over some of the delicate EVA tasks normally completed by astronauts during EVA. It will alleviate a great deal of the maintenance EVA load during the life of the ISS. The unit – commonly referred to as Dextre – will be equipped with twin arms, lights, video equipment, a tool platform, and four tool-holders.

Japanese Experiment Module Remote Manipulator System (JEMRMS)
This is the next robotic arm system that will be installed on the ISS, and will be located at the Japanese Pressurised Module. It will help support experiments being conducted on the JEM (known as Kibo), or for support maintenance tasks in that area of the station, thus further alleviating the astronauts' EVA work-load. Tests of a Manipulator Flight Demonstrator took place during STS-87 in August 1987, and basic technology demonstration experiments were conducted on the Japanese Engineering Test Satellite V-7 (Orihime-Hikoboshi), launched in November 1997. The control console for the arm will be launched after 2006, together with the pressurised section of the experiment logistics module onboard ISS-1J/A. The main arm will be launched a few months later on ISS-1J, and approximately a year later the 2J/A mission will deliver the small fine arm as part of the exposed experiment logistic module section. Both elements of the arm will have 6° of freedom, with the main arm measuring 9.9 m and the small fine arm nearly 1.7 m. The total payload-handling mass of the main arm is 7,000 kg, and the small fine arm has a capacity of 300 kg.

European Robotic Arm (ERA)
The European Robotic Arm is a cooperative venture between the European Space Agency (ESA) and Rosaviakosmos (RAKA) in Russia.[14] The original idea for the European RMS was derived from the European mini-shuttle Hermes during the 1980s. When that was cancelled, the studies were adapted to fly the arm on the

proposed Mir 2 space complex. When Russia joined the ISS project in 1993, cooperation with ESA continued, and the ERA became part of the Russian segment in July 1996. The primary European contractor is Fokker Space, of The Netherlands, funded by ESA. with the Russian industrial partner RSC-Energiya, under contract from RAKA. The setback for the arm was that it was to have been fitted to the Russian Science and Power Platform (SPP) during 2002 or 2003, but after much delay in development and delivery, it seems improbable that the SPP will be launched. The ERA was to have been launched as part of a Shuttle payload, and may still be launched for use across the Russian elements in support of payload and hardware handling and EVA maintenance tasks.

The arm is 11.3 m long, and has a mass of 630 kg. The 'inch-worm'-type function of the arm is similar to that on Canadarm 2, allowing the jointed arm to attach its end-effecter 'hands' to base-plates as it 'walks' across the station elements. With seven joints – of which six can operate at the same time – it will bring an added dimension to the Russian segment of operations.

Russian devices

During the 1990s, support crane devices were installed on Mir to support EVA operations. A pair of 14-m, 45-kg telescopic cranes, called Strela (Arrow), were

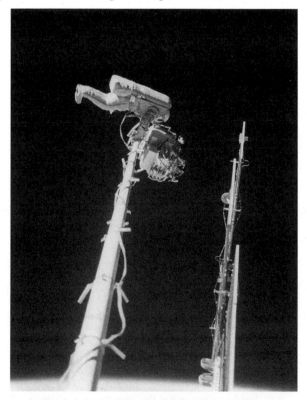

A cosmonaut rides the Strela arm on Mir.

installed on either side of the Mir base block during EVAs conducted in January 1991. A 2-m storage container housed the structure, which was anchored to the support that originally held the shroud covering the forward docking node of Mir during its launch. Each of the booms could support 700 kg, and they were used to move some of the large solar arrays and take cosmonauts from one point on the station to another. As it was fixed to the base block launch shrouds it was limited in reach, and it was also hand-operated. A cosmonaut would turn a pair of hand-cranks to move the arm above or below the base block. By rotating it in an outside arch away from the station, it could reach behind Kvant 1 at the aft end of the complex. The cosmonauts stored the Strela against the side of the Kvant 2 module, which housed the air-lock at its far end. In this way they translated hand-over-hand down the Strela to the area where the crank handles were located, and then went back up its structure to the air-lock hatch at the end of the EVA. It was a simple device, but was very effective in operation.

TOOL-KITS

The early EVAs revealed a need to provide the EVA crew-men with a facility to store, transport and restrain a growing array of tools and support equipment.

Tool carriers

Space Shuttle[15]
Mini Work Station (MWS) This is a mechanical device that attaches to the front of the EMU and provides the wearer with additional temporary stowage locations for EVA tools. It also incorporates an additional work tether to provide additional restraint at the place of work, and to prevent the drifting away of individual tools. Each tool is tethered to interchangeable tool caddies by 0.9-m retracting tethers.

Tool caddy Fabricated from stiffened fabric, and weighing 0.85 kg, this is designed to carry small hand tools. For additional temporary restraint, the two 0.9-m retractable tethers are supplemented with Velcro strips.

Cargo bay stowage assembly This was the stowage location for the Shuttle EVA tool suite (the tool-kit). It weighs 108.23 kg, including support assembly for mounting to the orbiter. The volume of the compartment, including dividers and cushions, was 0.52 m³. It could be used to store EVA tools and equipment up to a weight of 181.44 kg, and could be located on the sill longeron on either the port or starboard side, depending on mission requirements. Its location dictated the use of the weight support assembly, which was 53.07 kg lighter.

Provision stowage assembly This was a developmental system with a stowage volume of 0.4 m³ for EVA tools *under* the payload-bay liner. It was designed to carry up to 181.44 kg of tools and equipment, and was located within the forward 1.22 m of the bay, near the EVA hatch area.

Surface tools

For surface exploration, geological sampling and photodocumentation, a dedicated set of tools and equipment has to be provided. The only in-flight experience of using such tools was accrued during the Apollo programme during 1969–72. They proved very reliable, and were invaluable aids to the astronauts during their surface explorations. Although completed three decades ago, the Apollo surface activities continue to be re-evaluated as baseline experience for a return to the Moon and future Mars surface operations, as well as for the development of new tools for surface exploration in conjunction with the latest developments in technology and materials research.

Apollo Lunar Hand Tools (ALHT)

The need for adequate tools and implements for surface operations on the Moon was identified early in Apollo planning, and underwent numerous phases of evolution to arrive at what was actually carried on the lunar landing missions. As those missions evolved, so the tools themselves were adapted in concordance with the experiences of earlier landing crews.

Modularised Equipment Stowage Assembly (MESA) This was a pallet used as the stowage location for most of the surface tools. They were located in the folded MESA on Quad 4 (forward) of the LM. Descending the ladder for the first time, the Commander released the unit with a 'D' ring, and unfolded it for easier access from the surface. The unit also included fresh Portable Life Support System batteries and LiOH cartridges, a TV camera and cable, and lunar sample return containers. It also had a folding table on which to place the sample return containers for ease of packing, and the table also functioned as a bracket on which the astronaut could hang the transfer bag used in the transfer of Portable Life Support System batteries and cartridges up to the LM cabin.

The ALHTs were used mainly for geological sampling, but were also considered, by some, to be part of the ALSEP package. On Apollo 11 the astronauts worked close to the LM in a short EVA, and they therefore carried the required tools individually to the sample location. During the H missions (Apollo 12–14), a hand-tool carrier was devised to carry all the tools more easily on foot traverse. This was also attached to the Modularised Equipment Transporter on Apollo 14. On the three J missions (Apollo 15–17), a tool-rack was located on the back of the LRV. The tools were made of lightweight materials – including Teflon, stainless steel and aluminium – that would not contaminate the samples, and had attachments for stowage in the tool-racks, on the Portable Life Support System, or on the astronaut's suits:[16] The ALHT was not carried on Apollo 17.

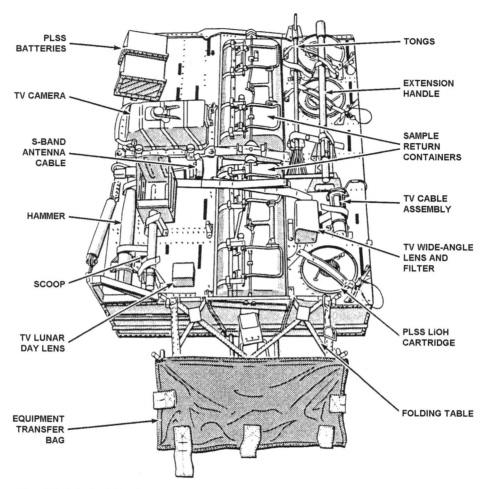

PLSS
BATTERIES

TV CAMERA

S-BAND
ANTENNA
CABLE

HAMMER

SCOOP

TV LUNAR
DAY LENS

EQUIPMENT
TRANSFER
BAG

TONGS

EXTENSION
HANDLE

SAMPLE
RETURN
CONTAINERS

TV CABLE
ASSEMBLY

TV WIDE-ANGLE
LENS AND
FILTER

PLSS LiOH
CARTRIDGE

FOLDING TABLE

The Modularised Equipment Stowage Assembly located on the Apollo LM Descent Module.

Flight	Mass	Main components
Apollo 11	22.9 kg	Large scoop; extension handle; tongs; hammer.
Apollo 12	29.2 kg	Scoops (large and small); hammer; extension handle; tongs; brush/scriber/hand lens/spring scale; tool carrier.
Apollo 13	34.0 kg	Similar to Apollo 12.
Apollo 14	34.1 kg	Similar to Apollo 12.
Apollo 15	50.3 kg	Extension handle; adjustable sampling scoop; hammer; tongs; spring scale; lunar rake.
Apollo 16	53.0 kg	Similar to Apollo 15.
Apollo 17	45.7 kg	Similar to Apollo 15.

Extension handle This was an aluminium alloy tubing approximately 76 cm long and 2.5 cm in diameter, with a malleable stainless steel cap designed for use as an anvil

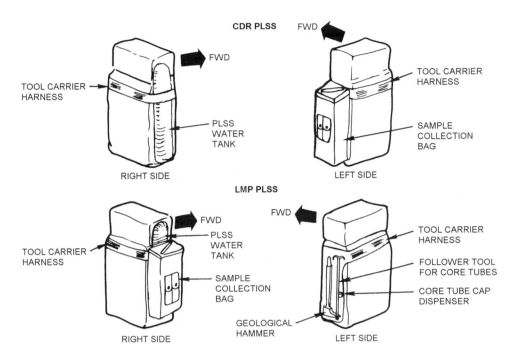

CDR PLSS FWD

FWD

TOOL CARRIER HARNESS

TOOL CARRIER HARNESS

PLSS WATER TANK

SAMPLE COLLECTION BAG

RIGHT SIDE   LEFT SIDE

LMP PLSS

FWD

FWD

TOOL CARRIER HARNESS

PLSS WATER TANK

SAMPLE COLLECTION BAG

GEOLOGICAL HAMMER

TOOL CARRIER HARNESS

FOLLOWER TOOL FOR CORE TUBES

CORE TUBE CAP DISPENSER

RIGHT SIDE   LEFT SIDE

Stowage on the life support back-pack for Apollo lunar traverses.

surface. It was primarily used as an extension for several other tools (scoop, rake, hammer, drive tubes) to allow their use without the astronaut having to bend down or kneel in the suit. Its lower end featured a quick-disconnect mount and lock designed to restrict, compress, tension or torsion, or a combination of these loads, while in use. The upper end was fitted with a sliding 'T' handle which allowed torqueing operations.

Sample scales These were used to weigh the collected samples in their bags, the sample return containers, and other items, to stay within the mass budget for the return to Earth. Samples were not collected by simply filling the rock boxes and hauling them into the LM, as the weight had to be carefully monitored against the LM ascent engine's capability to launch the astronauts off the Moon. The centre of mass had to be calculated to ensure a smooth and accurate ride, for equipment transfer from the LM to the CM for the flight home, and for stability and precision during the re-entry and parachute landing. The scale used on Apollo had graduated markings in increments of 5 lbs (2.26 kg), with a maximum capacity of 80 lbs (36.28 kg). This scale was stowed and used inside the LM.

Tongs This was a useful item that enabled the astronaut, while standing, to grab samples ranging from a small pebble to a rock the size of a clenched fist. The handle was made of aluminium, while the twin tines were made of stainless steel. The tool was operated by squeezing the T-bar grip at the top of the handle, which opened the tines. Once the sample was selected, the T-bar was released to capture the sample for

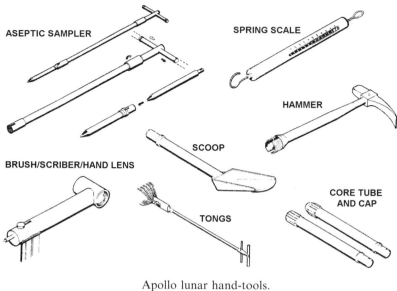

ASEPTIC SAMPLER

SPRING SCALE

HAMMER

SCOOP

BRUSH/SCRIBER/HAND LENS

CORE TUBE
AND CAP

TONGS

Apollo lunar hand-tools.

placing in the sample bag. This tool was also useful for retrieving items which had been inadvertently dropped.

Lunar rake This was used for the collection of discrete samples of rocks and rock chips between 1.3 cm and 2.5 cm in size. The handle, approximately 25 cm long, was attached to the extension handle for ease of collecting, and the rake was adjustable for ease of collecting and stowage of samples. The tines of the rake were formed in the shape of a scoop, and were made of stainless steel. As the astronaut used the rake, small particles and soil passed through the tines and back onto the surface, leaving the required collection of rocks in the scoop tines.

Adjustable scoop This tool was used when the sample was too small for the tongs or rake to retrieve, or when a sample of soil was required. The handle was also compatible with the extension handle, and the pan was adjustable from the horizontal to 55° and 90° for use in both scooping samples and in creating a trench to retrieve samples from just below the surface. The pan measured 5 × 11 × 15 cm, had a flat bottom with flanges on either side, and a partial top to prevent the loss of contents once collected.

Hammer This single tool actually served several functions. It was a sampling hammer used to chip or break large rocks, and it was used as a pick for leverage, as a hammer to drive in the core tubes, and for impact force on other items of hardware and tools as necessary. The head was manufactured from impact-resistant tooled steel, and featured a hammer face on one end and a broad flat blade on the other, with large hammering flats on either side. The 36-cm handle was made of aluminium, and had an attachment to fit into the extension handle when the hammer was used as a hoe.

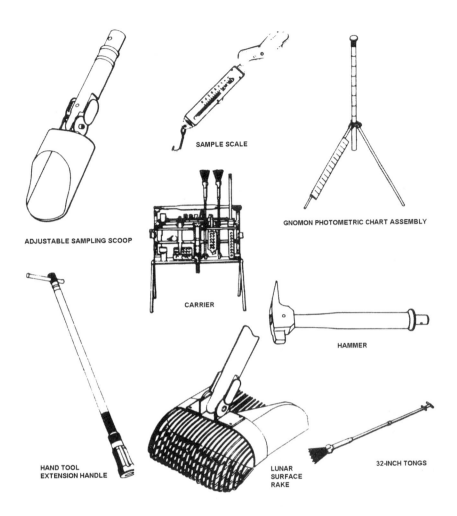

ADJUSTABLE SAMPLING SCOOP

SAMPLE SCALE

GNOMON PHOTOMETRIC CHART ASSEMBLY

CARRIER

HAMMER

HAND TOOL
EXTENSION HANDLE

LUNAR
SURFACE
RAKE

32-INCH TONGS

Additional Apollo lunar hand-tools with the hand-held tool carrier.

Surface sampler Used for the first time on Apollo 16, this tool was designed to capture a sample from the very top of the surface. Using the Universal Hand Tool (UHT) from the ALSEP suite (see below), the astronaut attached a 1.25-cm deep box measuring 11.25 × 12.5 cm. Inside this small box was a plate that 'floated' in channels. One plate was covered with deep-pile velvet cloth, and the other was covered by beta cloth (or perhaps a thin layer of grease). To use the device, the astronaut opened the spring-loaded door at the base of the box, and gently lowered it until the floating plates touched the surface. About 300–500 mg of material only 100–500 μm deep would be trapped in the fabric piles or the grease. As the box was lifted, the lid was closed, and the exposed box was placed in a numbered bag for return to Earth.

LRV soil sampler Used on Apollo 16 and 17, this was a scoop device, attached to the end of the UHT, for gathering surface soil and rock samples without the astronaut having to dismount from the LRV. The device was 25 cm long and 7.5 cm wide, and on the end featured a 7.5-cm-diameter ring and a five-wire stiffened cage which held twelve telescoped plastic cap-shaped bags. As each sample was retrieved, these bags were removed and sealed, revealing a fresh bag in the cage for the next sample.

ALSEP tools

To assist in deploying the Apollo Lunar Surface Experiment Package, several implements were provided. These included:

Universal hand tool Also termed a 'general purpose device', this tool resembled an elongated Allen wrench. Two of these tools were located in the second ALSEP sub-package. Each insertion end of the UHT fitted both the carrying sockets on the ALSEP instruments and the structural units and the bolt fasteners, and a spring-loaded ball lock at the head end of the tool retained the device in the carrying sockets. To release the lock, the astronaut used the trigger at the handle end. The UHT was used for handling and positioning the ALSEP units, transportation and emplacement of sub-systems, release of the bolt fasteners, the removal of pull-pins and release fasteners, and to initiate electrical power from the central station to the experiments by means of 'astronaut' switches on the central station. It was also used for extending the reach of the Surface Sampler and LRV Soil Sampler tools (see above).

RTG tools Two tools were designed to transfer the Radioisotope Thermal Generator fuel element from a fuel cask on the outside of the LM descent stage to the RTG unit. The dome removal tool had a temperature label on its shaft, and was used to remove the dome of the storage cask after the cask had been canted to ease retrieval. The fuel transfer tool was then used to extract the fuel element from the cask and transfer it to the RTG. This also had a temperature label on its shaft, and an engage/disengage knob. Once the fuel element had been safely transferred, neither of these tools were of further use, and were discarded.

Lunar equipment conveyor This consisted of a strap, 18.28 m long and 25.4 mm wide, looped through a support ring in the ascent stage. The two ends of the conveyor were joined by two hooks, which allowed one astronaut (normally the Commander) on the surface to attach items to be transferred up to the ascent stage crew compartment for removal by the LM Pilot inside the compartment, or in reverse for items to be moved to the surface. This simple system was, predictably, also termed the 'clothes line'.

Sample return containers

Contingency sample This device ensured that at least a small sample of undocumented surface material would be collected early in the first surface EVA, in case an early lunar lift-off was necessary. It was stowed in the LM cabin and, with

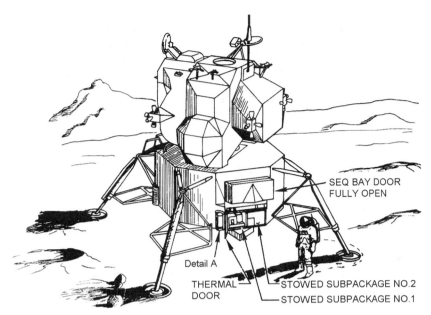

THERMAL
DOOR

SEQ BAY DOOR
FULLY OPEN

Detail A

STOWED SUBPACKAGE NO.2

STOWED SUBPACKAGE NO.1

ALSEP EQUIPMENT READY FOR REMOVAL FROM LM SEQ BAY

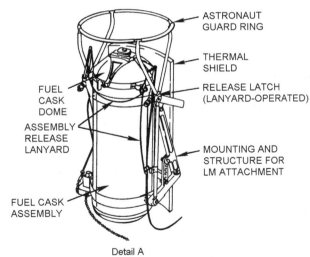

ASTRONAUT
GUARD RING

THERMAL
SHIELD

FUEL
CASK
DOME

RELEASE LATCH
(LANYARD-OPERATED)

ASSEMBLY
RELEASE
LANYARD

MOUNTING AND
STRUCTURE FOR
LM ATTACHMENT

FUEL CASK
ASSEMBLY

Detail A
RTG FUEL CASK STRUCTURE ASSEMBLY

The location of the ALSEP and RTG fuel cask on the LM descent stage.

a sample bag attached to an extension handle, was taken out by the Commander and used shortly after he stepped onto the Moon. The bag was then stowed in a pressure-suit pocket on his leg until he returned to the LM at the end of the first EVA.

Special environment samples Samples taken from a carefully selected area were sealed inside a container retaining a high internal vacuum. This was the Special

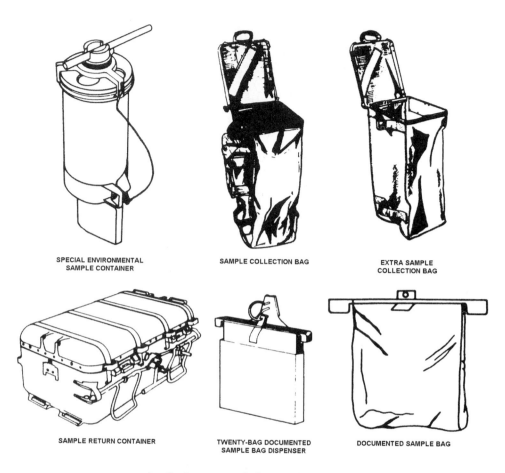

SPECIAL ENVIRONMENTAL
SAMPLE CONTAINER

SAMPLE COLLECTION BAG

EXTRA SAMPLE
COLLECTION BAG

SAMPLE RETURN CONTAINER

TWENTY-BAG DOCUMENTED
SAMPLE BAG DISPENSER

DOCUMENTED SAMPLE BAG

Apollo lunar sample bags and containers.

Environmental Container that protected both the sample material and the original environment from which it was gathered, it was sealed on the surface. It was not opened until it was inside the Lunar Receiving Laboratory at the Manned Spacecraft Center in Houston – a controlled environment facility where the sample and the environmental components could be studied in as near to their original state as possible.

Core tubes To obtain a layered sample of the surface to a depth of several centimetres, the crews had a number of hollow aluminium tubes – each 4 cm in diameter and about 41 cm long – designed to be driven into soft surfaces such as gravel, sand, or even soft rock. The top end of each tube could be sealed, and they were designed for use with the extension handle. The bottom open end featured a non-serrated cutting edge. To retain the sample inside the tube and prevent it falling out when extracted, an internal spring device held the retrieved material securely. A Teflon screw-on cap with a metal-to-metal crush seal replaced the cutting end and closed the tube. The caps were stowed in a small cap dispenser of approximately 5.7 cm. In the Lunar Receiving Laboratory, each core tube was split to reveal the layered sample inside. Apollo 12 took four core tubes, Apollo 13 and 14 carried six tubes each, and Apollo 15, 16 and 17 carried nine tubes each. A deep core sample was obtained by joining a series of two or three tubes.

Field sample bags Eighty 5 × 4-inch sample bags, in twenty-bag dispensers, were provided for collecting lunar samples. They were made of Teflon, and were stowed in the tool-kit or on the harness of the Portable Life Support System. On the later J missions, two sizes were provided, with a twenty-bag dispenser of 8 × 7.5-inch Teflon bags for documented (numbered) samples. Used bags were stowed in the sample return containers.

Sample collection bags These featured pockets used for stowing the core tubes, special environmental samples, and magnetic shield and gas analysis sample containers, and there was a special sample return bag capable of holding large unbagged samples to be stowed in the sample return containers.

Sample return containers Capable of returning up to 60 kg of bulk and documented samples, other items stowed in these suitcase-like containers for return to Earth

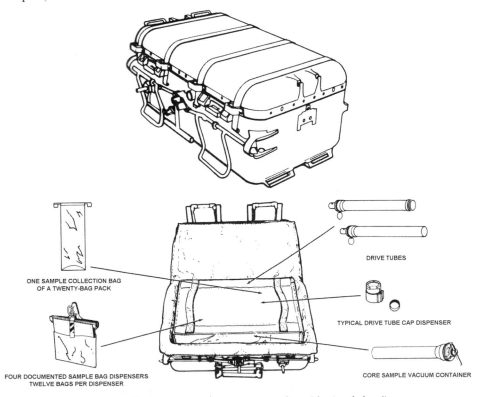

ONE SAMPLE COLLECTION BAG
OF A TWENTY-BAG PACK

FOUR DOCUMENTED SAMPLE BAG DISPENSERS
TWELVE BAGS PER DISPENSER

DRIVE TUBES

TYPICAL DRIVE TUBE CAP DISPENSER

CORE SAMPLE VACUUM CONTAINER

The Apollo lunar sample return container (the 'rock box').

included large and small sample bags, core tubes, and gas analysis and lunar environmental sample containers. All collected samples were documented and stowed in these boxes, which were then sealed and stowed inside the LM for return to Earth.

Photographic equipment

Some of the best images of the human exploration of space are photographs taken by the crew-members on EVA and by other members of the flight crew from inside the vehicle while monitoring the activities outside. However, these photographs are not mere holiday snaps. They constitute official photodocumentation for post-flight evaluation, although some of them are more celebratory. Various movie and TV cameras also record activities as they take place, and can also be used for additional monitoring for the rest of the crew or ground controllers to assist in the operations of the EVA crew. This photographic flight-deck crew support has been used to great effect on all Shuttle missions, using an array of payload bay cameras and the RMS mounted camera. In addition, the real-time viewing of the Apollo crew on the surface, by means of the LRV TV camera system, significantly assisted in the productive exploration of each geological sampling site. During the early years, cameras were primarily used for engineering documentation. Pictorial photography was difficult without stabilisation or restraint, and the results were frequently poor. Moreover, unsecured cameras were sometimes lost. As the programmes developed, however, so the photographic technology became more refined, while at the same time the skills of the astronauts and cosmonauts improved.

Voskhod Leonov carried a still camera – which he was not able to reach to operate, due to the stiffness of the suit – and a movie camera which he temporarily attached to the rim of the air-lock to record his historic EVA.

Gemini A 16-mm camera was installed on the hatch area to record EVAs, and a small hand-operated camera was installed on the HHMU used by Ed White. EVA cameras were also used by other Gemini astronauts, either automatically in the hatch area or by hand.

Soyuz A TV camera and still camera were used during the Soyuz transfer in 1969, but unfortunately the still camera was not tethered correctly and was lost.

Apollo A considerable amount of photographic equipment was used for documentation of crew activities, mission events and scientific objectives, both in-flight and on the lunar surface. Equipment included 16-mm sequence cameras, 70-mm and 35-mm still cameras, and associated lenses, film cartridges, brackets, filters, light meters and other accessories.

Space Shuttle The EMU helmet carries a real-time closed-circuit TV broadcast through the S-band frequency. Battery-powered spares are available for change-out after each EVA.

In-flight service tools

Shuttle baseline tool-kit Experience from earlier EVAs indicated that once properly restrained, an astronaut could perform manipulative tasks using modified off-the-shelf industrial or consumer tools. These could be modified to include tethers to

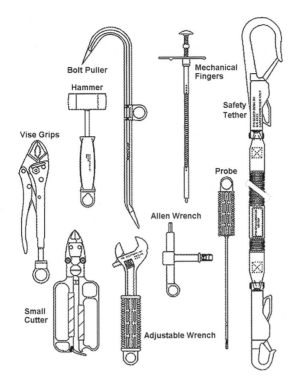

Bolt Puller

Hammer

Mechanical
Fingers

Safety
Tether

Vise Grips

Probe

Allen Wrench

Small
Cutter

Adjustable Wrench

A selection of Shuttle EVA tools.

prevent loss and extended handling, or provision for additional grip. In developing
Shuttle EVA capabilities, the specialists tools for certain tasks were supported by a
suite of tools designed to perform baseline tasks on the early Shuttle missions (1981–
86), which were adapted for later phases of the programme. Such tasks included the
removal of a jammed system, or the bypassing or disconnection of cargo bay
mechanical systems, such as the closing of the payload bay doors or the severing of a
faulty RMS. These tools are stowed in a cargo bay stowage box, and are
interchangeable if specific tools are required for a given mission.

Winch system It was determined that one of the most probable contingency
operations on a Shuttle flight was the need to manually close the twin payload bay
doors on the orbiter prior to re-entry. To assist this, a pair of manually operated
winches were installed on each bulkhead of the payload bay, although they could be
relocated on orbit. The forward winch weighed 10.9 kg, and the aft winch weighed
9.5 kg. The winch line (Kevlar rope) was 7.3 m long and had a load limit of 265.45
kg, but could be overridden if required. Load was transmitted to the reel and line by
means of a winch ratchet handle and a gear system. This load had a limit of 1,935 N
through a toque limiter that was part of the winching handle. There was also the
option to ratchet out the rope at this load. There was also a 24.38-m length of
additional Kevlar rope (3.67 kg) designed to extend the length of the cargo bay
winch, and a snatch block (quantity as determined by the mission) – an adaptation of

the common marine device that could route the cargo bay winch line to support payload or any orbiter contingency task required. During the early phases of the Shuttle flight programme, designers also considered other operations for which this system might be used in support of future EVAs.

The basic Shuttle EVA tool-kit, located cargo bay service area, consists of at least the following items for each mission:

Two *adjustable wrenches*, each weighing 0.45 kg, and with an approximate opening of 2.84 cm; *90-degree needle pliers*, weighing 0.27 kg, and with hand-grips modified for use with EVA gloves; a 0.27-kg *diagonal cutter*, with hand-grips modified for use with gloves; a 0.36-kg *bolt puller*, a crowbar modified for use on EVA, and with improved leverage in confined spaces; *vice grips*, weighing 0.55 kg; a *brass-headed hammer*, weighing 0.91 kg; a screwdriver-type *probe*, weighing 0.27 kg, 21.91 cm long, and with a 0.48-cm-diameter tip; a *lever wrench*, weighing 0.68 kg; a developmental *³/₈-inch ratchet drive*, weighing 0.54 kg, and including a large mushroom-shaped disc that can be turned by hand when overtorquing needs to be avoided or restricted and when obstruction restricts the use of the ratchet handle; a developmental *screwdriver*, weighing 0.18 kg shrouded, 22.86 cm long, and compatible with the ratchet drive; an *Allen wrench* No. 10, weighing 0.18 kg, 22.86 cm long, a developmental tool incorporating a moveable sleeve sized to capture No. 10 Allen head-cap screws, and also compatible with the ratchet drive; a developmental *power ratchet*, a 3.8-inch ratchet-drive electrically-powered tool for low-force and high-revolution tasks; a *⁷/₁₆-inch socket/4-inch extension/³/₈-inch ratchet drive* common tool with all three elements pinned together, weighing 0.45-kg; a *½-inch open-ended wrench*, weighing 0.41 kg; a *½-inch ratcheting box wrench*, weighing 0.45 kg; a ¼-inch hex Allen, a *19.5-inch drive extension*, weighing 0.45 kg, used with the ³/₈-inch drive for orbiter radiator drive disconnection; a *³/₈-inch ratchet drive*, weighing 0.36 kg, using ratchet forces only in counter-clockwise direction to loosen, and slips in the counter-clockwise direction; a *loop pin extractor*, weighing 0.36 kg, available for the removal of loop pins from any mechanism to be disconnected by bolt removal; a *pry bar*, weighing 1.45 kg, and 58.42 cm long, a tool designed specially for use on the centre-line latches of the orbiter payload bay; a pair of *forceps*, a medical instrument modified for use in removing foreign objects from a jammed mechanical system while on EVA; two *tube cutters*, weighing 0.77 kg, resembling a plumber's pipe-cutter, capable of cutting tubes between 1.27 cm and 2.54 cm, and used for the Inconel linkages of the door drive system; a *tape set*, weighing 0.38 kg, a tool caddy carrying twelve 15.24-cm strips of waterproof cloth duct tape, to be used for restraining any loose hardware or disconnect mechanisms; a *Velcro strap set*, weighing 0.82 kg, a tool caddy containing four 25.4-cm Velcro straps, used for restraining mechanical disconnecters or other loose items of hardware; a *trash container*, weighing 0.45 kg, that could be attached to a mini-work-station by securing points, and with a sealable lid to hold loose items; two *payload retention devices*, weighing 3.63 kg, and with a maximum pre-load capability of 10 kN, used to position and/or secure any unlatched articles to mounting points (maximum 4.52 m between points) in the payload bay.

Over the years there have been very few amendments to the basic Shuttle tool-kit, but in light of the loss of *Columbia* in February 2003, and subsequent

The power driver used on Shuttle EVAs.

recommendations from the accident investigation board, additional tile inspection and repair tools will be incorporated on future flights.

Space station EVA tool-kits
Each space station has carried an internal maintenance kit that includes spanners, pliers, hammers, hacksaws, nuts, bolts, a certain amount of spares, and ducting tape. These could be used for most short-term repairs, pending a subsequent longer-term repair if required. However, many of these tools are not adaptable for use with pressurised gloves when the astronaut is outside the vehicle, and they could float away. Over many years, numerous tools, tethers, cameras and support devices have been developed for space station missions. For use on the ISS they have evolved from tools and implements used on the Shuttle, or from those which were specially developed for particular tasks on Skylab and Salyut. The tool-kit for ISS EVAs is being continually developed, and will be expanded with the continuation of the life and enlargement of the station.

Zero-g aircraft tests of the lunar cycle design that was never taken to the Moon.

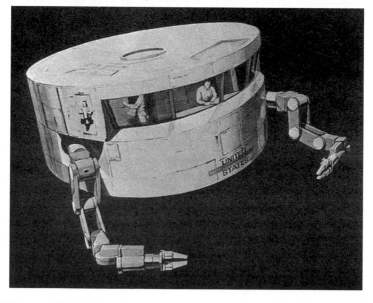

A late 1960s/early 1970s design for an orbital servicing/construction tug with twin RMS grapple arms.

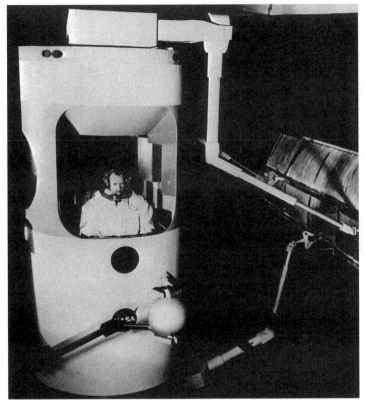

A 1970s design for a one-man pressurised EVA service vehicle with RMS.

TRANSPORTATION

The development of mobility units for EVA has not only focused on the MMUs described above. Designs for astronaut-assisted mobility units or completely robotic EVA assistants have been in the design stage for several decades. These have been depicted as complete pressurised free-flying servicing craft resembling telephone kiosks, with the shirtsleeved or spacesuited astronauts (depending on the design) flying around a large space structure, or servicing and repairing a satellite with small RMS units. Designs for smaller units have included the 'space scooter', on which the besuited astronaut sits astride a 'jet mobile' to quickly move around from one location to another. Fully robotic EVA aids have been studied for assistance in ISS operations, and it is clear that such vehicles will eventually be employed to assist and supplement future human EVA activity.

Apollo surface mobility aids
Studies into mobile lunar vehicles have included flying platforms, lunar motor-bikes, and a fully pressurised roving laboratory with an exploration capacity of fourteen

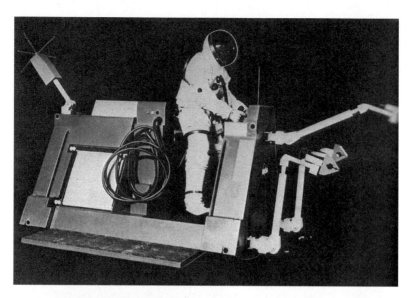

A one-man unpressurised EVA 'scooter' with RMS servicing aids.

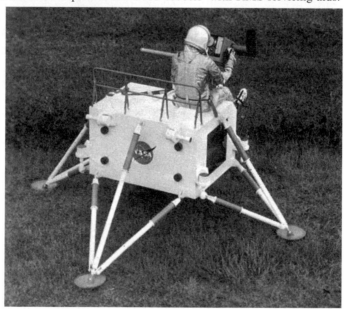

A two-man unpressurised lunar flying vehicle (1960s).

days away from the landing site. Although these never became flight units, the evaluation and development of the technology helped in the evolution of the two small craft that assisted in the Apollo missions. The Russians also studied mobility units for their exploration of the Moon, including a series of unmanned lunar orbital, sample return and automated roving vehicles (Lunokhod) to supplement the

A one-man lunar flying platform (1960s) derived from US Army flying platform studies.

An early 1980s hybrid design for an EVA suit and EVA service vehicle with RMS attachments.

Wernher von Braun drives a mock-up of the lunar rover (Mobile Laboratory, MOLAB) chassis.

manned lunar programme, though when the N1/L3 programme was terminated the automated Luna programme became the main tool with which the Soviets explored the Moon up to 1976.

Modularised Equipment Transporter (MET) The MET was designed to assist the Apollo 14 astronauts on their geological traverse before the advent of the LRV from Apollo 15. It was designed to obtain the maximum return of collected data and samples, and reduce the amount of equipment that had to be hand-carried on a Moon-walking traverse of some distance. This two-wheeled tubular aluminium cart weighed only 13.6 kg but was able to support a load of 163.29 kg, and was pulled by hand. It featured two special synthetic rubber tyres and inner tubes specially designed to operate in low temperatures (to −20° C), and was folded and carried in the MESA. The effort of pulling the transporter could prove very tiring for an astronaut in a pressurised suit, and the handle was therefore designed to alleviate strain on the hand and arm. This triangular handle was wide enough for the hand to be inserted, but the two sides were shorter than the pressure-suited hand, allowing insertion of the hand and movement towards the front of the handle to secure the MET, but without the need to enclose the fingers. It therefore provided control of the unit without need of a constant tight grip.

Lunar Roving Vehicle (LRV) The design and development of the Boeing-fabricated LRV, used on Apollo 15–17, was a marvel of engineering. The period of development was only seventeen months, from the awarding of the contract to the delivery of the first flight unit. This manually controlled, four-wheeled, electrically-powered vehicle extended the range and capabilities of the final three Apollo crews in their exploration of the landing area. It was designed to carry both suited astronauts and their equipment at approximately 16 kph on a 'smooth level surface', and at lower velocities could climb slopes of up to 25°. The design allowed operation by either astronaut, using the central T-shaped control handle on the central control and display column, moving the unit forward to go forward, and to the rear for reverse. Steering was achieved by moving the handle to the left or right, which could turn either or both of the two electronically driven rack-and-pinion assemblies on the front and rear sets of wheels. Braking was achieved by pivoting the hand-controller rearwards about the braking point. The two steering mechanisms were independent of each other, so that if a failure occurred the steering linkages could be disengaged and the wheels locked in the forward position to allow drive from the other system. Power was supplied by two 36-V silver–zinc batteries. The navigation system displayed data on pitch, roll, speed, heading, total distance travelled, range, and bearing back to the LM. The heading was determined by using Sun-aligned directional gyros, with wheel-rotation counters recording speed and distance, to compute both range and bearing. Alignment was achieved with Sun-angle readings and by relaying data to Earth, where initial heading angles were calculated and relayed back to the unit for calibration, with the gyro adjusted by using the torquing

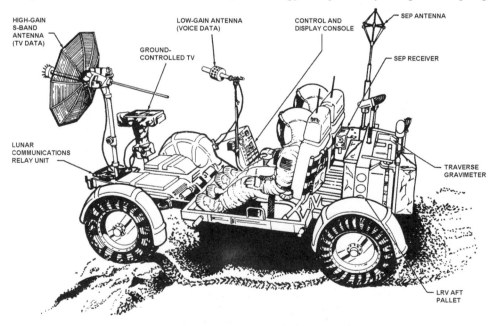

The Lunar Roving Vehicle.

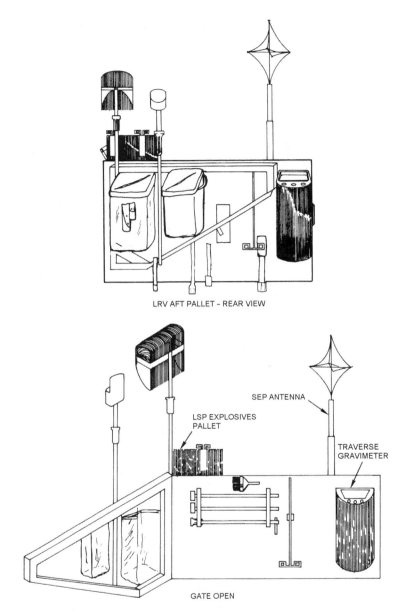

LRV AFT PALLET – REAR VIEW

SEP ANTENNA

LSP EXPLOSIVES
PALLET

TRAVERSE
GRAVIMETER

GATE OPEN

Stowage locations on the Apollo LRV.

switch to match the calculation value on the heading indicator. It was originally folded in Quadrant 1 of the LM descent stage for transportation to the Moon, and was released and deployed by a system of lanyards, pulleys and a lowering cradle to its full dimensions of 3 m long, 2.1 m wide, and 1.14 m high. The unladen vehicle had a mass of 204 kg, but when it was loaded with the pressurised crew and all the equipment the weight could vary between 694 kg and 725 kg. The four wheels

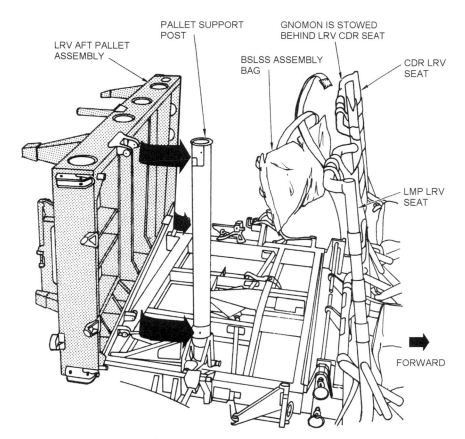

LRV AFT PALLET
ASSEMBLY

PALLET SUPPORT
POST

GNOMON IS STOWED
BEHIND LRV CDR SEAT

BSLSS ASSEMBLY
BAG

CDR LRV
SEAT

LMP LRV
SEAT

FORWARD

Further stowage locations on the LRV.

featured open-mesh tyres with a chevron tread, covering up to 50% of surface contact area. Separate traction drive allowed each wheel to be disconnected from the drive mechanism for free-wheeling. All three LRVs were left deployed at the landing sites of Apollo 15, 16 and 17, and the onboard TV camera system – remotely controlled from a console at Mission Control, Houston – was used to record not only the activities of the crew and the surrounding terrain at each geological station stop (it was not able to use the TV camera during traverses due to a requirement to align the antenna with Earth), but also the lift-off from the Moon, and poignant views of the unoccupied landing site after the astronauts had begun their trip home.

REFERENCES

1 'Footprints on the Moon', *National Geographic*, **125**, No. 3, March 1964, 380.
2 Kozloski, Lillian D., *US Space Gear: Outfitting the Astronaut*, Smithsonian, 1994.
3 Abramov, Isaak P. and Skoog, Å. Ingemaar, *Russian Spacesuits*, Springer–Praxis, 2003.

 4 Kennedy, Gregory P., 'Jet Shoes and Rocket Packs: The Development of Astronaut Manoeuvring Units', *Space World*, October 1984, 4–9.
 5 Shayler, David J., *Gemini: Steps to the Moon*, Springer–Praxis, 2001, p. 347.
 6 Abramov, Isaak P. and Skoog, Å. Ingemaar, *Russian Spacesuits*, Springer–Praxis, 2003, p. 193.
 7 Shayler, David J. and Hall, Rex D., *The Rocket Men*, Springer–Praxis, 2001, p. 259.
 8 Gaston, Nancy, 'Backpacking through Space', *The Citizen* (Harris and Galveston County newspaper), 12 June 1988, p. 1–2.
 9 *Skylab: A Chronology*, NASA SP-4011, 1977, pp. 97–98.
10 Belew, Leland F., *Skylab: a Guidebook*, NASA SP-107, 1973, p. 197.
11 *Skylab: A Chronology*, NASA SP-4011, 1977, pp. 107–108.
12 Whitsett, C.E., Status Briefing Notes, 15 November and 20 December 1983, NASA JSC History Archive, Space Shuttle Collection, University of Clear Lake.
13 Abramov, Isaak P. and Skoog, Å. Ingemaar, *Russian Spacesuits*, Springer–Praxis, 2003, p. 199.
14 'Reaching Out in Space: Europe's Robot Arm', *On Station*, No. 2, March 2000, 10–11; Hendrickx, Bart, 'From Mir 2 to the ISS Russian Segment', in Hall, Rex D. (*ed.*), *The ISS: From Imagination to Reality*, British Interplanetary Society, 2002, p. 24.
15 *STS EVA Description and Design Criteria*, NASA JSC-10615, Revision A, May 1983.
16 Sullivan, Thomas A., *Catalogue of Apollo Experiment Operations*, NASA Reference Publication 1317, January 1994; Apollo (LM) News Reference (1969) Grumman; Apollo 11 Press Kit (69-83K) 6 July 1969; Apollo 12 Press Kit (69-148) 5 November 1969; Apollo 13 Lunar Surface Procedures (Final), MSC, 16 March 1970; Apollo 13 Press Kit (70-50K) 2 April 1970; Apollo 14 Press Kit (71-3K); Apollo 15 Press Kit (71-119K) 15 July 1971; Apollo 16 Press Kit (72-64K) 6 April 1972; Apollo 17 Press Kit (72-220K) 26 November 1972; Apollo Mission J-3 (Apollo 17) Mission Science Planning Document (Final), MSC-05871, 31 July 1972.

Practice makes perfect

One of the major problems in training for any flight into space is that it is impossible to simulate microgravity for an extended period of time. Some training, with the astronaut either suited or unsuited, is carried out in a 1-g environment, but this restricts the training programme and cannot accurately simulate the conditions of the mission, whether in space, on the Moon or on Mars.

Since the beginning of the Space Age, specially padded aircraft have been used to fly parabolic curves of varying angles to recreate short bursts of near-zero-g or the gravity on the Moon or Mars, but from the early EVAs – particularly during the Gemini programme – it was found to be difficult to prepare an EVA crew for the challenge of working in a pressure garment in open space. It was not until the introduction of huge water tanks and suitably weighted and modified pressure garments that EVA crews could more accurately practice their EVA tasks over many cycles and for much longer periods than was possible during parabolic flights in aircraft.

For lunar excursions the training programme needed to focus on the reduction of the 1 g of Earth to the $^1/_6$ Earth-gravity of the Moon (and in the future will need to consider the $^1/_3$ Earth-gravity of Mars). The parabolic flights of the training aircraft were very useful, but did not provide training over an extended period. Neither was the water tank practical, and so elaborate devices had to be devised to simulate gravity on the surface of the Moon.

Over the past four decades, numerous engineers, scientists, students, technicians, astronauts and cosmonauts have developed, tested and trained on EVA operations and hardware intended for use in Earth orbit, deep space, and on the Moon or Mars. Over these years, a number of procedures or devices have become standard training requirements for EVA, and with the change from lunar to space station operations, the water tanks continue to play a leading role in EVA training forty years after their introduction. Numerous 1-g simulations of proposed martian EVA scenarios have been demonstrated, and continue to evolve towards the commitment to return to human space exploration beyond Earth orbit. In the 1990s, a new tool – virtual reality – became available to assist in EVA training, and this is being very quickly developed to meet the demands of ever more complex EVA goals. In addition, because of the extended duration of spaceflights, the time between training for an EVA and actually accomplishing it might be several weeks or months. 'Practice makes perfect' is never truer than for spaceflight training and EVA activities in particular, and the development of on-orbit training facilities and modes will become a growing field of crew preparation on the ISS and, eventually, on flights to Mars.

WHERE THERE'S A NEED ...

Over the four decades of human spaceflight and EVA operations, requirements for training have changed with succeeding generations of spacecraft and with different objectives and goals of the programmes. Although some aspects of EVA training have been retained, their applications have been refined as technology and flight experience provide an ever-expanding database of knowledge and understanding of the skills of performing EVA.

Voskhod and Soyuz: a new skill
The programme was relatively simple for the two early Soviet EVAs. On Voskhod 2 in 1965 – the first time an EVA was attempted – the plan was simply to allow the cosmonaut to leave the spacecraft and then re-enter it. Training focused on entering the air-lock, putting on the back-pack, opening and closing the hatches, and entering and exiting the air-lock structure. This was followed by a programme of zero-g flights and altitude chamber work, which was also later undertaken by the two Soyuz cosmonauts who were preparing for their in-flight EVA transfer in 1969. For this mission they had to exit one spacecraft and move hand-over-hand to a second docked Soyuz. No additional experiments or activities were planned in what was a simple demonstration of crew transfer or a potential rescue scenario.

Leonov decided to expand on his physical training programme for Voskhod 2 to an exhausting degree. In the twelve months prior to his EVA he ran more than 500 km, cycled more than 1,000 km, skied more than 300 km, and spent hundreds of hours in the gymnasium. The physical training of the Soyuz cosmonauts was probably similar, but less demanding. Not all EVA astronauts follow such a physically demanding routine, although they exercise and train to stay in condition, as an EVA can take its toll on even the most physically fit astronaut or cosmonaut. However, in contrast to Leonov, American astronaut Neil Armstrong once stated that in his opinion he was given a specific number of heartbeats in his life, and he was not about to waste any of then in undue exercise! He became the first to walk on the Moon.

Gemini: a learning curve
The early success of Ed White on his short EVA from Gemini 4 led to more adventurous planning for later EVAs, and the experiences on Gemini 9, 10 and 11 led to further evaluation of techniques of EVA restraint for the Gemini 12 flight. One of the new techniques made available during Gemini in 1966 was underwater training; and due to the training for the various manned manoeuvring units, new

types of training facilities had to be devised to simulate EVA while using the units in a 1-g environment.

To go to the Moon
Both the Americans and the Russians devised harness devices to suspend the occupant and counteract $^5/_6$ of their Earth weight, to approximate conditions on the Moon. Of course, when the missions finally reached the Moon, the skill of learning to walk (and stay upright) had to be learned anew, as there was no harness to catch a stumbling astronaut. Throughout the development of the Apollo missions and hardware, Apollo lunar surface activities evolved through simulated 1-g runs. This type of scenario is still being conducted in the current quest to evaluate the best methods for early surface exploration of Mars. During Apollo, zero-g aircraft flights and water immersion facilities were used to train the crew of Apollo 9 and the CM Pilots of Apollo 15–17 to retrieve film cassettes from the SIM-bay area of the SM; and because of the inclusion of the LRV on Apollo 15–17, an Earth-bound version of the LRV had to be developed for training runs.

Skylab: training for a rescue
On Skylab, American astronauts again ventured out of their spacecraft in Earth orbit. Considerable experience of surface EVA had been gained on Apollo on the Moon, but since Gemini, very little had been developed for orbital spacewalks. Apollo 9 had included a brief EVA to test the lunar suit in vacuum conditions, but the transfer from the LM to the CM had to be curtailed – due, in part, to the earlier illness of the LM Pilot. When the Skylab EVA programme began in 1973 it had been four years since an American had walked in space in Earth orbit. It is therefore all the more remarkable that the first EVAs on Skylab were so adventurous as to include an attempt to release the trapped solar array and deploy protective sunshades – although plans for these EVAs were not introduced at the last minute. Work on the OWS programme originated in the early 1960s, and evolved from studies into the use of Apollo hardware for other missions after the lunar missions. In parallel, there were also various studies for conducting EVA to support scientific and workshop activities. Training for all Skylab EVAs – including the more routine film canister exchanges – took place in the water tank at MSFC, as well as in several 1-g run-throughs. The two experiments flown inside the station (the Foot Controlled Manoeuvring Unit and the Manned Manoeuvring Unit) required improvements in the various mobility training devices used for Gemini, such as the air-breathing table.

Salyut and Mir: routine operations
Despite long-range planning studies, the Skylab programme constituted only one space station and three manned visits, and the performance and perfecting of EVA at space stations was developed more by the Soviets during their Salyut and Mir operations (1971–2000). Like the Americans, the Soviets recognised the advantages of training for EVA in water tanks together with the development of 1-g and zero-g training programmes, and most of the Salyut and Mir EVA crews trained in water

immersion facilities. The introduction of a manned manoeuvring unit for space station operations also required a new method of cosmonaut training.

The Space Shuttle: stretching the envelope
The development of EVA from the Shuttle also introduced a whole new range of training methods to the Astronaut Office. The water tanks were enlarged, special simulation devices were configured for contingency EVAs in the payload bay of the Shuttle, and mock-up satellites were fabricated for practicing repair and servicing techniques. The more advanced MMU required a new specialist facility at its prime contractor; and with the robotic arm being controlled by astronauts from within the Shuttle, each EVA became almost a full crew effort. The development of virtual-reality training techniques to prepare for more complex EVAs, including the servicing of the HST and construction of the ISS, began during the Shuttle missions of the late 1990s.

The International Space Station: a new challenge
Training for EVA was a challenge before the ISS, but it was nothing compared to what lay ahead in building the ISS from 1998. Not only were the EVA tasks mounting, but the training programmes also became more international, with crews visiting facilities across America, Canada, Europe, Russia, and eventually Japan. Special cadres of astronauts would therefore train for EVA on the Shuttle, or as part of long-duration training as resident crews, from either Russian or American elements, using hardware from either country. And because of the necessity to adapt to different training psychology, hardware, procedures and languages, EVA became even more complex on the ISS. In future, however, virtual reality and the development of robotic EVA support systems will begin to blend manned and unmanned space exploration to a level that will be extended to the Moon and planets.

Training for the future
For the foreseeable future, EVA will imply work at the ISS, as that is where most of the manned missions over the next ten years will be focused. Well before construction of the ISS is finally completed around 2010, and the Shuttle returns to flight, it will be evident how this changes the crew training schedule from construction to maintenance and repair. How, therefore, will EVA develop? Partly, the new systems, procedures and hardware developed for the ISS will be used as a training facility for later EVAs in deep space or around the Moon and Mars. The best place to train for spaceflight is in space itself, and as we venture further from our home planet, perhaps EVA training will move from Earth to the Moon and on to Mars, where actual EVAs will be performed to simulate more adventurous exploratory EVAs in remote and challenging terrains.

Whatever challenges might be encountered in future, it is clear that early training (particularly EVA training) for space begins here on Earth – in the classroom, and in the minds of those who plan the missions and build the hardware.

TYPES OF TRAINING

Most EVAs follow a particular format in planning and preparation once it has been established that EVA is required to achieve the desired results. The planning then examines what the EVA is designed to achieve, and whether it can be accomplished safely within the limitations of the mission flight plan, the hardware, the consumables, and the crew. Once the objectives of the EVA have been determined, the timeline is developed and merged into the mission flight plan, with contingency and back-up options. The required training programme is then prepared to support the development of techniques and hardware.

As with most aspects of a human spaceflight, the process can take anything from a few days or weeks to as long as several months or even years. In extreme cases it is possible to perform EVAs for which there has been no planning on Earth and which have not been included as part of the flight plan. Over the past twenty-five years or so, EVAs have come to be typified as planned, unplanned and contingency, and training has been adapted for each of these categories.

Planned A planned EVA is an intentional part of the flight plan, to meet a particular goal for an experiment, for the inspection, repair or deployment of hardware, or for the development of new equipment and procedures. As these procedures are linked to the success of the mission and the safety of the EVA crew, training for this type of EVA is the most protracted.

Unplanned An unplanned EVA is executed when little or no EVA training has been accomplished. It is normally carried out in response to mission events as they occur, and relies on the ingenuity of the crew, the flight controllers and the support teams to develop procedures to ensure success.

Contingency This is an EVA for which the crew can train to deal with a malfunction or rescue scenario, even though, in the course of a normal mission, such an operation is not expected. It is possible to train for procedures such as the manual closure of the Shuttle payload bay doors, the separation of a disabled RMS, and the deployment of spacecraft appendages, but training is limited to the time available after higher-priority training.

PLANNING

The inclusion of EVA in a mission is determined in the planning stages, and is guided by mission objectives, duration, crew activities and experience, and any impact on other aspects of the planned flight. There are also several key factors that have to be taken into account:[1]

Microgravity The capabilities of a crew-member under EVA conditions can be affected by the zero-gravity or partial-gravity environment in which they intend to work. Major factors affecting performance during EVA include the encumbrance of the pressure suit, restrictions in the working area, and inadequate restraints or mobility devices. In addition, mass handling and transfer is partially controlled by the agility of the EVA crew-member, the availability of mass-handling devices, the mass of the object to be moved, the path along which the object is to be transferred, and the time factor of the task and governing limitation of the overall time of the EVA and consumables. There is also the consideration of the thermal environment in which the crew-members are working, which in Earth orbit can be 116–394 K ($-250°$–$+250°$ F). This is well within the operating parameters of the EMUs, but extended orbital exposure to direct sunlight and indirect reflection, combined with several orbital night-time passes, can create EMU parameter infringements on EVA over 4.5–7 hours. In certain circumstances, focusing of the Sun in the work area can create flux levels that can rapidly exceed the parameters of the suit and place the crew-member in jeopardy. Careful planning and management of the vehicle's orientation, and the length of extended exposure to the Sun, is therefore of primary concern during EVA planning. With careful consideration of Sun/vehicle angles there should be no added constraints imposed in performing EVA over several day and night cycles. Another environmental factor affecting EVA performance is the limitation of working in a suit pressurised at around 3.1 psi, to protect the wearer from hypoxia, and the requirement to denitrogenate the blood to avoid the bends. An operational suit with higher pressure – which would alleviate this task by operating at around 55.2 kN/m^2 – is still under development.

Consumables and servicing The length of an EVA is affected by the availability of consumables and the servicing of the EVA equipment. On short missions it is also affected by the timeline and other mission objectives, and on longer missions such as those onboard space stations, EVA preparation, maintenance, servicing, and on-orbit training, can help pace a series of EVA tasks over a longer period of time. A supply of spare or replacement parts presents a certain amount of redundancy, but there remains a limit as to what could or should be repaired, replaced or maintained, because all of this affects the crew work-load and time available for other tasks. The improvement of systems and procedures to eliminate on-orbit servicing and repairs will ensure a reduced preparation time for EVA, so that more time will be spent in carrying out other mission objectives. State-of-the-art technology in EVA systems and hardware will certainly be a requirement for EVAs at Mars.

Pressure suits The wearing of a pressure suit brings its own limits, and many factors need to be considered. These include sizing, comfort and reliability; mobility, reach, and application of forces; the handling of mass; the dexterity of the gloves when gripping, manipulating, removing and replacing small items of hardware; the helmet's range of vision, including the checking of information on heads-up displays, cuff checklists and other displays; communications and data recording; and the translation or mobility rates, either by hand, by manoeuvring unit, or as affected in a partial gravity environment. Finally, safety and alternative back-up procedures have to be included and trained for in every EVA programme, although these are never expected to be called upon.

EVA task design In addition to the above issues there has to be a guideline for designing EVA interfaces, and accommodation in the task or payload. Experience in earlier programmes has always led to refinements in later programmes, and more accurate planning for future tasks, both in generic tasks and specific programme requirements. In considering any EVA, payload designers and task designers need to address the following: air-lock/hatch access path to and from the payload/ objective; translation aids to work-sites; crew equipment and safety; cargo transfer requirements; restraint provision at work-sites; visibility and lighting requirements; working volume requirements; EVA glove interfaces; EVA tool design; knobs, switches and actuators; access doors and panels; and deployment and set-up procedures.

During the early missions that included EVA (Voskhod, Gemini and Apollo), the duration of the mission was a limiting factor, and each EVA was meticulously timelined and practiced. With Apollo there was also the limitation (within the prescribed 8–12 day mission) of 1–3 days of lunar surface time available to complete the surface EVAs. With the advent of space station missions from 1971 (Salyut, Skylab, Mir and the ISS), certain EVAs could be trained for prior to the mission, but there also had to be a degree of flexibility for unplanned EVAs to be completed as the mission progressed. While certain space-station EVA tasks might not be completed during one EVA, they could still be reassigned to a later EVA or even to a subsequent crew, which helped in planning the work-load and pace of the EVA much more effectively.

Another factor affecting EVA planning was the health and adaptability (or lack of adaptability) of the crew to the conditions of spaceflight prior to EVA. To ensure that the main objective of the mission – the EVA – could be accomplished in the first

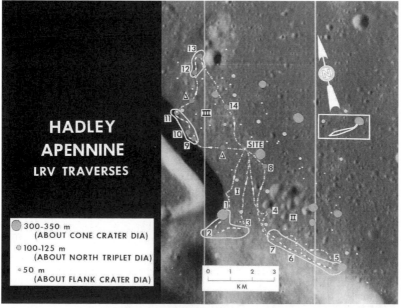

EVA traverse planning for Apollo 15.

few hours of the flight, it was placed early in the mission (at least on the first flights), so that it would have already been accomplished if the mission was subsequently terminated early. However, as Gemini 4 clearly demonstrated, due to other issues early in the mission, the EVA cannot necessarily be completed quickly. Over the years, crew illness and/or adaptation to spaceflight conditions has certainly affected more than one EVA, and with strict mission rules governing the health of all crewmen, especially when undertaking EVA in an enclosed pressure suit outside the vehicle, this is one of the more important factors to be considered. On the Shuttle, EVAs are not normally scheduled for the first two days of the flight unless there is a requirement to work on a payload or on safety equipment. The crew is therefore provided with sufficient time to adapt to orbital flight.

EVA INTEGRATION

The integration of EVA into a mission has changed considerably with the American Shuttle programme, in which the possibility of conducting an EVA is available on *every* mission – at least as a contingency, if not planned. The integration of EVA into the Apollo programme focused on geological and surface activities, while those at space stations have centred on maintenance and repair, or in expanding the resources and facilities of the station. For the Shuttle, with a wide variety of payloads to be carried, a broad range of EVA scenarios had to be integrated into a defined plan to allow for both fabrication of EVA support interfaces and system requirements, as well as crew training and mission safety. Using the STS EVA Mission Integration Plan as an example, the payload EVA requirements constituted part of the Payload Integration Plan (PIP), developed between the STS Program Office and Payload organisations. A scenario of EVA activities and interfaces was issued as the PIP EVA Annex to mission documentation, and established a requirement for the designer of the payload to incorporate these interface requirements into the development of the proposed payload. The integration of EVA requirements was planned at the earliest stages of payload design, and followed a sequence of formal integration responsibilities. The following table shows where the responsibilities lay for each activity (C, consultation role; FDF, Flight Data File; OSA, provided by the STS throughout the additional optional service agreement).

For the STS, a payload EVA was viewed as an optional service to the user, but its inclusion in the PIP initiated a series of EVA planning activities that can also be categorised for most EVA planning in any programme:

- Design reviews
- Procedures development
- Support hardware design
- Flight planning
- Crew training
- Flight performance

Activity	User	STS
Payload design for EVA task	x	C
Operational evaluation of design	x	x
EVA support hardware (tools, restraints, *etc.*) that are payload unique	x	OSA
EVA support hardware operational evaluations	x	x
EVA 1 g, WETF, KC-135, altitude chamber, *etc.*; mock-ups/trainers as required for crew training	x	OSA
Stowage provisions for payload EVA hardware that exceeds standard STS stowage provisions	x	OSA
EVA technique development	x	OSA
EVA technique evaluation	x	x
EVA procedures development	x	OSA
EVA procedures evaluation	x	x
FDF procedures development		x
FDF procedures verification	x	
EVA crew training		x
Flight performance of EVA tasks		x

1-G SIMULATIONS

Some phases of EVA training can be completed in a 1-g environment with little compromise in the value of the training, and are more convenient and cost effective than more elaborate phases of the training cycle. The type of 1-g environment training is dictated by the flight plan and by the facilities available at the time. The 1-g simulations have continued over the years, and form an introduction to the EVA being planned. The 1-g walk-throughs have included benchtests, altitude chamber runs, air-bearing platforms and body-harness training procedures, and have been accomplished in all of the EVA programmes undertaken.[2]

Bench reviews
Engineers, controllers and crew-members review items of EVA equipment, certain procedures and facilities in an inspection-style process, to familiarise themselves with components of the EVA and to understand their operation. This can be done at the contractors or at crew training facilities, and includes representatives of contractor staff, EVA support teams, training officials, and crew representatives or members of the flight crew, depending on the mission requirements.

1-g walk-through
A 1-g walk-through allows the crew to go through a detailed checklist of EVA procedures for either practice or training. This involves crew familiarisation with procedures, EVA-related hardware, and stowage, and allows them to develop

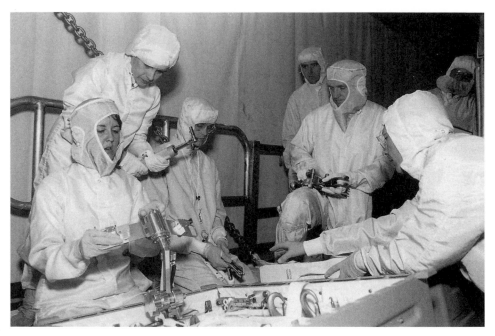

A 1-g walk-through of EVA tools and components.

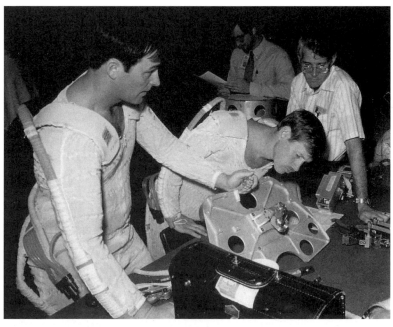

Wearing the liquid-cooling garments, a Shuttle EVA crew examines elements of the hardware to be used during the impending training session in the water tank.

procedures into a coordinated team effort. This formula began with Gemini and Voskhod, and has continued to the present day. Full-size mock-ups or partial mock-ups of the spacecraft hardware are used to familiarise crew-members with facilities relating to their planned EVA, and crew station mock-ups are fabricated and laid out to model as closely as possible the real spacecraft on which they will work. Initially, these walk-throughs are completed in shirtsleeve conditions, and are not timelined. Where equipment is stored in specific parts of the spacecraft, the walk-through allows the crew-members to become accustomed to the locations and restraints with which they will be working when wearing their pressure garments.

In some instances, 1-g walk-throughs were carried out in partially-suited or full-suited configuration. During Apollo, landing crews would practice the deployment of the mock-up ALSEP experiments – usually wearing only the EVA over-gloves for the unsuited phase, and full training pressure garments and Portable Life Support System for suited runs. For the early missions, geological training was undertaken in some of the mission simulation facilities at the Manned Spacecraft Center. For the J missions, a similar training facility was located at the Kennedy Space Center. However, photographs of spacesuited astronauts practicing lunar EVAs against

Suited 1-g lunar surface EVA training for Apollo J missions.

a backdrop of tropical palm trees never really captured the atmosphere of the desolate lunar terrain! With the introduction of the MET and LRV, more detailed 1-g lunar traverse training was added to the geological field trips.

In 1965, a full-scale LM mock-up was installed at the Lunar Topographical Simulation Area at MSC. It was called the 'the rock pile', and until 1970 was used for unsuited and suited simulations of ALSEP deployment, and training in the use of lunar tools and LM equipment deployment. During the mid-1960s, many of the astronauts were involved in suited evaluations of Apollo lunar surface techniques, exit and entry into and out of the LM, the deployment of experiments, and other exercises.

Russian 1-g training facilities
The Russians also had their own 1-g training facilities for the lunar programme, as well as various suspension rigs and devices for simulating $^1/_6$-g loads. Since the mid-1970s the majority of training for space station EVA has been completed in the Hydro Laboratory at Star City, and 1-g facilities have also been used for certain elements of pressure-suit familiarisation and training. A recent addition for station-based EVA training is the device called Vykhod 2 (Exit 2, or EVA 2). This is a 1-g facility, based at the back of the current Mir Hall at TsPK, where cosmonauts train

The Soviet 1-g training facility at TsPK. (Courtesy Bert Vis.)

The EVA handrails and attachment mock-up at TsPK. (Astro Info Service collection.)

to operate various hatches (although the American hatches on the ISS and on the Shuttle are represented by full-size photographs), and includes a full-size mock-up of the Pirs hatches. Two Orlan suits hang on cranes underneath booms that support the weight of the suits and occupants to simulate zero g. There is also a retention board on a wall for practice in attaching and removing experiments or sample cassettes, and foot-restraint mock-ups and tether hooking devices.

Altitude chamber tests

The use of altitude chambers in both the American and Russian programmes has allowed astronauts and cosmonauts to perform training with EVA equipment in simulated space environments, and to participate in pressure-suit tests, qualification and evaluation. In the Gemini programme, flight spacecraft were placed in the attitude chamber to provide the crews with experience of opening and closing the hatch in a vacuum. The training included checkout and donning of the ELSS, the ESP, and later, the AMU. The tests not only provided familiarity with EVA systems and procedures in a vacuum, but also increased confidence in the performance of flight-qualified hardware and procedures.

These are huge sealed facilities in which the internal pressure can be reduced to simulate the vacuum of space, and can accommodate anything from small experiments to suited astronauts to large complete spacecraft. They are produced in various sizes, and all the major space powers use them to test space hardware. They are particularly useful for EVA training and suit evaluation. Designers and test engineers evaluate new suits and hardware and procedures in these chambers, and they are also used by the astronauts and cosmonauts for familiarisation exercises in basic training and for mission simulations.

Rusty Schweickart prepares for an Apollo EMU test run in the altitude chamber prior to Apollo 9.

Air-bearing platforms

These facilities are used when some sort of manoeuvring equipment needs to be evaluated. They allow an astronaut to glide over a highly polished surface on an air cushion, with the manoeuvring unit providing propulsion, and with little or no resistance caused by contact with the ground. These devices simulate 3° of freedom, and allow the astronauts to experience handling of the manoeuvring devices and some of their handling characteristics.

These facilities were also used for MMU training for Skylab and the Shuttle. The Precision Air Bearing Facility (PABF) is currently located in the Space Vehicle Mock-up Facility in Building 9 at JSC. The principle is similar to that of an air-hockey table, allowing heavy objects to slide over it with minimal friction. It consists of a 10 × 7-m metal surface providing two-dimensional simulation (3° of freedom) of the weightless environment. The PABF has proven useful as an engineering tool and as a crew training facility in the development and teaching of mass handling techniques. This was important for handling large masses for the HST servicing missions, and in demonstrating the translation of an incapacitated astronaut between two points. The basic techniques of flying an MMU can also be practiced with the use of a chair system incorporating onboard thrusters and two hand-operated controllers.

Features of the air-bearing floor and simulated MMU controls on a training 'chair'.

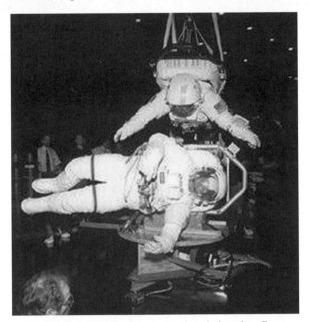

Simulation of crew rescue on the air-bearing floor.

Body harnesses

Various slings and body harnesses have been used for some exercises, by suspending the crew-member in a variety of rigs to simulate the reduced gravity of working in space, on the Moon or on Mars. They provide 4° of freedom for evaluation and training, but have very limited use.

The current American device at JSC is the Partial Gravity Simulator (also known as POGO), located in the Space Vehicle Mock-up Facility at Building 9. A 9-m tall A-frame supports a vertical servo and a horizontal air-bearing rail, with air and

Simulation of $^1/_6$ gravity on a 'Peter Pan' rig.

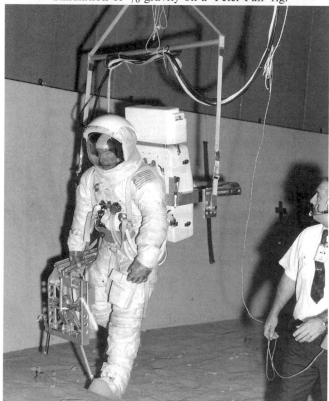

Suspended from a rig supporting $^5/_6$ of his weight, Apollo 13 astronaut Jim Lovell simulates a $^1/_6$ gravity traverse on Earth.

electrical overhead feed lines, and a seat and gimbal assembly supports the test subject during locomotion studies and training exercises. It is used for astronaut training and procedure evaluation, and can also be used to demonstrate EVA techniques in both microgravity and various partial gravity environments.

During the 1960s, several harness devices were used to simulate lunar gravity in the development of equipment and techniques for the Apollo programme. During the development of extended lunar activities for Apollo after the initial landings, various mobility devices were evaluated on test rigs, including a lunar motorcycle that was tested in the centrifuge building (29) prior to its conversion to the JSC WETF facility (see below). These rigs were also adapted for EVA simulations of ATM film retrieval in early AAP/Skylab simulations and have been used in early Shuttle EVA demonstrations and space station construction tests. They are currently being adapted for Mars-gravity mobility simulations.

Geological training

To support Apollo lunar surface activities the Americans began geological field training with the intake of the second group in 1962, and continued through 1972 with several trips to geological sites across the United States (including Hawaii),

John Young and Charlie Duke, wearing mock-up Apollo back-packs, undergo unsuited lunar traverse/geological training. The chest-mounted cameras were duplicates of those used on the Moon, but the head-gear was the choice of the individual astronauts.

Unsuited geological sampling using Apollo lunar surface tools.

Apollo 17 astronauts Jack Schmitt and Gene Cernan use the 1-g rover mock-up ('grover') during unsuited traverse training.

Iceland and the Canary Islands. Geological training was also carried on for Earth observations from Skylab and the Shuttle, but not to the extent that it was undertaken for Apollo. It is probable that the Russians also completed similar training through 1969. Classroom studies were divided into several courses over a number of years, and these were supplemented with field trips, with leading professional geologists, to sites that offered the best comparisons with expected terrain on the Moon. The astronauts would complete this type of training as part of their basic candidate training prior to flight-crew selection. After assignment to a flight crew, then more specific EVA simulations would be incorporated with geological field trips throughout the mission training programme, especially for the three J-class Apollo missions. Although having a piloting and engineering background, several astronauts enjoyed these geological field trips, while others tolerated them – and a few hated them. With the renewed interest in evaluating techniques required for Mars traverses, several privately funded and government-funded simulations are being conducted around the world. These are pioneering the techniques anew, but this time for missions to Mars, and not lunar missions.[3]

For the Apollo programme a total of 11,408 hours of lunar science studies – including briefings, geological field trips, lunar surface simulations and LRV training operations – was logged by all assigned landing crew-members. A further 526 hours (not all relating to EVA) were logged by the crews in zero-g training in the water tank, and 919 hours were logged by all crews in the checkout of the EVA hardware, excluding CM and LM systems familiarity training to support EVA operations. Each geological field trip lasted for between one and seven days, and the total number carried out by the landing crews, including the back-up crews and some support astronauts (such as CapComs and Mission Scientists), were as follows:

Apollo mission	No. of trips
11	1
12	4
13	7
14	7
15	12
16	18
17	13
Total	62

The following shows the total number of lunar surface simulations (the number of training sessions) for Apollo:

Neil Armstrong works at the MESA during suited 1-g simulation of the Apollo 11 EVA timeline at MSC, Houston.

Apollo mission	Surface operations	Operations before and after EVA	Total per mission	Actual lunar surface EVAs
11	20	10	30	1
12	31	4	35	2
13	42	11	53	(2 EVAs cancelled)
14	43	18	61	2
15	91	20	111	4 (including stand-up)
16	67	10	77	3
17	47	20	67	3
Totals	341	93	434	15

ZERO-G AIRCRAFT TRAINING

The strange condition of 'weightlessness' was first encountered during the aerial dog-fights of the Second World War, during which fighter pilots would dive on enemy planes from above and then pull up into a steep climb before conducting a further dive. The pull-up at the bottom of the dive resulted in positive g (more than normal), while at the top of the climb, negative g was encountered, with short periods of weightlessness during which the pull of gravity was counteracted. By 1951 the USAF had determined that brief periods of sub-gravity and so-called zero gravity could be

Aircraft simulations of Gemini EVA exit and entry techniques.

achieved by flying aircraft in a parabolic curve within the Earth's atmosphere, and a variety of experimental curves were subsequently flown by test pilots such as Chuck Yeager, Bill Bridgeman and Scott Crossfield. These studies continued into 1955 and 1956, when the School of Aerospace Medicine flew a Lockheed T-33A and also a F-49C Starfighter at altitudes around 5,200 m and 7,600 m to produce a period of zero g of 25–35 seconds. In one test in which an afterburner was used, an F-49C remained 'weightless' for 42 sec.[4]

NASA's KC-135 aircraft
NASA's Reduced Gravity Program, begun in 1959, used the Boeing military aircraft company model 367-80 (the basic design for the later commercial 707 and the KC-135 Stratotanker). The Mercury astronauts were the first to experience the phenomenon – during training flights in the rear seat of an F-100 in late 1959, in a padded rear compartment of a KC-131 transport aircraft in March 1960, and finally in a converted KC-130 cargo aircraft in September 1960. One of the pilots on these early flights was Ed White, who later became a NASA astronaut and the first American to walk in space. The addition of KC-135 aircraft to the NASA fleet has allowed the agency to conduct its own reduced gravity programme in support of

Aircraft $^1/_6$-g training for pulling the MET over simulated lunar terrain.

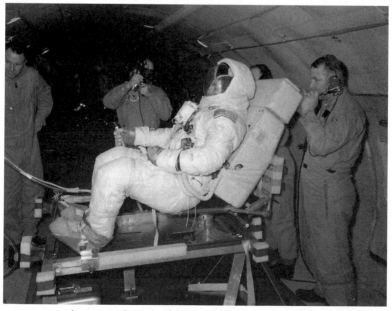

An astronaut practises procedures in the LRV mock-up crew seat in an aircraft flying a $^1/_6$-g profile.

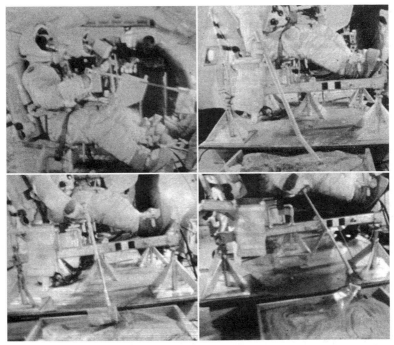

An astronaut practises the gathering of samples whilst seated on the LRV onboard an aircraft flying a $^1/_6$-g profile.

Mercury, Gemini, Apollo, Skylab, the Shuttle and the ISS. Because of the limit of only a few seconds to travel through each parabola, the technique is of minimal value for extended EVA operational training, but it has proven useful for training astronauts in certain aspects, procedures and details of EVA. By adjusting the flight path, lunar or martian gravity conditions can also be achieved. Since 1959, NASA has completed more than 80,000 parabolic flights in support of investigations into human and hardware reactions to varying degrees of gravity and weightlessness. The four-engine KC-135 carries a crew of five, a pilot, a co-pilot, a flight engineer, and two reduced gravity test directors.

A typical mission for this type of training lasts between two and three hours, and consists of thirty to forty parabolas, flown in succession or with short breaks between manoeuvres to reconfigure test equipment. On a single flight, 10–20 minutes of zero g can be experienced during these short bursts. Although the sensation can be very pleasant, repetition can be most uncomfortable for even the most hardened air-traveller; thus the aircraft is known as the 'vomit Comet'. The current KC-135 programmes flown out of Ellington Air Force Base, Houston, near the Johnson Space Center, take the aircraft to 7,300 m, where the nose of the aircraft is pitched up, typically to 45° (1.8 g), until approximately 10,000 m is reached, after which the aircraft pitches over to create the zero-g environment for about 20–25 seconds. After this, the aircraft dives at 45° (1.8 g) until it levels off at 7,300 m to repeat the process. The period from pull-up to pull-out is approximately 65 seconds. Some typical g-levels used on different tests are as follows:

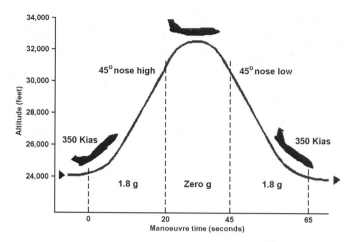

The flight profile of the 'reduced gravity' aircraft in simulating a range of gravity forces by adjusting its flight path to support Earth orbital (near zero-g), lunar ($^1/_6$ g) and martian ($^1/_3$ g) conditions.

Negative g	Approximately 15 sec
Zero g	Approximately 25 sec
Lunar g ($^1/_6$ Earth g)	Approximately 40 sec
Martian g ($^1/_3$ Earth g)	Approximately 30 sec

The cargo bay test area of the KC-135 is approximately 18.2 m long, 3 m wide and 2.1m high, and is equipped with electrical power, overboard venting/vacuum systems, and photographic lights. In addition to the air supply, nitrogen is available as liquid or gaseous supplies, and still and motion picture photography and/or video facilities are also provided. Most of the test equipment is bolted to the floor of the aircraft.

Russia's Flying Laboratory
The Russians have used similar aircraft in their cosmonaut training programme since 1960, and currently use an Ilyushin IL-76 MDK aircraft for up to twenty parabolas during a flight. Designated the Flying Laboratory, it is operated out of the Gagarin Cosmonaut Training Centre near Moscow, and is used in the training of cosmonauts to introduce them to weightlessness, aspects of entry and exit of EVA hatches, and crew transfer techniques. Early simulators of Russian EVA used the Tu-104 for the Soyuz 5/4 transfer. The aircraft carried partial mock-ups of the Soyuz OM hatch area for cosmonaut exit and entry training. In another section of the padded aircraft, the cosmonauts used mock-ups of the two linked spacecraft to rehearse the transfer from Soyuz 5 to Soyuz 4. In these simulations they encountered difficulties with the life support system, which caught in the hatch, and they had small problems with the mobility of the pressure suits. During the training process the prime crew performed entry and exit operations while the back-up crew practiced translation techniques in another part of the aircraft. The two teams then exchanged operations.[5]

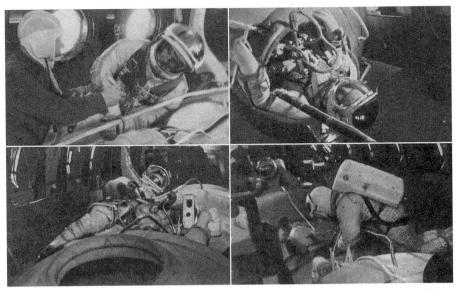

A Soyuz 5 cosmonaut practises EVA transfer in an aircraft flying a zero-g profile.

A Soviet cosmonaut undergoes zero-g training on the MMU.

Europe's A-300 Airbus programme

The European Space Agency began its programme of parabolic flights in 1984, and used the NASA KC-135 aircraft at Ellington Air Force Base six times. From 1988 the French space agency (CNES) subsidiary Novespace conducted more than fifty parabolic campaigns in the Caravelle aircraft, and between 1988 and 1995, fifteen of these campaigns were conducted by ESA. In 1994, ESA flew one campaign in the

Russian IL-76 MDK, and in 1996 used a NASA KC-135, flying out of Bordeaux. The following year, CNES purchased an Airbus A-300 and, under Novespace, operated it out of Bordeaux-Mérignac airport. It was used as a test aircraft for Airbus Industries, and had logged little flight time before transfer to the space agency. In 1988, it was used for training Jean-Loup Chrétien during preparations for his EVA on the Aragatz mission onboard Mir.

The Airbus cabin measures 20 × 5 × 2.3 m, and is the largest zero-g parabolic aircraft in service. With sub-systems and user services similar to the American and Russian aircraft, the ESA programme follows a similar flight profile, flying a 1-minute parabolic profile (20 sec at 1.8 g, 20 sec of weightlessness, and 20 seconds at 1.8 g) followed by a 2-minute 'rest period' at 1 g. After the tenth and twentieth parabolas, the rest interval is increased to 6 minutes. During all flights in these aircraft, the test crew is constantly informed of the campaign status, with indications of how many seconds remain before the next parabola, the approaching positive gravity, the number of minutes of rest period, the number of parabolas, and other data.

UNDERWATER TRAINING FACILITIES

One of the most effective methods of training crews for complex or long-duration EVAs has been the use of large water tanks, in which suitably weighted crew-members, supported by safety divers, work underwater on spacecraft mock-ups, and perform simulations of the tasks that they will carry out in space. This type of EVA training began in the mid-1960s, and is currently a major element of ISS EVA training. The pressure suits resemble the flight suits, but are specially adapted training suits for use underwater over longer periods, or for multiple submerges.

The general term for this type of training is 'neutral buoyancy'. Anything that is configured to be neutrally buoyant (achieved by a combination of flotation devices and weights) has an equal tendency to float as it does to sink, and an object prepared in such a way seems to 'hover' between the surface and the floor of a water tank. Large items can therefore be moved in much the same way as in orbit. Their 'weight' is controlled, but their mass remains.

There are, however, two significant differences between working in a water tank and working in space. First, the fact that the subject is not truly 'weightless' in the tank is not clear in film or photographs of the activity. Although neutrally buoyant, the test subject can still feel his suit around him, no matter what his orientation, and especially if he is upside down in the tank. On Earth, gravity is always present. It is essential that the suit fits comfortably, so that the test subject feels as comfortable as possible for the duration of the submerged test. The second factor is that the water creates a drag on movement, and this always hinders the motion of the body, arms, legs and hardware. Tasks are therefore generally much more difficult to accomplish in the tank than when in orbit, and this has to be recognised during planning and training.

Early studies
Studies of various Man-in-the-Sea programmes (including the Navy's Sealab programme) clearly demonstrated that the ability to work underwater in sub-aqua diving gear was comparable to an astronaut wearing a pressure garment in space. However, up to the mid-1960s the theories and applications of such techniques had not been fully explored. With the Gemini EVAs encountering difficulties that had not arisen in zero-g or 1-g training, however, the option of using water tanks for EVA training became more appealing.

Shortly after his Mercury flight in 1962, Scott Carpenter became involved in the evaluation of underwater training for space, and after discussions with the French oceanographer Jacques-Yves Cousteau he approached the United States Navy. He afterwards participated in the Sealab Man-in-the-Sea programme, and in 1965 spent 30 days in Sealab II. Carpenter had often stated that being in space was similar to being submerged in water, and when he returned to the Astronaut Office in 1966 he was placed in charge of its AAP Branch Office. Thereafter he supported underwater training for EVA, and later worked as a NASA liaison with the USN before returning to the Navy in 1967.[6]

According to one account,[7] in early 1966 the two directors of a small aerospace firm on the outskirts of Baltimore had lobbied several agencies (the Air Force, the Navy and NASA) to sponsor an experiment to determine whether underwater techniques could be adapted as a suitable EVA training method and in understanding body motion in space. The first demonstration was held in a rented swimming pool at a Baltimore Boys' School.

Simulations at Langley
The first major demonstration of underwater techniques as a possible EVA simulation was sponsored by NASA Langley Research Center during July 1966. The simulation was filmed and clips were reviewed. After this it was decided to pursue the technique and develop a programme to evaluate equipment and procedures for Gemini EVAs, and to investigate the use of underwater simulations for future programmes. As a result of this, test simulation services were contracted, full-scale Gemini mock-ups were fabricated for underwater use, and Gemini EVA training suits and support equipment was modified for the simulations.

The underwater EVA training facilities were adapted for later Gemini missions. For Gemini 10, authorisation to use the procedure came too late in the training cycle for the prime crew to participate in the underwater simulations. With less than a month left before launch, the crew was simply too busy to add underwater training to their schedule. They therefore approached the Gemini Program Manager Chuck Mathews, arguing that due to the simplicity of the equipment (which could be prepared prior to opening the hatch), EVA astronaut Mike Collins had only to attach the nitrogen line to the outlet supply on the spacecraft. It was located just outside the cockpit, there was a hand-rail providing adequate support, and the nitrogen was only required to supply the manoeuvring gun. Mathews agreed, but decided that a checkout of procedures should be undertaken by others in the water tank.[8] The partial task evaluation, simulated by the contractor of the Gemini 10

EVA, concluded that the tasks assigned to the flight were 'reasonable and feasible'. Collins later wrote that he was told that he 'might have trouble with the task, and then again, I might not. John and I received this news with straight faces ... sometimes I had trouble with it in the zero-g airplane, and other times I did not.'[9]

The second simulation was completed *after* Gemini 9 had returned home, and with the EVA Pilot (Gene Cernan) acting as test subject. This was a post-flight re-run of the proposed AMU EVA that had to be abandoned on the mission. By evaluating the accuracy of the underwater simulation compared with the actual orbital conditions which he had already experienced, Cernan was able to offer an objective analysis of the water tank in relation to what was actually experienced in orbit. He concluded that the demonstration had merit in the areas of suit dynamics and in continuation of a particular task, and also reported that the maintaining of body position and the associated fatigue strongly resembled the situation he had faced during Gemini 9. Cernan's comparisons of EVA to swimming in a pool help secure the use of underwater EVA training from then on.[10]

A third simulation was held prior to Gemini 11. This was a contractor evaluation of the planned EVA procedures and equipment to be used. Following the simulation the results were analysed, and after discussion with the prime crew (EVA Pilot Dick Gordon), the EVA plan was changed. Both the still and motion picture tasks were removed from the EVA in order to focus attention on the retrieval of the experiment cameras and the HHMU.

For Gemini 12, the water tank proved very useful in preparing the EVA astronaut (Buzz Aldrin) for his tasks in orbit. During one of the underwater simulations for the impending Gemini 12 EVA, Aldrin and Cernan were both underwater, following the EVA plan. Aldrin later recalled that as he followed the sequence of clipping, connecting, screwing and torquing, he 'suddenly felt absolutely ridiculous.' He realised that his pet monkey could probably have performed the same tasks, 'probably not underwater, but a least out in the garage. I let out a high-pitched screech.' A startled Cernan looked at his colleague and wondered whether there was a serious problem, only to be greeted by a glaring stare as he asked what was wrong. 'Shut up and pass me a banana', Aldrin replied. This response became a standing joke in the Astronaut Office, and for some time his desk was fully stocked with fresh bananas.[11]

The Neutral Buoyancy Simulator, MSFC
The Neutral Buoyancy Simulator at the Marshall Space Flight Center was completed in 1968. It was 22.8 m in diameter and 12 m deep, held 11 million litres of water, and had four levels to allow observation at different depths. Audio and video facilities were also installed, together with a fully equipped test control centre used for directing, controlling and monitoring activities in the tank. The facility was sited in the south-west corner of Building 4705. Initially it was built for design engineering and for developing new space hardware, but it later became a valuable crew training aid.

(Early in the AAP, MSFC had performed some neutral buoyancy design and development work in a small tank previously used for explosive forming.[12] Funding

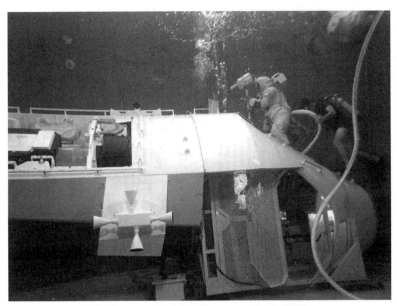

Underwater training for trans-Earth EVA.

for the facility – which was used extensively during AAP and Skylab and later on the Shuttle programme for EVA simulations – would have been provided by the Cost of Facilities budget. The funding, however, was denied, and so the resourceful MSFC management reclassified the facility as a tool and spent $1 million of the Research and Development budget. This led to a reprimand from the General Accounting Office for such 'creative movement of finances'.[13] Initially, the location of the NBS at MSFC caused friction at Houston. In August 1966, a meeting of the NASA Management Council of the Office of Manned Spaceflight had assigned the Manned Spacecraft Center (later the Johnson Spaceflight Center) the responsibility for the astronaut team and the EVA equipment which they would use. But MSFC was assigned responsibility for 'large structures' and for studies of 'EVA equipment and procedures which may by used to carry out operations on large space structures'. The rivalry went as far as MSC Director Robert Gilruth, who plainly stated that he thought the Marshall tank was a needless duplication of Houston's role and an attempt by MSFC to become a 'manned space centre'.)

The Water Emersion Training Facility, Houston
Building 29 at JSC originally housed a centrifuge for simulating launch and entry up to 15 g on Saturn rockets. For the Shuttle programme, however, the change in the flight profile and launch pressures resulted in its being able to pull only 3 g during certain phases of launch. It was therefore no longer required, and in the latter half of the 1970s it was disassembled, to be replaced with a Water Emersion* Training

---

* The correct word is 'immersion' – the entry of an object into a liquid, or the occultation or eclipse of a celestial body. 'Emersion' is the appearance, above the water, of a previously submerged object.

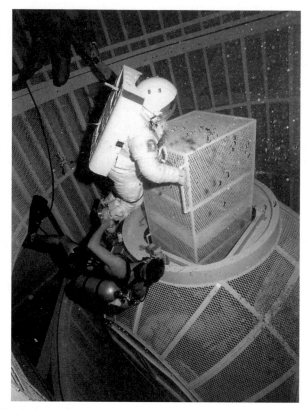

Water-tank simulation of passing a large-volume item through a space station (Freedom) node.

Facility (WETF), for training astronauts on EVA from the Shuttle, and for water egress training. The information below pertains to operational use of the facility during the 1970s to 1990s.[14]

The facility includes a rectangular pool instead of a circular pool, and was originally designed to hold a full-size mock-up of the Shuttle payload bay, where most of the Shuttle EVAs would to be performed during the early years of the programme. At this time (during the late 1970s) there was no firm commitment to a space station programme, let alone a final design, and it was therefore a balance of economics that governed the construction of the Building 29 WETF, although early space station mock-up elements were used in the pool. The facility is 24 m in length, 9.8 m in width, and 7.6 m deep, and it contains 1.82 million litres of water maintained at 31° C.

The JSC WETF facility has standard filtering, chlorinating and pumping systems, and additional sub-systems that support test activities including diving, electrical power systems, an environmental control system, a closed-circuit TV system, a communication system, overhead cranes, and a medical support facility. Air is supplied for the use of all air-powered tools.

Orbiter mock-up The primary facility used in the pool was a full-size orbiter payload bay mock-up complete with operating payload bay doors, with a full-size RMS mock-up to practice different EVA scenarios, including payload bay closure and the restraint or removal of a simulated faulty RMS.

Cranes A 4.5-tonne pneumatic hoist on a monorail and a 4.99-tonne circular pneumatic hoist are available for lifting mock-up hardware or EVA crews in and out of the facility.

Environmental control system This provides both air and water through umbilicals to the life support system of the Shuttle EMU that is worn during EVA simulations. The system is monitored and controlled at the ECS console and at the test director's console. The ECS contains a series of switches that monitor preset limits for airflow and pressure, and incorporates an alarm warning system which is activated if these limits are not maintained. A 30-minute back-up emergency air supply can support suited crew-members in the event of a failure in the primary system.

Communications system There are two-way communications between all members of the test team, except the underwater support divers. The communication system is housed in the test director's console, and transmits to the EVA test subject via an electric cable in the life support umbilical. It has a prime and a back-up channel in case of a primary mode failure, and as an additional precaution a battery-powered communication system is available should both primary and back-up channels fail, although this provides only one-way communication, through an underwater speaker, from the test director to the support divers.

Suit ballast system In order to simulate the microgravity of space as closely as possible, a system of weights is used to ballast each test subject as required. Front and back weight packs, two arm weight cuffs and two ankle weight cuffs can be attached to the exterior of the EMU, and are adjustable (by adding or removing weights) to achieve the required weight distribution for neutral buoyancy. The packs are designed so that in the event of an emergency situation they can be removed in 5–10 seconds.

Closed-circuit television This consists of two underwater pan-and-tilt cameras, two underwater hand-held cameras carried and operated by support divers, and TV monitors mounted on the test console.

Dressing rooms Separate dressing rooms are provided for male and female test members, and the divers have suiting, dressing and equipment storage rooms separate from those for the EMU/EVA test subjects.

Medical station This is manned by two trained medical technicians, with a medical doctor available for immediate response in the event of an emergency. During all WETF operations the team uses the TV to monitor activities, and an emergency vehicle is on close stand-by to transport the personnel to the hyperbaric chamber or other medical facilities as required.

The use of the tank in preparing astronauts for Shuttle EVAs was first demonstrated early in the programme, as payload bay door closure, stowing of the RMS, the Spacelab air-lock and exit, and entry of the middeck air-lock, were all simulated between 1978 and 1983, together with the first attempts at space vehicle servicing, repair and construction. After the first EVA was completed on the Shuttle (STS-6 in April 1983), Story Musgrave recalled the preparation for his tasks in the pool at Houston, and compared it to performing his first EVA: 'There's no viscosity [in space]. If you get [something] going you can keep it going. There's a reach right there I could make in the water tank that I can't make here,' he commented during the EVA, referring to his evaluation of hand-holds at the aft end of the payload bay. 'This is a little deeper pool than I've been used to working in,' he added, orbiting the Earth at about 300 km above the ocean.[15]

When the ISS mock-ups were projected as being too large to be submerged in the old WETF in Building 29, the size and design of the new facilty allowed all EVA and water egress training to be relocated there, thus rendering the old facility surplus to requirements. There were plans to have the WETF filled in and Building 29 converted to more office space, but the astronauts and their rehab specialists successfully lobbied for its new use as an ISS long-duration crew post-light rehab and personal fitness pool.

Sonny Carter Water Emersion Training Facility, Houston
With the advent of the ISS programme in the late 1980s, it became apparent that the WETF at JSC and the Neutral Buoyancy Simulator at MSFC would not be able to support the planned EVA training simulations required to prepare for space station construction. McDonnell Douglas was contracted to construct a new water

The platform for lowering an EVA crew into the WETF, or NBF, at Houston.

immersion facility for astronaut training at the Ellington Air Force Base, just north-west of the main JSC facility. The new facility was designated the Neutral Buoyancy Laboratory, and in 1995 was dedicated to astronaut Sonny Carter (1947–1991), who was instrumental in developing many of the current EVA techniques used by NASA. Construction was completed two weeks ahead of schedule on 20 December 1995, and after a year of system installation, tests and trial simulations, the first scheduled training operations began in January 1997. Training operations are managed, under contract, by Johnson Engineering (part of SpaceHab Inc). This facility is also used for water egress/recovery training of Shuttle crews.

The facility was designed to simultaneously cope with two activities, using mock-ups sufficiently large to produce an accurate and meaningful training content and EVA duration. The pool is 61.56 m long, 31.06 m wide and 12.18 m deep (6.09 m above ground level, and 6.09 m below ground level), and holds 28.18 million litres of water. However, even at this size it is unable to accommodate the full dimensions of the 106.6 m × 73.1-m ISS, and so only parts of the station can be submerged at one time. Two overhead bridge cranes, each capable of lifting 10 tons, and several jib cranes, each capable of lifting 1.6 tons, are used to change the configuration inside the NBL. The voice communication system includes a two-way communication link between the astronauts in the tank and the topside training facility test conductor, and can also be linked to the flight controllers in MCC Houston and the rest of the Shuttle crew in the Shuttle mission simulators. Divers again receive one-way communications through underwater speakers, although upgrades to this commu-nications system are under development.

Full-scale working models of the Shuttle RMS and SSRMS (Canadarm 2) are included, with hydraulic joints replacing the electrically-driven flight models. Because of the pressures encountered when working at these depths, the divers use oxygen-enriched breathing gas (Nitrox) to reduce the probability of the bends after long training sessions. Nitrox is also supplied to the EVA suits through umbilicals. The Portable Life Support System back-packs used underwater are mass models, and not working units.

The water in the facility is recycled every 19.6 hours, and is automatically monitored and regulated at 27–31° C in order to minimise the potential effects of hypothermia on support divers during a long diving session. The water is also chemically treated to eliminate the growth of contaminants in the water and to prevent deterioration of the mock-ups and equipment.

In addition to the main pool there are several support systems and teams on hand during suited operations. The main simulation and control area monitors resources for all aspects of the test programme, including safety, communications, video support, medical requirements, suit technicians, support divers, trainers and technical observers. The small medical team is always on station during the training sessions, with a hyperbaric chamber prepared to deal with unexpected decompres-sion and immediate evacuation to medical facilities on site or near the facility. There is also a support system, with personnel available for the design, manufacturing, assembly and testing of equipment for use in the NBL, as well as for maintenance, repositioning and storage of all NBL mock-ups. Hand-held and mounted video cameras provide visual coverage and data-recording links to topside consoles, to MCC at JSC, and, if desired, directly to NASA Select Cable TV channels.

Fabrication of the facility was initiated with the removal of the existing concrete floor of the building and the installation of steel castings for the support pilings. A pipe system was then added to reduce the effects of hydrostatic water pressure on the perimeter of the tank, and pumps were installed to prevent seepage into the evacuated area. The total amount of materials used in the building of this facility was gargantuan:

Excavation	641,088 cubic feet
Wood layering	1,782 timbers, each 10 × 9.5 × 3 feet
Reinforced steel floor	690 tons
Reinforced steel walls	645 tons
Steel wire	310 miles
Floor	6 feet thick; 143,100 cubic feet; 10,303 tons; 530 truck-loads of cement
Walls	5 feet thick below ground; 2.6 feet thick above ground; 94,500 cubic feet; 6,070 tons; 350 truck-loads of concrete
Water	200,000 gallons; 25,823 tons; filtered at 5,400 gallons per minute

The Hydro Laboratory, Moscow

The tank in the Hydro Laboratory at the Gagarin Training Centre near Moscow is 23 m in diameter and 12 m deep, and holds 5,000 m^3 of water with the temperature maintained at around 30° C. Every twelve to eighteen months the water is emptied

Cosmonauts perform EVA tasks underwater.

away, and the whole tank is cleaned.[16] The tank has been in use since the 1970s, and has supported EVA training and planning for Salyut, Mir and ISS operations.

During each test, a team of six or seven divers help guide the spacesuited test subjects to the desired work-station. In addition, two cameramen, inside the tank, photodocument the exercise. The back-packs are mock-ups, and oxygen is supplied via umbilicals. Test conductors monitor progress from poolside consoles or through several view ports. The main control panel is located on the side of the tank, with view ports provided for direct visual observation during the test procedure. A small team of specialists – including the test conductor, the chief doctor, and an EVA specialist – control and monitor the operation, and support teams monitor the parameters of the suit and the medical condition of the cosmonauts.

The test subjects wear suits with provision to add or remove weights on the wrists, ankles, waist, chest or back as required, to simulate weightlessness in space. The cosmonauts begin their training programme in a conference room or classroom, where the principles of the suit and the timeline of the EVA are discussed and evaluated. After a physical examination by a medical specialist, they then move to the suiting area near the poolside, and don the liquid-cooled undergarment, communication headgear and Orlan suit. The suits are supported in a frame attached to a hoist, and after an integrity check the subjects are hoisted up and over the pool by one of several overhead cranes, as they are unable to move without assistance. Following EVA simulation and removal from the pool area, the EVA crew takes a hot shower and then attends a debriefing session with doctors, EVA specialists, trainers and other cosmonauts.

The facility is used for training cosmonauts in EVA techniques, for testing and developing new procedures, equipment and hardware, and in support of EVAs being carried out in space to verify procedures and techniques for the cosmonauts in orbit.

The Russian hydro tank without water.

Inside the Orlan (underwater) training suit. (Astro Info Service collection.)

The Japanese water tank

As a major partner in the ISS programme, Japan has developed its own crew-training facilities, mainly focused around the Kibo experiment module. The facility, located at the Tsukuba Space Center (TKSC) in Ibaraki, Japan, includes a Weightless Environment Test (WET) building adjacent to the astronaut training facility. The circular tank is 15 m in diameter and 10.5 m deep, and can accommodate full-size mock-ups of the Kibo facility. As with the American and Russian facilities, there are several sub-systems for supporting test procedures. The water purification and heating system controls the quality and temperature of the water by using a system of filters, pH control equipment, pumps, and a boiler. Each EVA test suit is supplied, via an umbilical, with chilled water to cool the occupant. Breathable air is also supplied to the test subjects via umbilicals (the back-packs are mock-ups), and also to the support the divers' scuba tanks to prolong their submergence throughout protracted test. The facility also includes an adjacent control room with video and communication links, an emergency hyperbaric chamber and test support equipment where donning and doffing of pressure suits is accomplished, 1-g testing and training facilities, support equipment for the spacesuits and the scuba-diving gear, and a facility for drying the suits.

The Orlan underwater training suit. (Astro Info Service collection.)

SPECIAL TRAINING FACILITIES

Over the years, several additional facilities have been used in support of EVA training.

RMS EVA training

When new Shuttle astronauts are selected by NASA they are introduced to a range of speciality areas which they will expand upon, according to their ability and skills, during their careers as active astronauts.

While some astronauts become specialists in performing EVA (mainly, but not exclusively, Mission Specialists, as some Pilots are also trained in EVA operations), others master the Shuttle (or space station) RMS. The RMS has a critical support role in Shuttle-based EVAs, and a coordinated team effort of the EVA crew and the orbiter flight-deck crew ensures the success of the mission. As with all aspects of crew training on Shuttle missions, one crew-member takes the primary role, while a second or third astronaut is trained in a back-up RMS role (which on several occasions has included the mission Commander).

An EVA crew-man is lowered into the hydro tank for EVA training. (Courtesy ESA.)

RMS training using virtual-reality images displayed on a computer screen in a classroom, prior to moving to Shuttle/RMS mock-ups.

In 1977, Spar, the primary contractor for the Shuttle RMS, built a 30 × 18-m high-bay clean room for integration and testing of the RMS system. For astronaut training and simulations of RMS operations, a Simulation Facility (SIMFAC) was developed for both real-time and 'non-real-time' simulation programmes, in which the dynamics of the RMS, orbiter and planned payload could be evaluated. The SIMFAC reproduced the dynamics of the RMS and orbiter from data gathered in its early flights during the non-real-time programme, which was used for developing the design and verifying the performance of the RMS. This development was then used to verify the real-time facility to provide the operator with an accurate visual representation of the performance and behaviour of the arm, with the operator's reactions being realistic. It was completely enclosed in a replica of the RMS control station on the aft flight deck of the orbiter, and commands were generated through hand-ooperated controllers and the display and control panel. The display generated four real-time views of the outline of the payload bay, the payload and the manipulator arm on CRT monitors, using an internal data reference base to determine the shape of the payload structure and manipulator arm as it would be seen from the orbiter windows and with the closed-circuit TV camera.[17]

MMU simulators

MMU Space Operations Simulator (SOS) The complexity of the Shuttle MMU required an accurate duplication of its characteristics for astronaut training. To achieve this, Martin Marietta (the prime contractor) developed, at its Denver plant, a six-degree moving base simulator to develop flight procedures and train crews, to supplement air-bearing, harness rig and zero-g aircraft training at NASA. The facility – the Space Operations Simulator – featured a high-fidelity mock-up of the

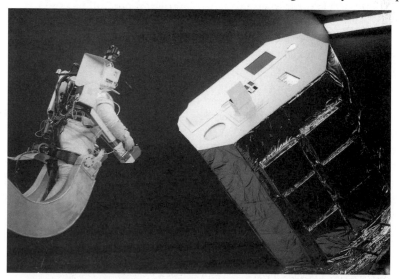

The MMU simulation facility at Martin Marietta, where astronauts/test subjects simulate approach to a target satellite (here Solar Max). (Courtesy Martin Marietta.)

Monitoring the MMU simulation run at the simulation facility. (Courtesy Martin Marietta.)

MMU mounted on a carriage. The unit was linked to computers that read the hand-control inputs from the Pilot, and modelled the affected MMU systems to calculate the appropriate drive response to move the carriage. This resulted in realistic high-fidelity representation of the MMU being flown in zero gravity, and the translational and rotational movement of the astronaut in the simulator. The Pilot could use the MMU simulator in either a suited or non-suited mode, and under simulated day or night conditions, to provide a broad range of training operations and scenarios. The SOS facility also included full-scale partial mock-ups of the Shuttle orbiter nose (aft crew compartment) and the forward section of the payload bay (containing the Flight Support Stations). Specific mission training utilised mock-ups of the Shuttle and related spacecraft (Solar Max, Palapa and Westar), set up in the simulator area to provide the flight crews with additional accurate training simulator sessions. Large-screen video displays enabled simulations of manoeuvring velocity to and from the vicinity of the orbiter. The SOS was used by the crews of only three Shuttle missions – STS-41B, 41C and 51A (six astronauts) – and for a number of other training simulations and demonstration exercises in the first half of the 1980s, before the MMU was retired from the programme following the *Challenger* accident. It was replaced by the simpler design of the SAFER unit, which did not require the SOS for training.

Soviet MMU training facilities The Russian MMU training and verification facilities featured a rotational test facility designed to verify angular rates of the axes of the unit in relation to the test facility table, with 3° of freedom to study the angular motion along two directions. This floor measured 7 × 8 m, and was machined to a precision of 20-μm smoothness and a slope of no more than 0.1 mm/m. Compressed air was supplied, through an umbilical, to the aerostatic support pads on the Pilot's support frame, and there was also a visual transfer simulator called Polosa.

SAFER facilities

Initial SAFER training was completed (for STS-64) at the WETF facility at JSC, where the astronauts practiced end-to-end choreography. Training focused on SAFER recharge, battery change-out, crew transfer, and set-up and clean-up operations. SAFER avionics and hardware was also used at the NASA Integrated Operations and Analysis Laboratory (IGOAL), which is similar to the Shuttle Avionics Integration Laboratory (SAIL), providing training on piloting techniques and procedures, and rehearsal of malfunction procedures. Suited operations gave the astronauts experience of the visual range from within the EMU helmet, and the limitations of using the suit while wearing the SAFER unit. The JSC air-bearing floor was also utilised. Finally, the Shuttle Engineering Simulator was used to practice SAFER rescue training, using a combination of the SAFER unit and Shuttle orbiter to effect the rescue of a crew-member.

ON-ORBIT TRAINING

It has often been stated that the best place to train for spaceflight is in space itself. On the early missions it was impossible to train for EVA operations, as the flights were so short and the EVAs themselves were experimental. Knowledge gained from the early Gemini and Apollo EVAs was utilised to prepare later activities to maximise their return. However, it was not until the Soviet Salyut programme that EVA techniques could be practiced in a microgravity environment before going outside to perform the actual operations. This was demonstrated during the Salyut 7 engine repair EVAs in 1984, during which a visiting crew supplied the crew with new training documents to read and videos to watch before they attempted their next EVA. This on-orbit preparation continued on Mir with repeated activities in pre- and post-EVA operations, and will probably become a feature of future ISS EVA service and maintenance operations.

VIRTUAL EVA

During the 1990s, computerised virtual reality training aids became more of a feature of spaceflight preparations as the graphics capabilities of visual displays become more refined. The integration of EVA crews with the RMS operation on Shuttle EVA training was one of the earliest uses of virtual reality training. The most valuable experience has been for the series of Shuttle service missions to the HST, during which, in the classroom, practice of instrument and system removal and replacement was possible wearing the VR helmet/goggles and sensitive gloves, before progressing to hardware mock-ups in the various water tanks.

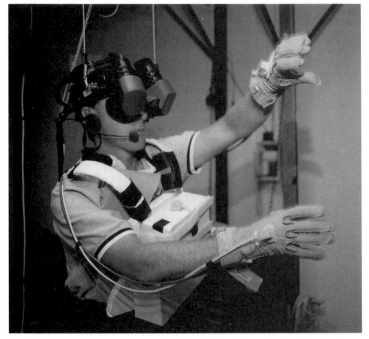

During the 1990s, EVA training introduced virtual-reality technology, with gloves and goggles linked to the computer to produce images in the goggles from movements of the head and hands. This type of training will be an increasingly familiar element in future EVA crew preparation.

CONCLUSION

It is very difficult to replicate zero g on Earth. Parabolic flights in aircraft can replicate varying degrees of gravity from negative g to positive g, but for only a few seconds. For EVA – which can occupy several hours – underwater facilities have proven the most beneficial. Various rigs and 1-g training devices have been used, but not to the extent of the water tanks, which for the space station have proven invaluable. Geological field trips across various terrains on Earth can help in surface exploration of the Moon and Mars, but only to a certain degree. More recently, the use of on-orbit training, computer-generated visual systems and virtual reality has provided an added dimension in preparing for spacewalks. Those who have flown in space have often commented that the best place to train for a spaceflight is in space; indeed Russians believe that only those who have flown in space should be designated cosmonauts. This type of 'in-flight' training will probably become part of future candidate astronaut and cosmonaut training, where a short flight to the ISS is included in gaining 'hands-on' training before assignment to a flight crew that will travel to a lunar base or to Mars; indeed, training for Mars may include developmental EVAs on the Moon. Whatever the method chosen, it has become clear that despite the increased training for spaceflight – and in particular, for EVA – nothing compares with the experience gained during an actual flight. After exiting the spacecraft, the view makes all the training worthwhile – provided there is sufficient time to admire and appreciate it.

REFERENCES

1 *STS EVA Description and Criteria*, NASA JSC, 10615 Rev A, May 1983. This focuses on EVA planning for STS missions, but the criteria are generally applicable to all types of EVA planning.
2 *Summary of Gemini EVA*, Section 7, EVA Training and Simulation, NASA SP-149, 1967.
3 Shayler, David J., Salmon, Andy, and Shayler, Michael, *Marswalk*, Springer–Praxis, (due in 2004).
4 Gantz, Lieut-Col. Kenneth F. (*ed.*), *Man in Space*, Duell, Sloan and Pearce, 1959, pp. 108–132.
5 *Transfer in Orbit*, Novosti Press Agency, 1969, pp. 69–73.
6 Carpenter, Scott, and Stoever, Kris, *For Spacious Skies*, Harcourt, 2002, pp. 313–328.
7 Aldrin, Buzz, *Return to Earth*, Random House, 1973, pp. 173–174.
8 Collins, Michael, *Carrying the Fire*, Farrar, Straus, Giroux, 1974, pp. 192–193.
9 *Ibid.*; *Summary of Gemini EVA*, NASA SP-149, 1967, pp. 7–34.
10 Cernan, Gene, *The Last Man on the Moon*, St Martins Press, pp. 156–157; Astro Info Service interview with Gene Cernan, Houston, Texas, August 1988.
11 Aldrin, Buzz, *Return to Earth*, Random House, 1973, pp. 174.
12 Compton, David, and Benson, Charles, *Living and Working in Space: A History of Skylab*, NASA SPP-4208, p. 170.
13 Dunar, Andrew, and Waring, Stephen, *Power to Explore: A History of MSFC, 1960–1990*, NASA SP-4313, pp. 187–188 and 648.
14 WETF information brochure, NASA, 1988; *Orbiter*, No. 80, Astro Info Service, October 1991, 18–21; Astro Info Service notes from personal tour of WETF, August 1988.
15 *EVA Report No. 2*, Astro Info Service Publications, 1983.
16 Astro Info Service notes from a personal tour of the Hydro Laboratory, June 2003.
17 STS-3 RMS Press Kit, Spar Aerospace, March 1982.

Surface exploration

Writers of science fiction have long had their space explorers visit both the known worlds of our Solar System and planets of their imagination, and this has continued into the twenty-first century in science fiction feature films and television series. In the eyes of many, true space exploration involves visiting and exploring 'strange new worlds'; but in the forty years of actual human space exploration, only one programme has managed to achieve that goal: Apollo. Although other programmes, both in America and in Russia, have proposed further exploration of the Moon and Mars, and even some of the moons of the other planets, no programmes have yet been fully authorised. So far, the history books record that only twelve humans have walked on a world other than our own Earth. Thirty years after the last of them left the dusty surface of the Moon, we are further from those events than we were in the 1950s, when science fiction began to turn into science fact.

The early missile race evolved into the space race and then the Moon race, and within a period of little more than twenty-five years we witnessed the first orbital flight of a small satellite, men walking on the Moon, probes to the distant planets, the first space stations, and the beginning of international cooperation. Over the last twenty-five years, however, planetary exploration has struggled along at a faltering pace, and although our spaceflight experience has increased with ever more complex space stations and orbital shuttle craft, no human has ventured out of Earth orbit since 1972. Of the twenty-four men who journeyed to the vicinity of the Moon we have lost six, including three of the twelve who walked on the lunar surface. It is sobering to think that many of us who witnessed those few small steps across the Moon might not be here to experience the thrill, satisfaction and sheer sense of awe of seeing the next human 'giant leap' in space and on to the red plains of Mars.

This section reviews what has been learned from surface-exploration EVAs, represented only by the experience gained by the Apollo astronauts, although there should also be recognition of the dedicated efforts of their colleagues in Russia as they tried to compete with the achievements of Apollo in the late 1960s and early 1970s.

' ... OF LANDING A MAN ON THE MOON'

President Kennedy's historic speech of 25 May 1961, committing the Americans to go to the Moon by the end of the decade, was delivered six weeks after the first manned orbital spaceflight by Yuri Gagarin, and less than three weeks after Alan Shepard's 15-minute sub-orbital flight. No person had yet spent an extended period of time in space, nor had anyone attempted to exit a vehicle in orbit or conduct any scientctic experiments. Yet here was the challenge to fly to the Moon and back in a mission that would have to last 7–10 days. It is interesting to note that Kennedy said that the goal was a landing of a man on the Moon and his safe return to Earth. There was no reference to walking on the surface, collecting rocks, or setting up experiments, and this argument between safety versus science, intertwined with 'why send men at all when robotic craft could be used?', would constantly plague Apollo from creation to cancellation.

The engineers and pilots wanted to prove the concept and 'get there and back' in one piece, saving the science for later missions. The scientific community argued that if men were to be sent, then the flight should include professional scientists who could take advantage of the uniqueness of each Apollo landing site to select the best samples for analysis on Earth. The inclusion of the first scientist-astronauts in the programme in 1965 was a step in that direction, but difficulties in developing the hardware, minimising the weight, and flight safety, constantly pushed the idea of

An early impression of an Apollo lunar surface EVA.

science on Apollo further and further down the flight manifest. As Deke Slayton
once said: 'It was far easier to train a pilot to pick up rocks than to prepare a scientist
to fly a spacecraft.' Indeed, although only one professional scientist (geologist Jack
Schmitt) reached the Moon, on the very last landing mission, it transpired that some
of the pilot-astronauts also happened to be good geologists.

Stories of the evolution of Apollo programme, the hardware, the mission
operations, and the frustrations of the Soviet programme in attempting to reach the
Moon, have been told countless times, and are referenced in several works (see
Bibliography). Here, however, the focus is on the development of the EVA aspects of
the programme, and the lessons learned from the surface activities that might have
an application for future programmes of surface exploration.

Early studies
One of the first post-Sputnik studies into human exploration across the surface of
the Moon was published by the USAF Ballistic Missile Division on 25 April 1958,
as a development plan for an Air Force programme of military manned space
systems, the overall objective of which was 'to land a man on the Moon and return
him safely to Earth'.[1] This high-priority effort was a four-phase programme
designed to place a man in space (Man in Space Soonest), to expand human
capabilities in space (Man in Space Sophisticated), to conduct unmanned
photographic and soft landing exploration (Lunar Reconnaissance), and for human
exploration of the Moon (Manned Lunar Landing and Return), during which
animal passengers would complete circumlunar missions, and human crews would
land. The programme, to be completed by December 1965, included an initial brief
surface exploration by astronauts. and later flights that would thoroughly explore
the lunar surface. The USAF Pioneer series of unmanned space probes would
support this effort.

From the creation of NASA in October 1958, to Kennedy's speech in May 1961
and the completion of contracts for the main lunar landing spacecraft in November
1962, there were dozens of studies, proposals, plans, ideas and suggestions for the
goals of the next US manned space programme after Mercury, and the role that the
US military might play in direct human spaceflight programmes. Plans to dispatch
unmanned robotic reconnaissance craft to the Moon evolved into what eventually
became the Ranger hard-landing probes and the Surveyor soft-landing probes,[2] and
the Soviet Union was also showing an interest in reaching the Moon with its early
Lunik (Luna) probes. During these formative years, most design and development
was devoted to methods of reaching the Moon, and the types of craft and crews.
Several diagrams of 'astronauts' exiting spacecraft and 'exploring the Moon' were
issued, but no details were revealed.

On 17 September 1960, an interesting paper was presented at a meeting of the
British Interplanetary Society, in London. The paper was a typical popular review of
what extended lunar exploration might be able to achieve by about 1970, and echoed
the hopes of the 1950s concerning the coming decade. Advanced roving vehicles,
unmanned robotic spacecraft and the development of a focal-point lunar base were
all part of the overall plan, which projected initial short reconnaissance landings

during 1968 and 1969, followed by an extensive surface exploration expedition over ten weeks (five complete lunar-day cycles) between 1970 and 1971. What was interesting was that the roving surface explorers would visit a variety of surface features, including craters, rilles, mountains and maria, and would place automatic experiments on the surface, collect rock and soil samples, conduct seismic and electric sensitivity experiments, and take core and drilling samples – all of which the Apollo astronauts later completed in the timeline suggested in the paper![3]

Meanwhile, NASA was beginning to consided what would be done when the first crews landed. During the first meeting of the Manned Lunar Landing Task Group on 9 January 1961, all that was detailed on surface exploration was a three-step projection of a manned landing on the Moon, with return to Earth, limited manned lunar exploration, and then a scientific lunar base. By 2 May 1961, NASA Associate Administrator Robert C. Seamans had established an Ad Hoc Task Group for a Manned Landing Study to identify tasks assigned to the manned lunar landing mission, time phasing of these tasks, and their estimated cost, shortfalls and manpower requirements. In this study, the estimated date of the first manned landing was 1967, and one of the objectives was to determine the time spent on the lunar surface and what the astronauts would do there. The report was limited in that the hardware design and mission profile was far from being selected, but the general

An early 1960s design for a lunar exploration suit and its associated equipment.

consensus was for a maximum 24-hour first landing, with a single-person excursion close to the landing vehicle to take samples of soil and rock and perhaps set up an instrument and take photographs.[4]

Introducing EVA to Apollo
Studies to define the main CSM in 1962 also began to include provision for EVA. Life support provision would allow two or more crew-members to perform EVA operations at the same time from the CM, but some of the ideas for EVA at this time, when no-one had actually performed EVA, verged on the adventurous and the downright dangerous. North American Aviation completed a preliminary require-ment outline for spacecraft rendezvous and docking, in which two Apollo spacecraft would rendezvous to within a few metres of each other, hold position at a relative velocity of less than 15 cm/s, and then steer to withing a few centimetres of each other. Then, a crew-man – 'assumed to be adequately protected against radiation and meteoroid bombardment' – would reach out and grasp the docking spacecraft and manoeuvre it to the docking seals for a hard docking! In June 1962, five NASA 'scientists' wearing pressure garments conducted a study, at the Rocketdyne facility, into the feasibility of using special maintenance tools and a variety of designs of pressure garment to repair or maintain components on the J-2 Saturn rocket engine while in space! Finally, in a study by NAA in July 1962, several docking methods were investigated, including 'a system in which a crew-man, secured by lanyard, would transfer into the open lunar module [by EVA]. Then a second crew-man in the open CM 'air-lock' [hatch] would reel in the lanyard and bring the modules together.'[5]

In July 1962, statements of work for the proposed Lunar Excursion Module were issued to industry. These included the capacity to separate the LEM from the CSM in independent flight for up to two days, including time spent on the Moon. The LEM had to support surface activity by the two-man crew, but would not have its own mobility system. Its pressurisation system would have to be capable of six complete cabin repressurisations, and the two Portable Life Support Systems for EVA operations would have to be recharged up to a maximum of six times, with a nominal operating time without recharge of up to four hours. The astronauts could therefore each perform up to three four-hour EVAs (including contingency EVAs).

The statement also indicated the scope of the scientific investigation package being considered at the time. The LEM would contain 'a scientific instrument system [consisting of] a lunar atmosphere analyser, gravitometer, magnetometer, radiation spectrometer, specimen return container, rock and soil analyser equipment, scientific equipment, soil temperature instruments and cameras.' During the subsequent months, several studies evaluated supplementary 'lunar logistics' concepts to deliver additional payload and consumables to the lunar surface.[6] These would form the core of what became the Apollo (lunar) Applications Program which, it was hoped, would follow the initial Apollo lunar landings.

On 7 November 1962, NASA announced that Grumman Aircraft Engineering Corporation had been selected to develop the two-man two-stage (ascent and descent stage) Lunar Excursion Module (LEM – later changed to Lunar Module, LM) for

Early exit methods from the LM, with a squared-off hatch and no forward strut ladder.

the manned lunar programme. By the end of 1962, many of the main elements of what was flown under the Apollo programme had been defined. The three-man mission would be launched by a Saturn V following the Lunar Orbital Rendezvous profile. Two men would land in the LEM, and the third would remain in the CSM in lunar orbit, with transfer via a mechanical docking system featuring a probe and drogue and internal transfer tunnel. After a period of surface exploration, the ascent stage of the LM would lift off and rendezvous with the CSM in orbit, and the three astronauts would make the return flight home to end a mission of about seven days.

Factors affecting Apollo EVAs
With the baseline mission plan defined, refinements to the design of the hardware, the mission and the objectives would occupy the next six years, and delays and consequences of the Apollo 1 pad fire (in January 1967) would affect the operational phase of Apollo. To detail all these developments is beyond the scope of this book, but summarised below are the significant features that affected the Apollo EVA development effort.

Hardware
Because of the changes to the docking location on the LM, from forwards and

Astronauts Frank Borman and Elliot See, wearing thermal coverall garment over early
Apollo pressure garments, conduct simulated EVA on mock-up lunar terrain.

overhead to just the overhead docking port, the shape of the forward hatch could be
changed from round to squared off, to better facilitate the astronaut's movement in
and out of the spacecraft while wearing the pressure suit and straight-sided Portable
Life Support System.

The crew stations accommodate the Commander on the left and the co-pilot (LM
Pilot) at right, duplicating aviation flight stations. The hinges of the hatch would
open inwards, towards the LM Pilot station, which meant that, unlike the early
planning documents in which the LM Pilot would go out and perform an EVA, in
the new design the Commander had to get out first to allow the hatch to be partially
closed so that the LM Pilot could move across to the now empty Commander station
and then exit the hatch. Equally, the LM Pilot had to re-enter the vehicle first, to
allow the Commander to enter without inference. This was a major factor in
sequencing who became the first of a crew to step onto the Moon, and the first of the
crew to leave the surface.

A nine-rung ladder was installed on the forward landing strut, with a small
platform – 'the porch' – on top of the ladder area near the forward hatch, for ease of
movement. This was a vast improvement over earlier so-called 'Peter Pan' jury-rigs
and even a knotted-rope system.

The safe landing angle was limited to a 30-degree tilt to the landing site for safe
lunar launch back to orbit, and easier exit and entry between the LM and the
surface. The specifications of the LM were changed to facilitate a shorter mission
with greater hover time to enable a safe landing on the first mission(s), plus a ten-
hour surface stay, six man-hours EVA capability, 32 kg of scientific payload, and 36-
kg lunar sample return capacity. The LM would retain the capacity to extend to full
capability on subsequent landings, based upon experiences from the first landing(s).

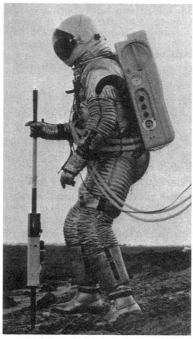

An astronaut test-subject uses a 'Jacob's staff' topographical/balancing instrument in early Apollo EVA traverse simulation.

This was adapted (in part) for the Apollo 11 mission, with five man-hours of EVA, less than twenty-four hours on the surface, an abbreviated scientific experiments package, and no extended geological traverse.

Suits

In November 1962, guidelines were issued on the preferred EVA suit design. It would include a Portable Life Support System supplying oxygen, pressure-controlled temperature and humidity, and shielding from extreme temperatures and solar radiation. The helmet would have an anti-solar glare faceplate that could be defrosted to improve visibility at extreme low temperatures, and an emergency oxygen supply system.

Liquid-cooled garments were selected in preference to gas-cooled, for a much-improved heat-load capability, providing the astronauts with better cooling when in training in a 1-g environment.

Sites

Early fears about the depth of dust on the surface focused initial landing efforts toward places where the dust would hopefully be at its thinnest. The early hard and soft lunar landings soon dispelled any fears that the lander would sink into a deep layer of dust, but there remained concerns about dust being kicked up by the descending LM and during traverses.

Walter Cunningham, wearing an early Apollo pressure garment and coverall, conducts a mobility exercise.

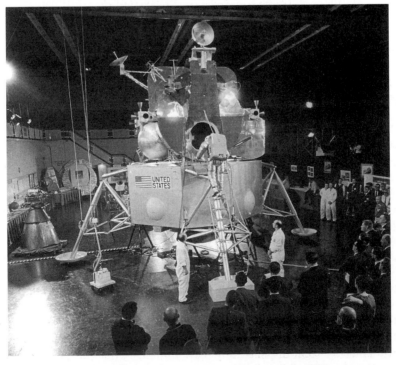

Early EVA operations near a mock-up of the LM.

By March 1963, preliminary studies had characterised a good landing site as being on a lunar sea (mare), 16 km from a continent ridge and 16 km from a main mare crater. This would permit the most scientific activity, thereby enabling planners to design future missions for even greater scientific return, and would define the nature and scope of sites for the most feasible landing

Candidate landing areas for Apollo were selected based on slopes, surface texture, strength, landmarks, isolated features, the size, shape and position of different areas, accessibility using the Apollo Saturn V/LOR profile, landing lighting conditions, and launch and splash-down constraints. The sites meeting these requirements were primarily in the central equatorial region of the Moon, restricted to $\pm 5°$ N/S and 50° E/W. Candidates sites would be reduced after further studies from Lunar Orbiter unmanned mapping missions, which would narrow target landing sites for Surveyor landing missions and lead to a final choice of manned landing sites based upon unmanned data and supplementary data from early manned lunar orbital Apollo missions. The primary concern of the first site selection board was safety, not science. That latter would ensue after the Apollo system had proven viable in supporting a manned landing mission.

Experiments

The first meetings of the Lunar Surface Experiments Panel were held at MSC between 24 February and 23 March 1963. The group was formed to study and evaluate the expected scientific returns of proposed lunar surface experiments, and how Surveyor and other unmanned probes could be adapted to support the manned lunar programme. General guidelines for conducting scientific investigations fell into three principle scientific areas; comprehensive observation of lunar phenomena, collection of representative samples, and placement of monitoring equipment.

The Apollo Experiment Ad Hoc Working Group proposals were incorporated into the final recommendations for Apollo's principle scientific objectives, including examination of both geological and physical aspect of the Moon in the area immediately surrounding the LEM, geological mapping, investigations of the Moon's interior, studies of the lunar atmosphere and radio astronomy from the surface.

The early definition for the size of an experiment delivered to the Moon was a package weighing 113 kg and measuring 0.3 m^3, with an initial return capability of 36 kg (later increased to 45 kg) and 0.06 m^3 (later increased to 0.09 m^3). For subsequent landings, the size and mass of experiments would fall within the guidelines for the safety of the crew, the LM, the objectives of the mission, and the overall success of the mission.

Contingency samples would first be collected in case of early termination of the excursion, and pending the collection of fully documented and environment samples.

Bendix Corporation would construct the Apollo Lunar Surface Experiment Package (ALSEP) and, following a review of deployment in 1968, an abbreviated experiment package for the first landing, called the Early Apollo Surface Experiment Package (EASEP).

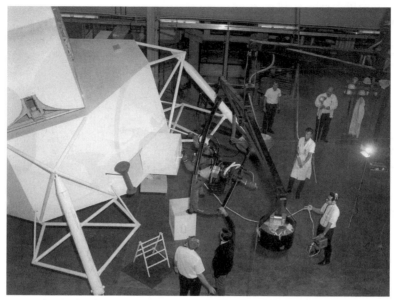

An early simulation of surface EVA using a 'Peter Pan' rig.

Excursions

On 27 March 1963 the Apollo Mission Planning Panel set two requirements for the landing mission. Firstly, both crew-men should be able to work on the surface at the same time, for productivity and safety. Secondly, based on the 48-hour LM-independent operational requirement, time on the surface was set between the two guidelines of 24 hours in flight and 24 hours on the surface, or 3 hours in flight and 45 hours on the surface. These guidelines would be incorporated in the first landing missions, up to Apollo 14 in 1971. The capability to conduct EVA from the CM during translunar/trans-Earth trajectory was included, beginning with Apollo 15.

In March 1966, Bendix Corporation was contracted for construction of the Apollo Lunar Surface Experiment Package (ALSEP), and in 1969, Boeing was selected to design, construct and deliver the two-man LRV, which was achieved within seventeen months of the contract agreement.

The duration of the later Apollo missions (the J series) was extended, increasing mission duration, surface stay, payload delivery and return capability. Surface EVAs were increased from two to three, and a scientific instrument package was added to the SM, requiring a deep-space EVA.

Safety

During a meeting of the Apollo Docking Interface Panel in November 1963, the probe–drogue system was recommended. The two methods for astronaut transfer thus became the primary internal docking tunnel, and EVA transfer from the LM to the CM.

In February 1964, EVA transfer from the LM to the CM was selected as the back-up method of crew transfer, to be used only in the event of inner transfer hatch faults or pressurisation failure. North American Aviation had not designed the Block I CM for EVA, and the Block II transfer tunnel was not designed for an astronaut in a pressurised suit. To transfer to the CM without docking was not possible, because the LM was not controllable without a crew on board.

COMPETITION: THE SOVIET MANNED LUNAR PROGRAMME

Although no Soviet cosmonauts ventured to the Moon when Apollo was operational, they did participate in testing procedures and hardware that could have been applied to lunar operations. The Soviet manned lunar programme was shrouded in mystery for many years, and at first was even denied by the Soviets to avoid embarrassment after the success of the Americans. Gradually, however, information about the troubled programme has emerged. During the mid-1960s, the leading Soviet design bureaux of OKB-1 (Korolyov) and OKB-52 (Chelomei) competed for the authority to head the Soviet manned effort to send cosmonauts to the Moon. It was a conflict of interests that would also surface in the production of the first space stations in the 1970s.[7]

The Soviet lunar programme consisted of a circumlunar effort (L1 programme of two cosmonauts launched by Proton rocket and based on the 'Soyuz' vehicle) and the manned lunar landing programme (the L3 programme of two cosmonauts launched on the huge three-stage N1 launch vehicle). From 1968 to 1970 the L1 profile was flown on a number of unmanned Zond missions, but in 1970 was cancelled before ever supporting a manned mission. The L3 programme did not complete any significant missions, and in 1974 was also cancelled without a manned launch.

The L3 mission required two spacesuits: one for lunar orbital EVA operations by the Commander, and one for surface operation by the Pilot. On the landing missions, the Pilot would have transferred by EVA to the lander, and then repeated this EVA in reverse after returning from the Moon. It was therefore decided to simulate this EVA transfer on Soyuz in Earth orbit, initially in April 1967 as part of the Soyuz 1/2 mission. But the in-flight difficulties of Soyuz 1 led to the cancellation of Soyuz 2, and a 21-month delay in the operation before it finally flew as Soyuz 4/5 in January 1969. By then it was a useful demonstration of the technique of EVA transfer, but due to several serious delays in the N1/L3 programme it had no further application in the Soviet lunar programme.

Soyuz EVA transfer, January 1969
During the autumn of 1966, the first crews were formulated from the Soyuz cosmonaut training group for the planed Soyuz 1 and Soyuz 2 docking and EVA transfer mission the following spring. Cosmonauts Alexei Yeliseyev and Yevgeny Khrunov (who had been a back-up to Leonov on Voskhod 2) were selected to make the EVA transfer, backed up by Valeri Kubasov and Viktor Gorbatko. Soyuz 1, with one cosmonaut (Commander Vladimir Komarov), was to launch the day before Soyuz 2, carrying three cosmonauts (Commander Valeri Bykovsky; Flight Engineer Yeliseyev; Research Engineer Khrunov). On the third day of Soyuz 1 and second day of Soyuz 2, the two spacecraft would dock, and the EVA cosmonauts would transfer

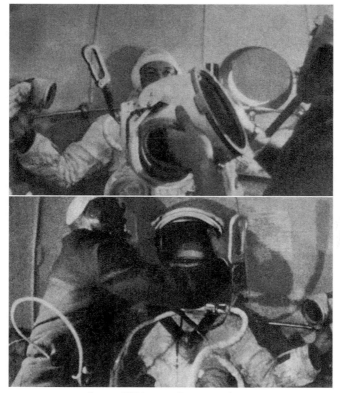

Soyuz EVA transfer operations.

from the OM of Soyuz 2 to the OM of Soyuz 1. The following day, Soyuz 1 would return to Earth, and Bykovsky alone would bring home Soyuz 2 the day after.[8]

However, the difficulties encountered by Komarov on Soyuz 1, almost as soon as he entered orbit, led to his tragic death, and the launch of Soyuz 2 was cancelled. After several unmanned missions flown later in 1967 and during 1968, the first manned Soyuz – Soyuz 3 – was launched with a lone cosmonaut on board. But this was a solo test flight designed to dock with the previously launched unmanned Soyuz 2. Georgi Beregovoi failed to achieve the docking, but succeeded in his rendezvous with the other Soyuz and thus qualified the spacecraft for further missions. The next flight was that of Soyuz 4, with Vladimir Shatalov on board. Two days later, Yeliseyev and Khrunov took off in Soyuz 5, but with a new commander, Boris Volynov. A day later they achieved the docking with Soyuz 4 and, after more than two-and-a-half years of training, finally prepared for their EVA.[9]

Volynov assisted his colleagues in donning their Yastreb EVA suits and then retired to the DM and sealed the interconnecting hatch. In the OM of Soyuz 5, Yeliseyev and Khrunov, wearing their full pressure garments, depressurised the module and prepared to exit through the hatch on the side of the module (the hatch used to enter the Soyuz on the launch pad). During simulations it had been determined that they could not easily pass through the hatch when wearing the EVA life support system on their backs, and although a larger hatch was being planned for later Soyuz craft it would not be ready for their mission. The back-packs were therefore transferred to their legs. The EVA lasted only 37 minutes, with Khrunov and then Yeliseyev transferring across to Soyuz 4, each monitoring the other during transfer. Once both were safely inside the OM, the hatches were sealed and, after repressurising Soyuz 4, the Commander opened the interconnecting hatch to welcome his new crew-members, who toasted each other with blackcurrant juice.

A successful L3 demonstration?

The EVA was the first and, to date, the only EVA accomplished through the OM of a Soyuz spacecraft, and the only one associated with the Soviet manned lunar programme. The EVA was only the second time cosmonauts had stepped outside their vehicle, and it was the last time that a Soviet/Russian EVA would be conducted on a non-space station mission. But was anything learned from the exercise? The primary purpose was to evaluate the technique for the L3 manned lunar programme and as a secondary demonstration of the possibility of EVA transfer in a hypothetical crew rescue situation. Expressed in these terms, the activity was successful, in that both cosmonauts (not just one on L3) transferred from one spacecraft to another. If such a rescue scenario was possible, then the two cosmonauts proved it could be done; but it depended on a variety of situations being favourable before such an EVA would be attempted as a rescue.

L3 transfer

In L3, the two cosmonauts would have transferred from the DM to the OM for exit into space. The suits would be stored in the OM, and would be donned before sealing the hatches between the modules (with no-one in the DM) and opening the outer hatch. The lunar explorer, wearing a Krechet suit with autonomous life support, would have moved hand over hand to the descent vehicle, with the Commander wearing an Orlan suit and a short umbilical, standing in or near the OM hatch to monitor and film his colleague, and remaining ready to go out and assist him (or rescue him) if required. This was the format used on Soyuz 5, with Khrunov making the trip first watched by Yeliseyev; except that on this mission, Yeliseyev followed instead of returning to the Soyuz 5 DM.

On Soyuz, the EVA was delayed by about 11 minutes when Khrunov had to return to the Soyuz 5 OM because he had erroneously connected the ventilation umbilical to his colleague's suit instead of his own. This was probably due to lighting conditions inside the OM, the similar appearance of the connections, the lack of clear identification marks, and an eagerness to proceed with the EVA. Such an operation was not planned for L3, but would probably have led to the better identification of components, better lighting, and careful timing. There would probably have been precious little time to extend or delay such an EVA without having a serious impact on lighting conditions at the landing site.

To move from Soyuz to Soyuz, a series of EVA hand-rails was provided to assist in moving across the surface of the vehicle. In the early post-mission reports, the cosmonauts were typically reported as saying that everything went according to plan.

Khrunov reported that he had had no difficulty in exiting the Soyuz, and that 'it was most practical to use our hands and hold on to the firm hand-rails as a means of support.'[10] More recently, reports have indicated difficulty in grasping such hand-rails, and fatigue resulting from the effort to hang on.[11] These hand-rails – which were also adopted across the family of Salyut/Mir space stations – would probably have been incorporated in the L3 mission hardware.

The portable camera was lost on the EVA, and with it, photographs of the event (a fate that had also befallen Gemini 10 in 1966). The tethers, apparently, were not secured. The cosmonauts also had difficulty in resealing the hatch of Soyuz 4, because floating straps and tethers were becoming caught in the hatch mechanism and seal. This again would have delayed events on L3 because there was only one cosmonaut, but on Soyuz there were two cosmonauts to solve the problems.

It is clear that the Soyuz 5/4 transfer provided the Soviets with valuable flight experience in the system developed for their lunar programme, and although that programme was far from operational, the experience would have given them time to incorporate any changes and revisions to procedures had the programme continued.

Rescue
One of the claims for the achievements of Soyuz 5/4 was that it was a demonstration of rescue operations of a stranded crew. But the problem with this is that on this mission the docking and transfer was preplanned, and the cosmonauts had trained for it. Contingency training can be completed to simulate probable emergency situations, but reality is almost certainly not like the simulations – if only in the fact that no matter how realistic the emergency being trained for, it will always be just a simulation, without the fear or stress. The other factor is the incorporation of adequate systems to allow a rescue by EVA. The docking systems have to be compatible, and the EVA would have to be viewed as a more viable option than an internal transfer via the pressurised tunnel.

In 1969 both the Americans and the Russians were operating a probe–drogue system, either without an internal transfer hatch (Soyuz 5/4) or with one (Apollo, and later Soyuz/Salyut missions). If a problem were to occur with either system, any rescue spacecraft would have to have a compatible docking system, assuming that the stranded vehicle was not also tumbling. It was this mission, and discussions between American and Soviet space officials over rescue and docking compatibility, which led to the 1975 Apollo–Soyuz Test Project and eventually to Shuttle–Mir and the ISS cooperative project. However, true space rescue capability has still not been fully developed.

The other factor is that to perform a rescue, EVA suits are required for every crew-member. On Soyuz 5/4, Shatalov and Volynov did not wear suits, and it was only after the loss of the Soyuz 11 cosmonauts in 1971 that Sokol suits were worn for IVA emergency requirements on all Soyuz launches. The early Shuttle programme also developed a 'theoretical' EVA rescue system for all crew-members, but it was not adopted for flight operations.

In summary, Soyuz 5/4 proved that a crew, given suitable conditions and equipment, could transfer from one craft to another. This demonstrated a theoretical space rescue system, but the work requied to put such a procedure into operational use was never carried out.

Lunar cosmonauts
The use of the Yastreb suit on the Soyuz 5/4 transfer reflects what the EVA cosmonaut on the L3 lunar missions would have to have completed, but in lunar orbit, not Earth orbit. The Pilot in the Krechet suit would probably have taken the same amount of time (approx 40 minutes), and the EVA would have been conducted on the near-side of the lunar orbit for direct communication with Earth. It is assumed that the second EVA would have taken a little longer due to the need to transfer lunar sample boxes, films and other items to be returned to Earth. This would have resembled the Apollo trans-Earth EVAs to retrieve SIM-bay film cassettes, and could have taken 45–60 minutes, again on the Earth-side of the orbit. The Orlan suit developed for L3 has been stated as being capable of supporting two 2.5-hour operational EVAs.[12] The Commander could therefore have been outside (or in a vacuum) for a complete lunar orbit (approx two hours, with a 30-minute emergency reserve) for each EVA. The Soviet lunar surface suit had a stated capacity of up to ten hours during a 52-hour stay on the Moon.

The Krechet design evolved into the Krechet-94, and by 1972 the L3 programme became L3-M, which would have had a three-person lunar mission with a two-person landing crew (similar to the Apollo profile). For this scenario it was suggested that modifications to the Krechet-94, with refilling and servicing of the suits while on the surface, could have supported six surface EVAs over a five-day stay. In some diagrams, small one-man lunar rovers were depicted as mobility systems available for surface activities, although they would probably not have been used on the first Soviet manned landing.

Experience of the Orlan suit on the later space station programme revealed to Zvezda that the projections for L3 EVA operations, at least in lunar orbit, were over-ambitious. The fabrication, testing and preparation of suits of different designs and objectives was complicated and prone to delay and there was no compatibility between them. The EVA transfer from one craft to a second was complicated, risky and time-consuming, and the length of time that the cosmonauts were required to remain in the suits was uncomfortable – even before they were due to walk on the Moon. Experience from Apollo demonstrated that lunar exploration had to be considered as at least a two-man operation, and the work-load on the single cosmonaut and suit system would have been an additional burden.[13]

MAGNIFICENT DESOLATION

The Soyuz EVA transfer was the first and last in-flight demonstration of a Soviet lunar EVA procedure, and would also be the final Soviet EVA for eight years as the development of manned space stations and internal transfer systems replaced plans to send cosmonauts to the Moon. During this time (1969–77), America flew and completed both the Apollo lunar programme and Skylab, which featured EVA as an operational activity, and developed plans for the Space Shuttle that would take EVA to a new level.

However, before sending spacecraft to the lunar surface, NASA planned a test of its own lunar suit and emergency crew transfer that was much the same as used on Soyuz 5/4. The first Apollo EVA therefore did not take place on the Moon, but from Apollo 9 in Earth orbit

Red Rover's Golden Slippers

Because two spacecraft were utilised, radio call-signs were employed for the CSM and the LM. On Apollo 9 these were Gumdrop (CSM) and Spider (LM). It was also decided that since Rusty Schweickart would be testing the self-contained Apollo EMU outside and would essentially become a separate 'spacecraft', he should have a separate call-sign: Red Rover. The transfer would be filmed from the open hatch of the CM by CM Pilot Dave Scott, while Commander Jim McDivitt would remain in the LM cabin in his Apollo EVA suit. Various explanations for the choice of call-signs have been put forward over the years, varying from the rusty colour of Schweickart's red hair, to having him 'rove' from the LM to the CM, to the red colour of his EVA helmet (the only time it was coloured red – probably because of the photodocumentation that was to be conducted).

The plan was for Schweickart to wear the EVA back-pack that would be used on the Moon, and to evaluate the exit to the front porch of the LM. He would 'stand up', place his feet in secured foot-restraints on the platform (painted gold – hence

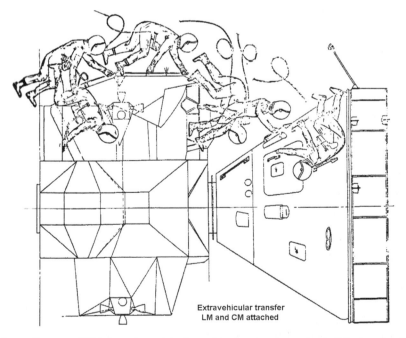

Extravehicular transfer
LM and CM attached

A contingency EVA traverse route for returning crews, evaluated during Apollo 9.

Golden Slippers), and, after taking 16-mm and 70-mm photographs of Scott emerging from the CM hatch and collecting thermal samples attached to the exterior of the LM, would translate toward the CM using the EVA hand-rails mounted along the front of the LM to enter the CM hatch as Scott photographed him. Early in the mission, the EVA was threatened with cancellation when Schweickart, on his first spaceflight, suffered bouts of motion sickness. McDivitt felt that this might seriously impact his performance outside the spacecraft, and recommended that the exercise be limited to only a cabin depressurisation. By the fourth day (EVA day) of the flight, however, Schweickart felt much better, and it was agreed that the EVA could continue. McDivitt therefore relented, and allowed his LM Pilot to at least go out and 'stand on the porch'.

The EVA was supposed to have lasted about 2 hours 15 minutes, but actually lasted a little over an hour. Considerably more effort was required to open the forward LM hatch than expected; but they finally did so, and Schweickart floated out – feet first and face out (instead of feet first and face down, as astronauts would do on the Moon) – for easier egress in a zero-g environment. He was attached by a 7.6-m nylon tether to prevent his floating away, since he was using the Portable Life Support System of the Apollo EVA suit and not umbilicals, which Scott and McDivitt were using. While Schweickart was on the 'front porch', restrained on the Golden Slippers, Scott reached out and retrieved thermal samples from the exterior of the spacecraft, finding those on the CM missing but successfully retrieving those from the forward edge of the SM near the open hatch. Schweickart managed to complete an abbreviated translation evaluation along the hand-rail in front of the LM, but not across to the CM. As he moved he evaluated mobility in the suit and the body attitude control to ensure that he did not strike parts of the spacecraft such as antennae, or tear the pressure suit as he moved. Scott filmed part of his transfer; but then the film jammed, allowing Schweickart to spend a few minutes just floating, hanging on to the LM and admiring the view of Earth and the stars. Schweickart executed an uneventful translation back to the LM cabin, and Scott closed the CM hatch to complete the first American two-person EVA operation.

Schweickart's demonstration increased confidence in the use of the suit on the surface of the Moon and, as with Soyuz 4/5, demonstrated that an LM crew could, of necessary, traverse from the lander across to the CM. Schweickart's performance on the porch after recovering from his illness convinced McDivitt that the hand-rail evaluation should be attempted, and in his evaluation Schweickart found that the hand-rail was easier to use, even one-handed, than in pre-mission simulations, including those carried out underwater. On a lunar flight, however, moving the return sample containers and other equipment back to the CM might have been a little more demanding; but such was the success of the EVA that it was decided to not include a planned similar exercise on Apollo 10. Scott's evaluation demonstrated that EVA from the CM was also a relatively simple exercise, increasing confidence in plans for future EVAs from the main spacecraft in the later scientific missions in the AAP. The next time an American would leave the spacecraft, he would step onto the Moon.

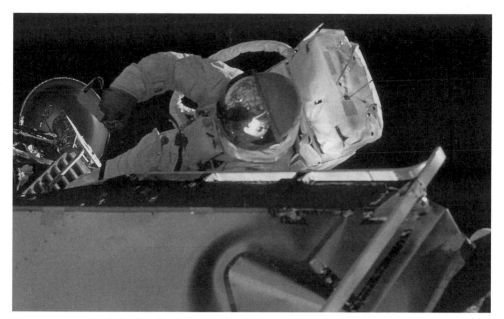

The Apollo 9 EVA test of the lunar surface EVA suit.

The Moon landings, July 1969–December 1972
History books record that the first men (Neil Armstrong and Buzz Aldrin) landed on the Moon on 20 July 1969, at a place called Tranquillity Base on the Sea of Tranquillity. They remained there for less than a day, and completed a single EVA. This was the first of an originally planned ten lunar landings that would have taken place by 1974. Unfortunately, three missions from the original programme (Apollo 15, 19 and 20) were cancelled, and due to redesignation, Apollo would terminate with Apollo 17 and not with the originally planned Apollo 20. Because of budget restrictions, the lack of available hardware, and a desire to move on to other programmes, there would be no Apollo follow-on exploration missions under the AAP. Accrued surface EVA experience during flown Apollo lunar missions was as follows:

Flight	Landing site	EVAs	Duration hrs, min		Surface stay hrs, min	
Apollo 11	Sea of Tranquillity	1	2	23	21	36
Apollo 12	Ocean of Storms	2	7	45	31	31
Apollo 13	Fra Mauro	(2 planned)				
Apollo 14	Fra Mauro	2	9	22	33	30
Apollo 15	Hadley–Apennine	4	18	34	66	54
Apollo 16	Descartes	3	20	14	71	2
Apollo 17	Taurus–Littrow	3	22	3	74	59
Totals		15	80	21	299	32

The duration is the period from cabin depressurisation to repressurisation. Apollo 15 also included a stand-up EVA (by Dave Scott) from the overhead LM hatch. There were also three deep-space EVAs (the first), performed on the final three Apollo missions to collect Scientific Instrument Module film cassettes: Apollo 15, 39 min; Apollo 16, 1 hr 23 min; and Apollo 17, 1 hr 5 min.

The lost Apollos
The loss of the three Apollo missions in 1970, and the abort of Apollo 13 on the way to the Moon, resulted in lost opportunities for further EVA experience. Two EVAs were planned for Apollo 13, for 4 hours each during a 34-hour stay, but from the flight records of Apollo 14 (which landed at the site intended for Apollo 13), an estimated total of 9–9.5 hours EVA would have been conducted during the two planned EVAs in a stay of approximately 33.5 hours. In addition, each of the cancelled Apollo missions (Apollo 18–20) would have completed at least three EVAs, each using the LRVs, which could have seen durations of around 21 hours (3 × 7 hours) during a 72-hour surface stay.[14] In recalling the achievements of Apollo, the lost opportunities and the potential for further knowledge and experience should not be forgotten, and perhaps serves as a lesson learned for future exploration of the Moon and beyond.

LEARNING FROM THE MOON: THE APOLLO EXPERIENCE, 1969–72

The first EVA on the Moon, by Apollo 11 astronauts Neil Armstrong and Buzz Aldrin on 20 July 1969, lasted just over 2 hrs 31 min. Although the astronauts

Geologist-astronaut and Apollo 17 LM Pilot Jack Schmitt (right) in discussion with Mission Scientist Bob Parker (left) during a break in training. The ground-based Mission Scientist and EVA CapCom were integral members of the EVA 'team' to support and liase between the crew, flight controllers and backroom support staff and scientists.

travelled about 1 km, most of this travel was confined to the vicinity of the LM – primarily for safety reasons, and to ensure that the majority of the activities were recorded by the TV camera. Towards the end of the EVA, Neil Armstrong ventured 60 m from the LM to inspect a 33-m diameter crater, and during the whole EVA, the two astronauts gathered 21.55 kg of samples. The lunar exploration phase of Apollo had had a great start, but just forty-one months later, Apollo 17 astronauts Gene Cernan and Jack Schmitt completed their third excursion – the programme's final excursion – on the surface. On that day, 13 December 1972, they spent more than 7 hrs 15 min on the surface, during which they drove their LRV 12.1 km for an accumulated time of 1 hr 31 min, and gathered 62 kg of samples. But all too soon, Apollo lunar exploration was over.

Between these two events, twelve astronauts left their boot-prints in the dusty soil of the Moon. More than thirty years after the last of those adventurers left the Moon, the footprints are still there – but no new prints have been added. So, what was learned in EVA techniques, developed during those few relatively short visits to our nearest neighbour, that might have application when we return to the Moon, and for safe and successful adventures to places farther afield?[15]

Planning

For reasons of safety, the first EVA on the Moon was planned to be of short duration. The objective was to meet the goals of Apollo by reaching the surface of the Moon, conducting a brief excursion, collecting limited samples, taking a few photographs, deploying a few experiments, and then safely returning to Earth. The basic LM and surface systems were enhanced for a second EVA, ALSEP, and a geological traverse for the next series of Apollo missions (Apollo 12–14), with Apollo 14 also using the MET. The next three Apollo missions (Apollo 15–17) were more scientifically orientated, with surface operations extended to three separate excursions, the LRV, and more geological sampling. Coordinated with orbital science operations, this provided a baseline of valuable data. Before the budget axe fell, these would have been followed by a further three missions of the same type (Apollo 18–20), to provide a firmer foothold on the Moon, to develop even more extended operations, and to establish one or more lunar research stations, eventually leading to a permanent lunar base over a twenty-year period. When the missions began to be restricted, serious rethinking was undertaken about where to target the remaining missions for the maximum scientific return each time. It paid off with a remarkably successful series of missions and a demonstration of surface activity planning and implementation that built upon the triumph and historical significance of the first landings.

Future application It is very difficult to determine the application of Apollo in mission planning without clarification of future programming. The advancement of hardware and sub-systems, the experience of several decades of spaceflight operations, and the knowledge that 'we've been there before', will lead to an approach different from that of Apollo. It is to be hoped that when we next go to the Moon it will be with the view to stay, and the planning of the exploration and exploitation of the Moon can be a more coordinated and comprehensive programme, with both immediate and long-term returns as part of a long-awaited space infrastructure.

Training

One of the most challenging human training and development programmes has been for human spaceflight, and in particular, EVA. As discussed in Chapter 4, ingenious methods have been devised to duplicate different levels of gravity to overcome our natural environment on Earth. As with space-borne EVAs, during Apollo it was clearly demonstrated that 'practice makes perfect', and training programmes evolved as each crew progressed through the lunar training cycle. Post-flight debriefings of early flight crews also helped to develop better training programmes for later crews, and the results are evident from the productivity of subsequent crews as they completed their missions.

Future application Does the training for the Apollo Moon missions have an application for our eventual return to the Moon? Had we followed Apollo with the extended lunar surface explorations and a lunar research station (early Moon-base), then perhaps there would have been a direct application, because the programmes would have been only a few years apart. But since we are looking at another decade or so at least before we return to the Moon, 40–50 years after Apollo, preparations will be quite different, due to advances in technology and changes in objectives. It is still difficult to simulate $^{1}/_{6}$ g, and so reduced gravity aircraft flights and suspension rigs will still be used for some aspects, if only in the development of new equipment. Will the next generation of lunar explorers dust off the files of Apollo? Probably. The archives of Skylab, Salyut and Mir have had direct application to the ISS, and certain Apollo technology has been reviewed during Shuttle–Mir (docking systems) and ISS (rescue vehicles), so it is reasonable to predict that Apollo will be revisited when we commit ourselves to a return to the Moon.

EVA preparations

It has been frequently noted that EVA preparations took longer than in the simulations; but this was because during the simulations the LM cockpit was relatively clean, whereas an operational LM sitting on the Moon had been occupied for some hours, and was cluttered with checklists, data, food packages, stowage pouches and other miscellaneous items, all of which impacted on the timeline and checklist to exit the LM. Despite numerous run-throughs of the egress, it always differed on the flight. The simulations were of great value, and the checklists provided a good guide; but there were always details, previously not considered, that arose when on the Moon, because it was several days into a mission and not solely an egress simulation. Moreover, deviations from the checklists, with the hope of making up time, in fact *lost* time and caused more delays. What was learned about completing several EVAs was that the crews became smarter, in that things that needed to be attended to on the first EVA were not required on subsequent EVAs – which focused attention on items that needed to be attended to only on the second EVA, or that could be left to the third.

Future application Development of clear procedures and checklists will benefit from the advanced suits under development. A programme of EVAs with multiple-use suits will be supported by systems capable of maintenance, repair and service on the lunar surface. Simplification of the checklists, through advances in monitoring technology for suits, will eliminate the steps that the Apollo astronauts tried to bypass to save time, but which actually cost them time.

Egress and ingress

With the the first use of the LM porch on the Moon, Armstrong found that simulations in the zero-g aircraft and water tank were reasonably accurate. This allowed him to adequately position his body, with Aldrin helping out visually from a different perspective. Inside the pressure garment, of course, it was very difficult to see behind and estimate clearances and foot-holds. With one man on the surface it was much easier to guide the second man out and down, because his view was not limited by the dimensions of the LM cabin. But there were still problems. When Conrad exited the LM, his Portable Life Support System tore a 15-cm hole in the hatch insulation. On Apollo 14, Mitchell did not deploy his EVA whip antenna atop the Portable Life Support System for fear of breaking it in such a confined place, and Shepard therefore deployed it for him when he reached the surface. Jack Schmitt found egress and ingress much more difficult on the flight than in simulations, and he snagged his suit pockets as he left or returned to the LM cabin.

Future applications Exit and ingress generally proceeded according to the timeline, but more room, better vision and a larger platform area may help future explorers when descending any ladder that may be used (if direct access through an air-lock is not provided on future landers). Having suits in an air-lock or service area, and using

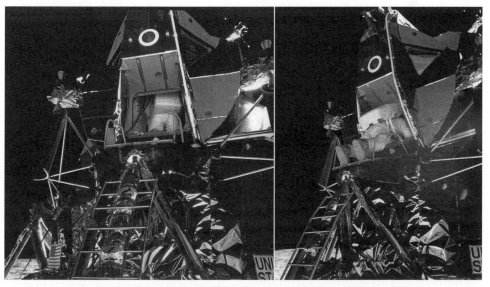

The tight squeeze during backwards exit from the Apollo LM.

a rear-door/hatch design similar to the Orlan, may assist in reducing the time taken to leave and re-enter the pressurised compartment.

Equipment deployment and transfer
The deployment lanyards generally worked well, although it was occasionally necessary to bypass the D-ring and pull directly on the cables to lower the MESA. The LEC also had a tendency to stick. On Apollo 14, Shepard was trying to remove a thermal cover on the stowed MET when he pulled the D-ring straight off, and he had to grab the blanket and tug on that. On Apollo 16, the crew asked that during the deployment of the MESA it should not touch the surface because of the potential of dust contamination; but when it was deployed it sat on the ground, so that the removal of some of the tools was quite difficult. Young reasoned that it might be an advantage to future crews to have an adjustment system inside the lander, so that when they landed, for example, on a slope, real-time readjustment could be carried out more easily than from a ladder, and with one hand. This was especially noticeable during deployment of the LRV. It was angled upwards, and its deployment, and the locking of the appendages, caused some concern. Some time was wasted in ensuring that the procedure was carried out correctly, because the angle at which the LRV was deployed appeared to differ from that during training, and it was assumed that it was incorrect.

Future application On future landings it may be advantageous to have a vehicle-levelling system incorporated, so that all deployments can be completed as per design, and almost regardless of the local surface terrain. Having equipment delivered on unmanned landers prior to crew arrival was an option considered during the 1960s for post-Apollo activities in the 1970s and 1980s, but it would entail more unloading and deployment. Having a fully configured roving vehicle

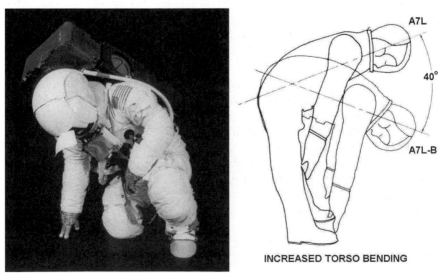

Mobility improvements on the Apollo lunar surface suit for later landing missions.

(pressurised or unpressurised) would save time, as would automatic deployment techniques (as with the Russian Lunokhod and Pathfinder rovers), but it would also increase the payload envelope and reduce the volume available for other equipment. It will be interesting to see how future hardware will be sent to the Moon, and the nature of any trade-offs. For some years, Mars has been the focal point of exploration by humans, and this has resulted in a variety of concepts for surface delivery of payload, consumables, exploration aids, and the crew. Perhaps if this development continues (and there are indications that it will do so), the result will be a 'spin-off' application directed towards the Moon, with the other option of developing Mars technology by first sending it on flight-tests to the Moon.

Mobility
All of the astronauts reported no difficulty in adapting to the $^1/_6$ gravity environment and moving around in the suit on the surface. Aldrin was the first to evaluate the mobility, and found that it seemed quite natural. All that had to be remembered was the mass of the suit and the Portable Life Support System on the back. A rear-mounted Portable Life Support System was not detrimental to moving around – except when losing balance and falling over, in which case the mass of the Portable Life Support System caused a couple of astronauts to flounder as they tumbled. On the TV it appeared like slow-motion falling, and the $^1/_6$ Earth-gravity seemed to give them time to react and prepare for the impact with the surface. Fortunately, no damage was done during these tumbles.

The MET on Apollo 14 was found to be more stable than expected, and it was able to traverse at various speeds without any difficulty. The pneumatic tyres were smooth and did not kick up any dust, and there was also no appreciable adhesion of soil to the tyres or wheel mechanisms. The MET bounced across level surfaces and down hill at relatively higher speed, but it was less than noticed during 1-g simulations and training on Earth. The main problem was in climbing steep gradients, and it was much easier for the two astronauts to carry the unit between them as they approached Cone Crater. It was also found that the MET served as a very useful workbench, and its ability to carry additional equipment and samples alleviated the fatigue and limitations of transportation by hand, as encountered on Apollo 12.

The operational constraints placed on the three LRVs on Apollo was that the astronauts had to remain close enough to the LM to enable them to walk back in case of LRV failure at any time during the traverse. In planning the LRV traverses, the astronauts would drive to the furthest objective early in the EVA, to maximise life support systems available, and then gradually work their way back towards the LM as the length of the EVA increased and their consumables were used. As an additional precaution, a Buddy system of secondary life support was carried on the LRV, for sharing of coolant water from one Portable Life Support System to the second in the event of a failure.

Following deployment, each Commander took the LRV around the LM to the MESA, where it was loaded with equipment. One of the items frequently mentioned

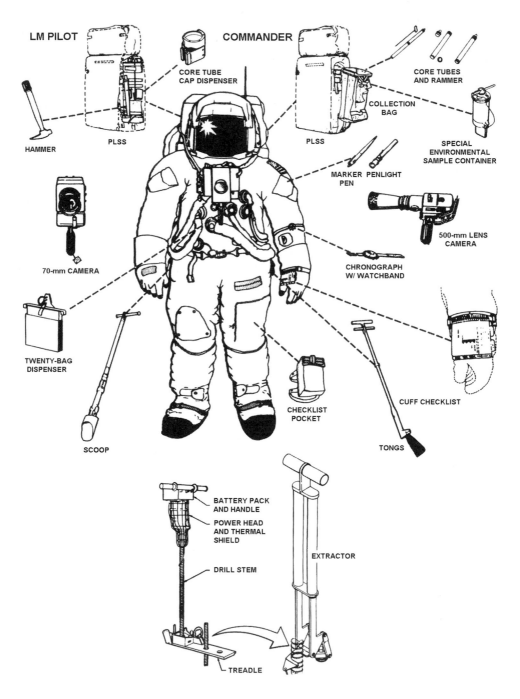

The attire of a well-dressed lunar explorer, with details of the Apollo lunar surface drill and extraction system.

during post-flight debriefing of the three crews who drove the LRV was the dust that was kicked up when the fenders were lost or broken. Even the repairs using EVA maps, ducting tape and clamps were not totally successful on Apollo 17, because the ever-present dust affected the adhesion of the tape. On Apollo 16, sprayed dust covered everything; but on Apollo 15, the fenders remained in place and prevented excessive amounts of dust covering the vehicle, although frequent cleaning of the lunar communications relay unit was required in order to prevent overheating of the TV circuits.

Adjustments to the seat belts, incorporated after training in 1 g, did not take into account the reduced gravity and the pressurisation of the suits, and on Apollo 15 the belts were too short. This was corrected and evaluated on the KC-135 for Apollo 16 and 17. On Apollo 16, the astronauts found that the map-holder on the LRV was 'useless', and jammed the map between the two driving positions between a camera and a support staff. Driving across a slope left the downward side occupant in a precarious position, and it was suggested that better seat restraint systems or a 'kiddie'-type safety bar should be incorporated in future open rovers. This was originally planned for the LRV, but was removed because it exceeded overall weight limits. Improved fenders would help to limit the dust, and more robust fittings would prevent them from falling off or being inadvertently broken by a crew-member. Having to reorientate the LCRU high-gain antenna for TV while stationary was also time-consuming, and was restricted by a dim viewfinder and the helmet visor. On future vehicles, an automatic orientation system would be an improvement – again offering constant quality communications and TV pictures of the traverse, and saving the crew time that could be allocated to other objectives and activities at each station stop.

Future application It is clear that the next generation of lunar explorers will not use Apollo-type suit technology, but more 'user friendly' pressure suits with better mobility and increased comfort, carrying advanced systems, microcomputers, information displays, and possibly integrated mobility-assistance systems for ascent/descent in craters, rilles and inclines. LRVs would probably be both unpressurised local survey vehicles and large pressurised 'roving laboratories' for extended long-distance exploration, of the Mobile Laboratory (MOLAB) type envisaged for post-Apollo exploration. Small lunar flying vehicles carrying one or two astronauts were planned for post-Apollo operations, but were never fully developed, although they may be reinvestigated for a lunar research station. In the science fiction TV series of the 1960s and 1970s, several feasible 'advanced lunar transportation vessels' were depicted 'hopping' across the surface between lunar bases. These would certainly be large vehicles that would be used to carry personnel and logistics between research stations or to aid in the construction of remote sites, thereby saving hundred of hours in EVA transportation and construction techniques.

Soil mechanics
Included in the objectives were soil mechanics experiments for determining the

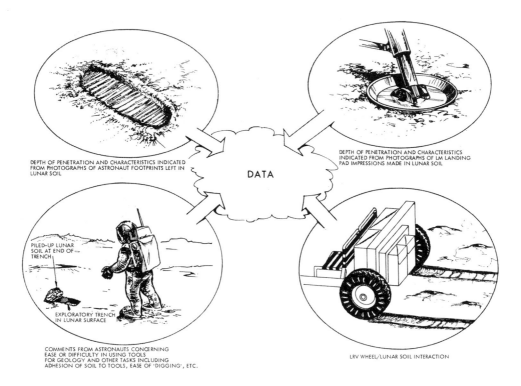

DEPTH OF PENETRATION AND CHARACTERISTICS INDICATED FROM PHOTOGRAPHS OF ASTRONAUT FOOTPRINTS LEFT IN LUNAR SOIL

DATA

DEPTH OF PENETRATION AND CHARACTERISTICS INDICATED FROM PHOTOGRAPHS OF LM LANDING PAD IMPRESSIONS MADE IN LUNAR SOIL

PILED-UP LUNAR SOIL AT END OF TRENCH

EXPLORATORY TRENCH IN LUNAR SURFACE

COMMENTS FROM ASTRONAUTS CONCERNING EASE OR DIFFICULTY IN USING TOOLS FOR GEOLOGY AND OTHER TASKS INCLUDING ADHESION OF SOIL TO TOOLS, EASE OF 'DIGGING', ETC.

LRV WHEEL/LUNAR SOIL INTERACTION

Apollo soil mechanics experiments.

characteristics of the lunar material. This had relevance to the LM footpads on the surface, the effects of the descent engine exhaust on surface material, the deployment of experiments, the digging of trenches, observations of the reaction of the soil, installation of equipment (antennae, flag poles and solar wind composition experiment) into the surface, the use of the lunar drill and core tubes, evaluation of walking, pulling the MET, driving the LRV, the impact of discarded objects, the depth of tread from the vehicles, and footprints.

Future application All of these had application in the later Apollo missions and potential in post-Apollo programmes. In situ experiments were supplemented by post-flight debriefing and analysis of samples, photographs, films, and the TV transmissions during each EVA.

Work-load
The need to improve working capabilities on the surface was recognised from the first excursion, when the Apollo 11 crew recommended that further crews should be able to kneel and bend to reach the surface, and that surface mobility vehicles would improve the range of their activities and hence the return from each EVA. This was confirmed by the next crew (Apollo 12), who estimated that a 20–30% increase in productivity, especially in retrieving samples, could be achieved if the suit allowed

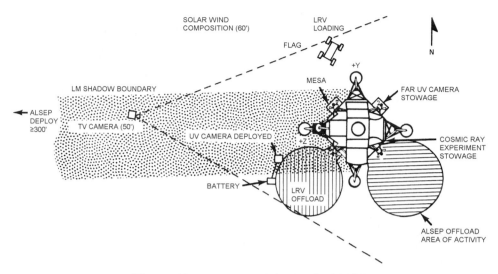

The working area around the Apollo 16 LM.

for more bending at the waist. On the final missions, improvements to the suit design provided more flexibility in the waist, knees and arms, and the LRV reduced the physical exertion of carrying equipment, walking, and carrying samples. This mobility also helped recovery from a fall, in that one man could manoeuvre to right himself instead of relying on the second astronaut for assistance.

On every mission it was found that hand fatigue was a common factor, because the glove was restricted by its design and the inner pressure, as well as the pressure required for a good grip on any object carried. It was not sufficient to rely on grip strength alone to grapple an object, or to hold it or manipulate it on a frequent or continuous basis. It would have been very tiring to manipulate objects throughout the EVA, and then do it again the next day, and again on the third day, if improvements to the suit and gloves had not been implemented. The Apollo 11 crew reported that although it was tiring to manipulate the sample return containers, they still had capacity for greater physical exertion. On Apollo 12, the two astronauts reported that although they were working at the maximum practical level required for their assigned tasks during each 4-hour EVA, they felt that each EVA could perhaps be extended to 8 hours without resulting in excessive fatigue, providing support equipment – such as suit-mounted tool-carriers instead of hand-held carriers – could be used.

On Apollo 11, twenty trips were required to collect samples and fill one sample return container on the MESA worktable. After Apollo 12, mounting the tools and sample bags on the suit was recommended, and although the use of the MET on Apollo 14 significantly increased the workload, walking everywhere still affected the astronauts' efficiency. It was not until Apollo 15 used the first LRV that workload and efficiency both improved, and planners could develop lunar EVAs for the scientific series of missions (from Apollo 15) of up to 7 hours duration. On the final mission, the three EVA periods resulted in more than 22 hours of activity without any deterioration to the working efficiency or well-being of the crew.

Future application Planning the EVAs for the return to the Moon has been the subject of numerous reports and studies, and will continue to be debated until the programme is defined. The question of what to do on each EVA is tied to the long-term objectives of that programme. When we go back to the Moon, will it be to work towards a permanent base, or for another series of short visits? If it is for short visits (of perhaps up to two weeks), multiple EVAs will have to be planned carefully so that the crew does not become fatigued, and the objectives will have to be made very clear. For a longer programme of many months, and the creation of a lunar base, multiple EVAs change in nature from crammed fast-paced activities to more leisurely and prolonged multiple EVAs to complete one or more objectives. The analogy to this has occurred on EVAs in Earth orbit, in which construction at the ISS or servicing of the HST has involved teams of EVA crew-members and several EVAs over one or more missions.

ALSEP/EASEP

For the first landing, two separate EASEP units were carried from the LM to the deployment site a few metres away. EVA duration and limits on the operational area away from the LM on this first surface activity restricted deployment location and the size of the package. The ALSEP packages on later missions were unloaded by lanyards and guided by booms to assist the crew in unloading them. On Apollo 15, the deployment booms were not required because of the slope of the landing, and on Apollo 17, at the request of the crew the pulleys were removed before the flight.

It generally took around 8–9 minutes to unload the packages, and although small craters hindered the unloading operation there was adequate space in which to work and for transferrence of the RTG fuel casket from the storage flask to the ALSEP package. The units were then transported on foot, using the antenna mast as a 'barbell'. On all missions, the ALSEP package had to be carried some 100 m west of the LM to ensure that it was not in its shadow. This was carried out by each LM Pilot, all of whom commented that the bouncing sub-pallets caused some difficulty on the traverse. It was also clear from the crews that holding the barbell was tiring on the hands or unconformable on the arms. On Apollo 15, Irwin decided to carry the bar in the crook of his elbow. This was found to be much easier, but even in the reduced gravity of the Moon, carrying a 20–27-kg mass was tiring and difficult.

It was also difficult to find a level site that would not affect the thermal control of the unit, and also be sufficiently distant from the LM that the instruments would not record the thermal leaking of the structure, venting of tanks, and debris from the ascent. Once a suitable site was located, the deployment could begin. Most crews had difficulty in erecting the flimsy antenna, and it was also found that using the central mast as a carrying bar created extra flexing of the mast, which was then difficult to set up in its final position. Some of the crews found that removal of the RTG from its cask occupied several minutes, and they had to use the geological hammer to entice it.

Each experiment was linked by telemetry/power cables to the central station. On Apollo 15, Scott tripped over one of the wires while moving the electrical box of the jet flow experiment. During training it was noticed that these cables retained their

Working at the ALSEP station on the Apollo LM.

Buzz Aldrin carries experiments across Tranquillity Base.

Jim Lovell carries ALSEP packages during training for Apollo 13.

stored shape and would not always lie flat, but nothing was done to correct it before Apollo 16. On that mission, Young inadvertently snagged his lunar boot in the cable and pulled it out of the experiment, thus rendering it unusable. Because of the restricted mobility of the suit and his limited vision he could not feel or see the cable that his boot had snagged. As these cables tended to lay about the surface, the astronauts had to 'jump' over them to avoid contact – if they could see them. Most crews found that it took longer to perform deployment tasks on the Moon than in training, and this instigated a time and motion study of Apollo 16 versus Apollo 15, which found a small statistical difference but generally indicated that it took 50% longer to perform an ALSEP deployment on the Moon than during training. To prevent further contamination of the ALSEP from debris left from the EVAs, and to prevent possible kick-up from the ascending LM, loose gear and trash (packaging) was kicked under the LM descent stage during the close-out of the final surface EVA. Above all, the most reported and recurring problem was the characteristics of the lunar dust, which dissipated everywhere.

Future application Because of the experience derived from Apollo, a small experiment package will probably not be flown on the return to the Moon, although it depends on what objectives are planned for the seventh manned lunar landing, and, of course, which country or organisation performs it. If it is the Chinese, then perhaps a small experiment package will be deployed, as on Apollo 11, with larger networks planned for subsequent flights. An American, or more probably an international, return to the Moon will probably focus initially on hardware delivery for a lunar research station before science (as on the ISS), but far more advanced experiments would be included for a one-off landing at a new site. During Apollo, the deployment and performance of each experiment was unique to ALSEP, but there were certain activities and procedures relating to general deployment that have clear application in future surface experiment deployment on the Moon. Future surface experiments will probably not be carried by hand to the deployment site, and advances in technology will probably be such that the troublesome cables would not be required, and so the situation of a crew-member tripping up and terminating the experiment would be avoided. Integrated power supplies, lighter materials, and miniaturisation, would reduce the mass of the package, and the main antenna would not need to be used as a carrying handle.

Cameras

During Apollo, various cameras and lenses were used for a variety of photodocumentation activities, such as documenting locations before and after sampling, deployment and installation of experiments, recording the local vertical in relation to the gnomon, close-up stereo photographs of the lunar surface in soil mechanics experiments, and general mission photography. Data acquisition cameras were used on the LRV, which also carried the remotely operated (from Earth) TV camera. Generally, the cameras worked well, but on all flights the equipment – including the camera lenses – became extremely dusty. Lens cleaning brushes were

Photodocumentation of a lunar sampling area.

not included on Apollo until the final flight, and dust clogged up several moving parts of film-advance mechanisms, filter installation recesses and mounting mechanisms. Dust also hindered viewing of the displayed frame settings, and the Apollo 15 crew commented that because the camera was installed on the chest area, where the crew found it easier to hold and seal sample bags, dust was inevitably transferred to the camera. They suggested that a dust cover on the camera might help when it was not in use. John Young suggested mounting a TV camera on the helmet, which could document the scene and operations without the crew having to do anything, thus saving them the time it took to photodocument each operation.

Future application Cameras have advanced significantly and rapidly since the days of Apollo, and miniature and digital cameras and webcams have all become standard photographic equipment around the world; and with the Shuttle and the ISS, this technology has also moved into space. Young's idea of a helmet-mounted camera was adopted for the Shuttle EMU, and remotely operated cameras are part of the RMS/Shuttle payload-bay equipment supporting EVAs, operated either by the crew on the flight deck or automatically from Mission Control. This type of application will undoubtedly be used in future activities and exploration.

Geology

Each of the planned Apollo missions featured geological field work of varying degrees and intensity. On Apollo 11, few hand-tools were carried, and the few samples that were collected were gathered over a very short distance and during a single EVA. Apollo 12–14 offered the opportunity to split the EVA objectives between local activity around the LM (such as deployment of the ALSEP) and an extended geological traverse. The provision of the MET on Apollo 14 also offered the chance to extend the capabilities of the second traverse (1.4 km) over that of Apollo 12's second EVA (500 m). From Apollo 15 onwards, the LRV greatly increased the geological sampling opportunities (up to 20 km) for the last three missions, within the safety perimeter limit.

New skills were developed for transferring samples from the collection device to the sample bag. It was difficult to gather a scoop of material in one hand and then transfer it to a sample bag in the other hand, but with practice the astronauts learned to collect the sample and then work their hand down the collector's handle so that it was close to the scoop end. The next feat was to open the bag with the other hand while not spilling the collected sample. Bagging the sample close to the chest was a preferred option, but this also transferred dust to the suit and camera. A two-man team effort worked best, with one astronaut taking the sample while the other photographed the operation and then acted as the 'bag man' to hold the sample bag while the soil or rock was inserted. The Apollo 12 crew noted that the identification of material types was more difficult on the Moon than during training on Earth because of the lack of distinct colour.

As the traverses increased, a protocol was required to fully document not only where each sample was retrieved, but also when and in what bag it was located. Photographing larger samples in place before removing them (and then the area after

Soil-sampling near a lunar boulder.

removal) was also completed, with a general photograph of the area for reference
and identification of when during the EVA the sample was collected, and in which
numbered bag it was placed. This was then cross-referenced in post-flight debriefing
and analysis. The detailed sequence for two-man sampling was generally as follows.
They would select the general area for sample collection, place the gnomon, and then
verbally describe the surrounding area to help locate the sampling position. As one
astronaut took cross-Sun stereo photographs, the other would select the tool to be
used and then pass it to his colleague. He would then take 'Sun down' photographs
before making ready the sample bag, record the bag number, and hold the bag for
the other to fill using the selected sampling implement (rake, scoop or tongs). After
this he would close the bag, seal it, stow it in a collection bag, and take 'after
sampling' photographs, while his colleague took locater photographs of the LRV or
background landmarks, as well as a 1-kg sample of fine material from a pristine local
area. The sequence for documented samples was similar.

Having the sample bags packed close to each other in the dispenser caused the
Apollo 14 crew some difficulty. When extracting one, two or three others would
become detached and float down to the soil, from where they were too difficult to
retrieve. The crew of Apollo 16 experienced much difficulty when their twenty-bag
dispenser repeatedly fell off its stowage location on the LRV, which slowed down
their activities. The Apollo 12 crew determined that perhaps extended and more
detailed geological training simulations would benefit later crews, and by Apollo 15,
numerous geological field trips to terrestrial sites over a longer period of training
time had helped the crews work with the equipment more easily. Improved suit

Using the lunar rake on the Moon.

mobility also helped these later crews, and the LRV carried the additional equipment needed to retrieve, document and store the samples during the traverse. From Apollo 14, geological tasks also included trenching, raking, core tube collection, and drilling.

The trenching activity on each mission was sometimes affected by other problems affecting the timeline, resulting in just one of the crew performing the activity while the second astronaut addressed the specific problem elsewhere. A few minutes was initially allocated for trenching, but with the addition of soil mechanics this increased to around ten minutes. Raking was achieved one-handed because of the mobility of the later suits. Early core tube collection was difficult because of the compacted regolith, and the tubes did not always retain the core sample when removed. Later missions used thinner-walled core tubes, which worked well. A single core tube took about five minutes to retrieve (eleven minutes for a double tube), but again this was a two-person activity, with photodocumentation, site description, gnomon activity, insertion (by hammer) and extraction, storage of the core sample, and stowage of gear, all having to be completed before moving on to the next site. Drilling for insertion of heat flow probes was achieved on Apollo 15, 16 and 17. On the first mission, high torque in the chuck-stem interface bound up the stems, and in one case a stem had to be destroyed to free it from the drill chuck. It proved difficult to remove the remains of the stem from the sample hole and it was therefore left in until other tasks were completed. It took the combined effort of both Scott and Irwin to extract the drill stem, and it physically exhausted them. Before Apollo 16 the system was redesigned, with a treadle provided for drill stem removal and a change in the design of the stems to allow clearance of dense soil from the drilled hole. These improvements worked well.

One of the frequent problems in surface exploration – particularly during foot excursions – was in recognising landmarks, so that navigation across a lunar surface void of distinctive features was quite challenging. Unexpected terrain features not indicated on relief maps often led to confusion; and equally, landmarks indicated on the maps were not so clear when viewed from the surface. The LRV navigation systems and the distinctive features of the more rugged terrain (backed up by the onboard TV at geological sampling station stops) clearly helped, and made for more productive excursions on Apollo 15–17. Perception of distance was also a problem, because haze was not present on the airless Moon; and use of the LM as a known height reference was not always possible on the undulating surface, as it often disappeared from view.

Stowing collected samples in the LM was also a balancing act to ensure that the sample return containers each weighed 20.4 kg for the trip home, to balance the additional weight for the ascent and the re-entry profile. During Apollo 11, samples were weighed on the surface, but all the other crews accomplished this in the repressurised LM cabin. Some shifting of samples between boxes was required to even them out, but this resulted in the spillage of loose dust from the exterior of the sample bags inside the LM. After reporting the weights to Earth, the crew waited for analysis to determine whether everything could be returned. If not, the excess weight

The LRV tool-carrier deployed on the lunar surface.

would have to be dumped overboard before ascent. This was an important crew safety issue, but was also frustrating given the effort of collecting and bagging all the samples and heaving them back into the LM cabin, only to throw some of them out again. Even the failed Moon landing caused a problem with mass calculations. During Apollo 13, several items of hardware had to be transferred from the LM cabin to the CM to compensate for the missing mass of the uncollected samples, so that the balance was correct for return and landing.

Future application The surface crews found that attempts to gather a sample without a field tool could be very awkward. The process of documenting the collected samples was occasionally skipped as they were pressed for time, although generally it worked well. In future surface exploration programmes, during which a prolonged surface stay is projected, a change in the surface protocol would be required. Both the crews and the Principle Investigators considered that the strict timeline on Apollo was a hindrance to true geological sampling, with no freedom to explore nor time for personal investigation and choice of which samples to collect. But the documentation protocol will probably remain the same on any return, to accurately record the location from where the samples are retrieved, and to provide comprehensive off-site analysis. Whether fewer samples would be accepted as a trade-off in the timeline was an open and debated question for Apollo. John Young suggested that instead of attaching sample collection bags to the sides of the Portable Life Support System, future explorers might benefit from a sack that rested on the ground and could be carried 'shopping-bag style' when required over short distances. Due to the restricted mobility of the Apollo suit, stowing of samples in the bags was a two-person activity, with one astronaut filling the other's bag. Weighing samples for the flight home would not be an issue for a lunar base, with extended expeditions that could perform on-site analysis and sample examination in a 'roving' laboratory, to bring back fewer but better selected samples to the central lunar base, and in turn, more selected extracts for return to Earth for even more detailed investigation.

Habitability
The success of any EVA depends on the crew being comfortable when carrying out their tasks, and, with multiple EVAs planned, in allowing for proper rest between the excursions. Due to the nature of the hardware used on the Apollo programme there were no luxuries, and Apollo activities were more akin to a camping trip. The cramped confines of the LM were designed for practical, weight/mission/safety purposes, and not as a scientific research station or long-term crew quarters. As a result, the habitability issues have not been as widely reported as the scientific return from Apollo, but they remain as important for future application. The Apollo 16 crew found that inadvertent activation of the in-suit drink bag sent orange juice squirting over them and the cabin and into the suit microphone, which had to be cleaned before use. The orange juice jammed up the helmet seals, and the crew thought that they would have to sleep in the suits between the second and third EVA. During the EVA this was not a problem, but during preparations their repeated bending activated the feed tube by entangling the microphone lead, or, when

pressing their chest against the bag dispenser, forced juice through and out of the tubes. At one point, Young thought that Duke had used orange juice as a shampoo.

The food system and waste collection system operated much easier in the $^1/_6$ gravity environment, and caused no real problems; but the Apollo 11 crew reported that they had little sleep in the LM cabin because of the light seeping into it, an uncomfortably low temperature, and a noisy spacecraft. The Apollo 12 crew reported that they slept in the hammocks, and although they noted the cabin noise, it was not enough to prevent sleep. On Apollo 14 the crew were uncomfortable in their suits, were aware of the sounds of the spacecraft, and had little sleep. Shepard stated that he thought the noise was caused by the spacecraft slipping over, which at one point caused both men to look out through the windows, only to find that they had not moved. Realisation of where they were, and the fragility of the spacecraft, certainly contributed to a restless night for the first three landings, before encroaching tiredness caused by the exertions of EVA. On the last three landings, with the spacecraft having been proven, the suits were not worn for sleep, and the crews were therefore fresher and more relaxed. Velcro worked well in general, but its performance was affected once the lunar dust entered the cabin. The crews tried to clean themselves by brushing and vacuuming the dust off each other, but it was really a lost cause. Gordon, the CM Pilot on Apollo 12, was so worried about dust dirtying his nice, clean CM that he ordered Conrad and Bean to re-vacuum themselves and then strip before they transferred to his CM.

Future application Habitability will be an important issue on long-term and multi-EVA activities. Having a living area separate from EVA preparation areas will enhance the well-being of the crew and the reliability of the equipment and systems, and the incorporation of rest periods by using more than one team of EVA crew-members will result in a more prolonged cycle of EVAs. This has already been incorporated on multiple EVAs from the Shuttle. The use and supply of spare clean and fresh pressure garments, replacement consumables, hardware and equipment will have to be weighed against the cost of logistics and payload mass limits.

Dust

Dust is pervasive on the Moon. It is unavoidable, and can cause equipment to clog up. It can affect static as well as moving parts, and will cover experiments, vehicles and astronauts alike. Comparing today's pristine EVA suits, used in Earth orbit, with those of the moonwalkers, it is clear that working on the Moon is a grubby experience. No matter how careful the astronauts were, dust entered the lunar cabin, and after removal of their suits for the night's rest, found its way inside their pressure garments and their underwear, and into their hair. In some cases it even found its way under their nails and into the pores of their skin, and it took some time to remove it, even after the return to Earth.

Future application With the decision to return to the Moon, dust will remain an important issue. Clearly, air-lock systems and changes in internal air pressures should prevent transport of dust from outside to the inside living quarters, but it will be difficult to remove all of it. For several years, work at a lunar base will

While working on the lunar surface, the accumulation of lunar dust is inevitable.

certainly be affected by the characteristics and presence of lunar dust particles in the low gravity, until an adequate isolated environment can be constructed.

Surveyor 3

On Apollo 12, the crew landed near, and retrieved parts of, the unmanned Surveyor 3 spacecraft that had stood on the Moon for thirty months. It had endured thirty lunar-day cycles (thirty cycles of fourteen continuous Earth-days of sunlight followed by fourteen continuous Earth-days of darkness). The astronauts retrieved specific samples from the spacecraft, and stowed them in a sample bag for return to Earth for post-flight analysis. These included the TV camera, a cable, the head of the soil-sampling scoop, a painted tube, an unpainted tube, and a sample of soil next to the scoop. The astronauts also photographed and described the condition of the spacecraft, wiped dust from its surfaces, and photodocumented the results.

Future application The purpose of this operation was to return samples of a vehicle that had remained on the surface for a known period of time, and to determine the degradation of materials after prolonged exposure to solar heating, radiation, cooling, solar wind, and lunar dust. An Earth-orbital experiment – the Long Duration Exposure Facility – was deployed on STS-41C in 1984 and retrieved on STS-32 in 1990, providing data on six years of exposure. That same year, the Hubble Space Telescope was launched, and it is hoped that some time in 2010 it will be retrieved for the gathering of information on the effects of twenty years of exposure in Earth orbit. Similar exposure of materials was conducted on several space stations and recovered by EVA; and there is an even longer experiment that began on Apollo 17, in which selected hardware was photodocumented and left in situ for an eventual return to the site to gather samples and compare them with similar material left in long-term storage on Earth. This is currently running at thirty and more years of continuing 'static' experiment. It is highly probable that when we return to the Moon, some of the early landings will include visits to former Apollo landing sites to retrieve parts for examination. All of this is important in the evaluation of components for future spacecraft and sub-systems, and their operation, degradation and performance under extended periods of vacuum, solar cycles, radiation and bombardment. This will have direct application in the construction of larger bases on the Moon and Mars, and deeper into the Solar System.

Ceremonies

Each landing included a plaque on the forward strut of the LM, marking the achievement of the landing. The American flag was unfurled to recognise the nation's achievement, statements were included to indicate the importance of the event, and mementoes were sometimes carried and left on the surface to mark past achievements and sacrifices. Apollo was an engineering and technical marvel, but it was also a very human achievement, and should be remembered as such. It is important to recognise the contributions of thousands of workers across the United States and elsewhere who were part of the overall accomplishment of the effort to reach the Moon, and to acknowledge the use of former Apollo hardware and operation of ALSEP experiments through 1977. Much of the work continues in studying the results and samples. The EVAs that made Apollo what it was – the first human exploration of a world other than our own – demanded appropriate and recognitive ceremonies.

Future application Times change and the world moves on, but the human spirit remains. Hopefully, around 2015 we will return to the Moon – and that will once again be cause for celebration, either national, international, or global.

Deep-space EVA

During the return leg of each of the final three Apollo missions, the CM Pilot completed short-duration EVAs from the open hatch to the SIM-bay area to retrieve cassettes of exposed film. The first, on Apollo 15, was conducted, some 273,600 km from Earth, by Al Worden, and was televised by a camera mounted on top of a long boom extending from the door of the open hatch. In his 41-minute EVA, Worden made three trips to the SIM-bay and back, evaluating the use of hand-holds and foot-holds, as well as retrieving the cassettes. He also inspected instruments that had failed. This was the first EVA beyond the protective envelope of the Earth's magnetosphere, usually designated a 'deep-space EVA' or 'cis-space EVA'. During these operations the CM Pilot was supported by the LM Pilot on stand-up EVA activity, guiding the 8.3-m tether and assisting in taking the cassettes back into the

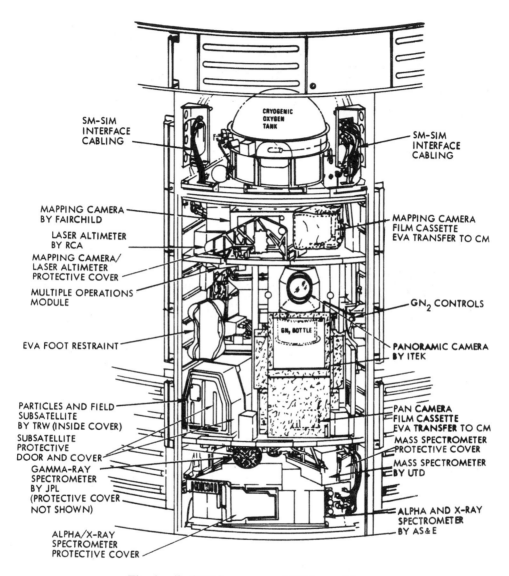

The Apollo SIM-bay area and EVA provisions.

CM. During the second deep-space EVA, on Apollo 16, CM Pilot Ken Mattingly made two leisurely trips to the SIM-bay and completed an excursion of 1 hr 24 min, admiring the view, inspecting the exterior of the spacecraft, and exposing a micro-ecological evaluation device for ten minutes. The final EVA of this type, and the final EVA of Apollo, was conducted by Apollo 17 CM Pilot Ron Evans in a three-trip excursion in which he logged 1 hr 7 min in deep space.

These three EVAs are often overlooked in the achievements of Apollo, but they are milestones for the future of EVA operations away from Earth and between the

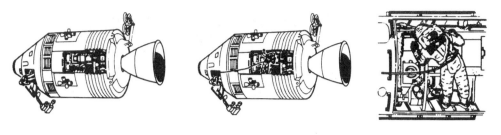

• START IN-FLIGHT EVA, EGRESS	• INGRESS FOR MEED/ATTITUDE MANOEUVRE	
• RETRIEVE PAN CAMERA CASSETTE	• EGRESS/ACTIVATE MEED	• INGRESS, CLOSEOUT
• RETRIEVE MAPPING CAMERA CASSETTE	• DEACTIVATE MEED	

The timeline and path for trans-Earth EVA.

Zero-g aircraft training for trans-Earth EVA.

planets. The trip to Mars will be a long one, and at some point it will probably be necessary to go outside and inspect, service or repair the spacecraft, as we have seen on space station missions. Whether this will be solely robotic (due to increased radiation levels) or a combination of human and robotic operations remains to be seen.

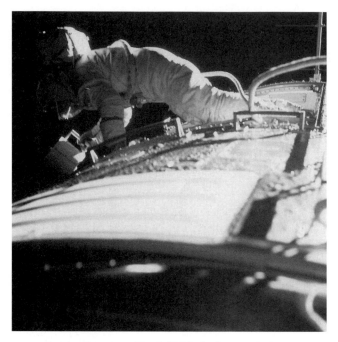

Conducting trans-Earth EVA during a mission.

BACK TO THE MOON?

Apollo was a momentous achievement, and a unique experience for those involved or for those lucky enough to witness events as they unfolded. It was the first time that EVA was used in a truly operational sense and as a major objective for each of the landing missions. Indeed, the objective of Apollo was human exploration across the surface of the Moon, and by implication, EVA was necessary to achieve this. Some difficulties and small setbacks featured on every EVA – but the overall success of Apollo and the relative ease of exploration also led to its downfall in the eyes of the public. Once Apollo 11 had achieved the goal of the programme set in 1961, the public attitude of 'been there, done that' became very evident by the time the last three missions flew. The achievements of those three missions were truly remarkable, given that they were only the fourth, fifth and sixth landings, and the only extended EVA experience prior to Apollo 11 was gained during a few short excursions from four Gemini spacecraft and an abbreviated trip outside the Apollo 9 craft.

Lessons learned from Apollo 11, 12 and 14 were applied to Apollo 15, 16 and 17, and could have been further expanded had Apollo 18, 19 and 20 been allowed to complete their missions. The hardware was built, the crews were short-listed, the landing sites were discussed, and experiments were selected, and clearly the desire of many in the programme was to continue the flights to the Moon. But others considered that we should stop before a crew was lost, and felt that the programme should be redirected towards supporting more earthly concerns then attracting publicity. These various pressures to reach new goals on Earth and in space before Apollo had fully been exploited have resulted in no humans walking across the surface of the Moon for almost two generations. We have lost the skills that have taken us to the Moon, and it will be very difficult to quickly regain them. In January 2004, President George W. Bush announced plans for a return to the Moon, to resume what Apollo began. If these plans are fulfilled, then the true legacy of Apollo will be realised.

REFERENCES

1 Ertel, Ivan, and Morse, Mary, *The Apollo Spacecraft: A Chronology*, Vol. 1, NASA SP-4009, 1969, p. 11.
2 Shayler, David J., *Apollo: The Lost and Forgotten Missions*, Springer–Praxis, 2002, pp. 156–164.
3 Stewart, Peter A.E., 'Surface Exploration of the Moon', *Spaceflight*, **3**, No. 2, March 1961, 34–48.
4 Hendrickson, Walter B., Reach for the Moon, Ian Allan, 1962; Branley, Franklyn M., *Exploration of the Moon*, Scientific Book Club, 1963.
5 Ertel, Ivan, and Morse, Mary, *The Apollo Spacecraft: A Chronology*, Vol. 1, NASA SP-4009, 1969, pp. 166 and 178.
6 Shayler, David J., *Apollo: The Lost and Forgotten Missions*, Springer–Praxis, 2002, pp. 197–229.
7 Shayler, David J. and Hall, Rex D., *Soyuz: A Universal Spacecraft*, Springer–Praxis, 2003, pp. 20–35.
8 *Ibid.*, pp. 120–138.
9 *Ibid.*, pp. 138–156.
10 *Transfer in Orbit*, Novosti, Moscow, 1969, pp. 135–137.
11 Abramov, Isaak P. and Skoog, Å. Ingemaar, *Russian Spacesuits*, Springer–Praxis, 2003, p. 96.
12 *Ibid.*, p. 115.
13 *Ibid.*; information supplied by Zvezda to Astro Info Service, 16 June 2003.
14 Shayler, David J., *Apollo: The Lost and Forgotten Missions*, Springer–Praxis, 2002.
15 Apollo Program Summary Report, NASA JSC, 09423, April 1975, Lunar Surface Operations, pp. 6-8–6-12; Orloff, Richard, *Apollo by the Numbers*, NASA SP-2000-4029; Sullivan, Thomas A., *Catalog of Apollo Experiment Operations*, NASA Reference Publication 1317, January 1994; Astro Info Service interviews with John Young (1992), Gene Cernan (1988 and 1989); Apollo Post-Flight Technical Crew Debriefs (listed in the Bibliography).

Station support

To keep a manned space station in orbit for any length of time it is essential to have the infrastructure to support a number of basic requirements. These include the means to boost the station to a higher orbit (because atmospheric drag gradually pulls it down towards the upper fringes of the atmosphere); the ability to exchange resident crews at regular intervals to conduct gradually longer-duration missions, and to provide a permanent presence in orbit; to provide fresh supplies and to dispose of unwanted trash; and the capability of exterior survey, maintenance and repair of station components by either human or robotic means. Between 1971 and 2001, eight space stations (seven Soviet/Russian and one American) operated in Earth orbit and developed these techniques, and the Skylab, Salyut and Mir programmes first demonstrated and mastered the techniques of conducting EVA at a space station.

The space station era began in 1971, and the first EVAs from a space station took place two years later. The world's first manned space station, Salyut, was launched (unmanned, as with all large space station modules) by the Soviets on 19 April 1971, and was followed by further Salyut stations operating between 1972 and 1976, with varying degrees of success in terms of manning and operations. Although EVA capability from these stations was rumoured, it was not until Salyut 6 in 1977 that cosmonauts finally left their spacecraft in orbit – for the first time since the Soyuz 5/4 transfer in January 1969. It was therefore the Americans who performed the first series of space station EVAs, from Skylab in 1973–74; but although they led the way, these would also be the last American space station EVAs for more than twenty years. When they resumed it would be from a Russian space station (Mir), by which time cosmonauts had gained considerably more experience in external maintenance and expansion of a permanent orbital base. This cooperative space station EVA programme would be a precursor (with a programme of Shuttle developmental EVAs) to the creation of the International Space Station.

SPACEWALKS FROM A SPACE STATION

Ironically, the first space station-related EVA, on 25 May 1973, was not executed

An early impression of space station EVA during the Apollo Applications Program.

from the space station. Standing in the hatch of the Apollo CM, Skylab 2 Pilot Paul Weitz attempted to pry loose the stuck solar array of the OWS during a fly-around of the station. This was the first American EVA in Earth orbit since Apollo 9 in March 1969, and it demonstrated a high level of confidence in performing suited operations, with so little prior experience, outside a spacecraft. Before Skylab, most of America's EVA experience had been gained on the Moon. Only three short deep-space EVAs (to retrieve film cassettes from the SM science package of the last three Apollo missions) and the pioneering Apollo 9 EVA had been performed in space since the Gemini programme. But though orbital EVA experience was lacking, hundreds of hours of studies and simulations had taken place over the previous decade as the OWS programme developed.

Working outside the vehicle
At the same time that President Kennedy challenged America to place a man on the Moon, studies into adapting Apollo lunar hardware for other missions were already underway. During May 1961 the Apollo 'A' design for Earth-orbiting scientific missions featured a cylindrical section between the Saturn second stage and the Apollo CSM, including an integrated section for use as a laboratory. In support of these studies, Smith J. DeFrance, Director of the Ames Research Center, proposed a number of experiments that might be performed in such a laboratory. These included, 'testing man's ability for work outside the vehicle'.[1] Between 1961 and 1965, further studies of Apollo lunar hardware for possible extended missions in Earth orbit included the development of a large manned orbiting telescope. Although many of these studies featured internal transfer systems between the Apollo craft and the 'space laboratory', some illustrations depicted crews on EVA in

a man-tending, servicing or repair mode. It was agreed by officials in the Saturn Apollo Applications (SAA) programme that astronauts would conduct tethered EVA operations whenever feasible – a requirement that continued throughout the lifetime of the programme as it developed into Skylab.

EVA from the Manned Orbiting Laboratory?
While studies for extended missions and space stations were based on both Apollo hardware and on future conceptual designs, plans for the first Americans to leave their spacecraft in orbit were being refined for the forthcoming Gemini programme. Meanwhile, the USAF was examining its own military space station in a programme of broad research, primarily under the Manned Orbiting Laboratory (MOL) programme. On 10 December 1963, the MOL replaced the X-20 Dyna Soar project as the primary USAF manned spacecraft programme. It would feature a series of Gemini/orbital laboratory combinations launched by Titan III launch vehicles, but would carry a range of military-orientated experiments and research objectives, one of which would be the development of EVA techniques by USAF MOL astronauts. Four decades have passed since MOL was conceived, and much of the history and development of that programme, together with hardware, experiments, and flight operations, remains classified. However, from the information that has been released over the years it is possible to gain an insight into what the MOL EVA programme may have included.

By 1964 there were four concepts for transferring the two-man crew to the laboratory behind the Gemini spacecraft: the use of a hatch in the aft heat shield (which became the desired option); an inflatable tunnel from the hatch alongside the Gemini Adapter Module and into the laboratory; rotation of the entire spacecraft for direct transfer from the Gemini crew compartment into the laboratory module; or EVA. The EVA transfer method was by far the easiest to accommodate, as it required little design change. In some designs of the MOL laboratory that have been released, there was an aft equipment section with a transfer/access tunnel to the rear of the module. This could have been used as an air-lock to open space for EVA activities, as an alternative to the Gemini spacecraft crew hatch.

MOL experiment P6: EVA
During the summer of 1964, technical development plans for the MOL also detailed a programme of twelve primary and eighteen secondary experiments, including at least one primary experiment related to EVA.[2] Experiment P6 – Extravehicular Activity – had the objective of evaluating an astronaut's ability to perform EVA, with a view to 'future military operations, including external spacecraft maintenance'.

A basic set of equipment would have been provided for EVA operations from the MOL. Planning documents mentioned an air-lock, but EVA could also have been performed from the hatch of the attached Gemini spacecraft, in much the same way as the NASA Gemini EVAs had been accomplished. There would be a selection of tether lines available for restraint, a 'strap-on' manoeuvring and control unit (the Gemini AMU, and an environmental control system. A set of basic hand-tools would also have been provided, together with task-specific equipment for replacement (for example, solar cells), connection (such as joining two fuel lines), patching (the application of an external pressure patch to seal a small leak in a pressurised compartment) and retrieval (during which an ejected target would be recovered).

The objective of these tests was to investigate the locomotion of the astronaut in both powered and non-powered tests. To minimise the hazards to the EVA crewmen it was important to determine how effectively both tethered and untethered EVA objectives could be completed, and so an independent environmental control system and a 'suitably designed spacesuit' would be required. Measurement of the difficulty of each task was also essential, so that the life support and environmental control systems could control and manage higher thermal outputs as the astronaut worked.

Initially, each task would be performed inside the laboratory, to gather baseline life support and performance information in a relatively safe (pressurised) environment. Operations would be conducted with and without pressure suits, varying suit pressure levels and environmental system flow rates, running tests with and without the EVA ECS and manoeuvring unit (or units as stated in the report), and completing a variety of alternative procedures.

These IVA tests would have been repeated outside, on EVA. Similar data would be recorded, and TV pictures and still photographs would be taken. The astronauts would then 'proceed through the air-lock, move over the exterior surface, operate [the TV camera] while tethered, and perform tasks with the manoeuvring unit, both tethered and untethered.' The evaluation report emphasised the importance of such tests: 'EVA operations are important to future military space missions, to provide external maintenance of spacecraft which are resupplied and stay on orbit for long periods of time. In case of failures that preclude direct access through internal tunnels to the recovery vehicle, EVA capability may be needed for the safety of the astronauts. Also, EVA may prove valuable for logistics reasons, to allow the transfer of equipment and for rescue operations.'

It was estimated that EVA equipment for the MOL would weigh approximately 181.4 kg, and would require a volume of approximately 0.39 m^3, as well as about 1 kW of electrical power for the lights, TV and associated systems. EVA support equipment would have included a tether-line reel and exterior hand-holds and foot-holds (to evaluate the difficulty of body restraint in space), and boom-mounted TV equipment, which would be returned to the inside of the laboratory after completion of the EVA (as happened on the Apollo trans-Earth EVAs in the early 1970s).

MOL experiment P7

In experiment P7, an astronaut inside the MOL was tasked to control a Remote Manoeuvring Unit (RMU), which would be sent to a target satellite. This package contained its own propulsion and attitude control system, and would be evaluated as an inspection scout and as a potential resupply bus for other spacecraft. Two decades before NASA astronauts used an MMU from the Shuttle, and before discussions concerning the potential of unmanned teleoperator systems in aiding astronauts on EVA, the USAF was already studying the potential of such vehicles in addition to the development of the Astronaut Manoeuvring Unit for Gemini 9 and 12 under a cooperative NASA programme.[3] According to the Technical Development Plan, the RMU had 'a fundamental capability of remote manoeuvring [that] may be extremely important in the logistics problem of resupplying military vehicles.'

The RMU experiment required about 0.08 m^3 volume, and had a mass of approximately 136.08 kg. The internal display and control console and associated data recording equipment required approximately 0.11 m^3, and had a mass of 22.76 kg and a power requirement of 500 W during the test flights. The vehicle would have been stabilised to within 1°, and during RMU observation and operation, laboratory stabilisation of 0°.01 per second would have been required. This was one of the forerunners of the automated teleoperated and robotic EVA systems developed during the early planning stages of Space Station Freedom and the more recent AERCam Sprint and EVA Retriever proposals for the ISS.

MOL EVA training
Each of the MOL Pilots completed an EVA training programme similar to that of the NASA Gemini astronauts, including 1-g simulations and bench-tests, underwater training, and flights on zero-g training aircraft. They would also have undergone the additional in-orbit training inside the laboratory before venturing outside the vehicle. The exact timeline and number of EVAs would have been developed from the training, testing and simulations. All equipment tested on the ground would have been put through simulations inside the laboratory before being qualified for use in the vacuum of space. It is reasonable to suggest that with a planned five-flight programme, each of the two astronauts on a crew would have had the opportunity to perform one EVA for direct comparison, so had MOL flown we may well have seen ten EVAs under the programme, although we may not have known about them given the classified nature of the programme.

There were also plans for the MOL to incorporate an EVA air-lock for exit, and it is possible that the training was used to evaluate such a system for use in the flight vehicles. Unfortunately, none of the MOL astronauts would be presented with the opportunity to put this training into practise, although Don Peterson performed one EVA during STS-6 in April 1983.

The USAF MOL was delayed several times, with the early flights – originally planned for the late 1960s – slipping into the early 1970s. At a time when NASA was focusing on following Apollo with its own orbital laboratory, MOL was finally cancelled in June 1969. Some of its technological development moved over to the NASA space station programme, including the further development of manned manoeuvring units.

Applications from Apollo
While the USAF was pursuing its own space laboratory, NASA was reviewing the possibility of its Apollo hardware supporting extended-duration research capabilities other than at the Moon. Variously called Apollo A, Apollo Extension System (AES)

and Apollo X, these centred on the use of the Apollo CSM, with experiments in a vacant bay of the SM, in joint flight with small cylindrical research modules, and even in adapted Lunar Modules (without the lunar landing radar and landing gear). These laboratory modules were either based on modified LM ascent stages, or were dedicated cylindrical structures to be flown in conjunction with the Apollo CM (Command Module Laboratories (COMLAB)). These would not require EVA for habitation, but there were facilities to support excursions from the vehicles using an attached air-lock.[4] In support of these proposed AES missions there was a short-listed programme of experiments, issued in May 1965,[5] including a range of EVA activities designed to support and develop new pressure suit designs, the techniques of working in space, and the rescue of stranded crew-members. It was a bold plan, at a time when no American and only one Russian had left a vehicle in orbit. It included the following experiments:

1501 – Advanced Spacesuit Investigations
1502 – Manned Locomotion and Manoeuvring Investigation
1503 – Emergency Rescue Operations Experiment
1504 – Personnel and Cargo Transfer
1505 – Maintenance and Repair Techniques
1506 – Propellant Handling Techniques
1507 – Extravehicular Assembly Operation

Although developed during the early 1960s, because of changes in the programme and restrictions in the budget they could not be carried out under the AES programme, but these research fields have remained at the forefront of mainline EVA operations to the present day.

At the time of these studies, Boeing was one of several organisations working on independent studies for NASA contracts,[6] and examining the potential of large pressurised vessels in which to conduct extended-duration scientific research in Earth orbit. In providing a pressurised vessel, the need for EVA for normal laboratory operation would be eliminated, and would be required only for external inspection and maintenance and for testing new equipment and procedures.

In 1965, some studies within NASA were considering the possibility of converting a Saturn S-IVB stage to support a cluster of modules and facilities around a common docking module. The Saturn-based concept also featured plans to conduct a series of suited internal EVAs to convert the spent stage to a habitable workshop once its role as a rocket stage had been completed. This became known as the Wet Workshop configuration, and in August 1965 the programme became known as the AAP. The Saturn S-IVB stage remained the focal point of AAP operations for the next eight years, as a programme of EVAs was devised to support tests of new equipment, procedures and planned experiments during the series of manned visits. Drawings of the concept, released at the time, clearly show the idea of internal and external EVA operations in and around the S-IVB, using both Apollo and Gemini spacecraft as crew and logistics ferries.

At the same time, there was a growing interest in flying the Apollo Telescope

Mount hardware to the S-IVB station, instead of on the Apollo SM as envisaged under the Apollo X and AES programmes.

EVA PLANNING FOR THE ORBITAL WORKSHOP

Under the original plans, the Apollo Telescope Mount would have flown in Sector I of a Block II (lunar capability) CSM in Earth orbit for 14 days. As the telescopes were located in a deployable facility in the SM, the film cassettes could be retrieved only by EVA, and as the film cassettes would not hold all the film required in one unit, exchange EVAs would also be required. Therefore, two EVAs were included in the mission definition studies for around the seventh flight day and towards the end of the mission.[7] The format of the EVA was similar to that used on later EVAs conducted on the three J-series Apollo missions in 1971–72. The first EVA would have seen the astronaut exit the spacecraft through the CM hatch, carrying replacement cassettes and retrieving the exposed cassettes, and probably supported (and photographed) by a second astronaut in the open hatch, while the third controlled the CSM. The second EVA would have retrieved the final cassettes for return to Earth, with the ATM being ejected prior to separation of the SM. At one time, a short stand-up EVA or umbilical EVA (similar to those conducted on Gemini missions) was considered, to observe the separation of the hardware from the SM. Had it ever progressed to flight, the Astronaut Office would have certainly objected to this risky operation, with the potential for debris to strike the EVA astronaut during an unnecessary operation prior to re-entry.

In April 1966 a letter from John H. Discher, Saturn Apollo Applications Deputy Director, to Jerry McCall, MSFC Deputy Director for Research and Development Operations, suggested that the MSFC should provide cost and scheduling estimates for incorporating the ATM in the modified descent stage of an LM laboratory instead of the SM. This led directly to the inclusion of such a design into the S-IVB workshop programme. For several months, discussions continued over mounting the ATM on the SM, the LM, or directly to the S-IVB, with the LM carrier becoming favourite.

As these discussions continued, the role of EVA in the AAP increased. By September 1966 the experiences of the Gemini EVA astronauts prompted AAP Deputy Project Manager Kenneth Kleinknecht to put forward recommendations to redesign the forward dome hatch in the S-IVB hydrogen tank, so that it could be easily removed. Kleinknecht argued for a flexible type of air-lock seal to be installed prior to the launch of the stage, which would seriously reduce the work-load of the astronauts once the vehicle was activated on orbit. The following month a meeting at Lake Logan, North Carolina, revised the division of effort between MSC in Houston and MSFC in Huntsville for future studies and responsibilities. Once again, drawing upon recent flight experiences during Gemini as well as a review of the assigned advanced activities, plans focused on a thorough analysis of the potential and capabilities of both the astronaut and the equipment, and on evolving a long-range plan for the development and use of EVA hardware and procedures to further expand man's usefulness in space. Gemini had clearly demonstrated the need for a very careful assessment of EVA requirements as directed by the objectives of each mission, as well as the creation of specific EVA hardware and procedures. It also ensured that with the resources available, the astronauts were capable of performing the various tasks assigned to them.

In the revised responsibilities, MSC would be responsible for the study, testing and development of the EVA equipment and procedures, which also involved the astronaut group. MSFC would be responsible for the development and testing of 'large structures in space that might require astronaut EVA for assembly, activation and maintenance, or repair.' This division of labour between the centres would create further arguments over the coming years, as each organisation promoted itself as a centre for manned spaceflight.

In early November 1966, the recommendation to attach the ATM to the ascent stage of an Apollo LM was accepted, and the configuration was assigned to the S-IVB workshop programme from early December 1966. This introduced the 'cluster concept' into the OWS programme, and highlighted a number of EVA issues on which the Astronaut Office in Houston had been working for some months – particularly with regard to retrieving and replacing the ATM film cassettes. EVA demonstrations of a manoeuvring unit were being evaluated as an experiment, but this new objective was for operational spacewalks, and with only the Gemini experience to work from at this point, it required further evaluation.

Early Apollo Applications Program EVA planning
The Apollo Telescope Mount was the largest item of scientific equipment on the OWS, and a major contributor to the whole scientific research programme. Therefore, to depend on EVA data retrieval a major study was required to support and validate the development of EVA methods.

Once the ATM was assigned to the OWS in December 1966, the objective of the EVAs focused on the replacement and retrieval of the film cassettes at intervals throughout the mission. As the AAP/Skylab programme developed over the next seven years, the EVA programme was expanded to include EVAs related to two corollary experiments (D024 and S230), the equipment for which would be placed on the exterior of the station so that astronauts on an ATM film retrieval operation would not have to divert far to gain access to the two experiments, thus saving time and additional EVAs. Although these EVAs could be pre-planned and trained for, there was always provision for unplanned EVAs ('off nominal' situations) that could be inserted into the EVA programme. After the launch of the Skylab workshop in 1973, the EVA programme had to be amended to include salvage and repair operations. Additional experiments and observations were also developed from the discovery of comet Kohoutek.[8]

As the EVA programme developed, it also had to reflect the changes in the configuration of the OWS cluster. As a result, three different systems were developed in response to significant cluster changes – the most notable being the change from Wet to Dry workshop configuration. In each of the changes, the task of transferring equipment and the decision on where to place the astronaut work-stations was re-evaluated. Many designs changed as a result, but there were several that remained unchanged and survived each of the configuration changes from inception in 1966 to flight in 1973. These constant design changes helped strengthen the Skylab EVA programme, and contributed to the success of the astronauts, who became familiar with tried and tested equipment in simulations long before they met the flight hardware in orbit.

This recognition that EVA was required on the OWS generated new studies, which in turn identified other areas of study and design. These included the number of EVAs, how often the crews should go outside and for how long, what they would do outside, and how they would achieve the task. It was also necessary to define exactly which crew would accomplish each task and at the same time develop a workable set of EVA design guidelines, constraints and safety procedures. When all this was decided there remained the question of integrating the EVA with the hardware sub-systems to achieve a capability that would also satisfy the objectives of the EVA and the overall mission.

These early design guidelines defined AAP EVAs that would begin at the air-lock module of the main OWS, with an Apollo-type pressure garment and an 18.28-m umbilical line feeding life support. A smaller portable life support back-pack would be carried as a back-up option only. Every fourteen days or so, the astronauts would venture out and exchange the ATM cassettes. On the three planned missions – 28 days, one EVA; 56 days, three EVAs; 56 days, three EVAs – there would be at least seven planned ATM EVAs for retrieval and replacement. The system remained flexible to react to real-time requirements and work on the ATM console.

Apollo Telescope Mount EVA concept studies
Early concept studies incorporated a 1-g part task mock-up with a mechanical simulator that was both low tech and inexpensive. This was used to review access to the ATM canister, pinpoint the best locations and configuration for the crew work-station to allow them to remove and replace the film canisters, evaluate several designs of astronaut translation from the air-lock module to the work-station, and determine exactly how the canisters could be carried.

In the early design of the ATM, located in a rack under the LM ascent stage (replacing the LM descent stage), seven instruments or experiments required film cassette replacement. The design of four of them required film exchanges from the side of the canister, and the other three would require film exchange at the outermost Sun end. It was clear that to have an astronaut move to each camera in turn would be both time-consuming and exhausting, and it was therefore decided that the astronaut would remain at a fixed work-station and bring each camera's access point towards him. Two LM/ATM work-stations were developed – one to provide access to cameras from the side (the LM-end work-station), and the other at the Sun end (the Sun-end work-station). During these evaluations it was realised that instead of emerging from the air-lock module and translating around the LM structure, it would be sensible, and more economic, if the astronaut were to use the forward hatch of the LM (the hatch used to enter and exit the cabin when on the Moon). The LM cabin would be reconfigured to include the ATM controls and displays, and

EVA activities from the Skylab space station.

could be isolated from the main station, with the overhead LM hatch and access tunnel providing air-lock capabilities (although care was needed when stowing or restraining important items to prevent them floating out of the hatch during EVA – as happened a couple of times on Gemini!). The film cassettes could therefore be stowed inside the LM ascent cabin instead of the air-lock module, and a new work-station – the LM hatch work-station – was developed for use just outside the forward LM hatch.

Despite the relocation of the EVA exit from the air-lock module to the LM forward hatch, there still remained a considerable distance to traverse to the work-stations. To reduce this, the designers located the LM hatch work-station 45° from the hatch area, and sited the Sun-end work-station directly underneath the LM hatch work-station. These locations would remain throughout the duration of the Wet Workshop design.

In evaluating these work-station designs in December 1966, there was some discussion between the Astronaut Office (CB), the Structure and Mechanics Division (EC), and Grumman (prime contractors of the LM) to evaluate possible access through the 55.88-cm diameter opening above the ascent stage engine (which was removed in the LM/ATM configuration) instead of a traverse outside the vehicle. With this internal EVA it was found that the top of the experiment package was 88.9 cm below the opening. In simulations, Grumman found that an 80-percentile man in an unpressurised suit just barely passed through their mock-up hatch and into a tunnel. It was clear that this configuration had neither a large enough opening through which to pass, nor a gap sufficiently small to reach across. Lack of lighting was also a problem in the confined workspace. Structural modifications could be made to enlarge the ascent engine cover hatch to about 73.66 cm diameter, but this had to be carried out at Grumman, and could not be accommodated on LMs already delivered to the Cape. In the evaluations it was assumed that no back-packs would be used, and that life support would be supplied directly from LM systems via an umbilical. In the event it was decided that such an operation would be severely restrictive in movement and in visual clues for the sole astronaut who could have conducted the operation, and so the idea was abandoned in favour of an exterior EVA operation.[9]

Development of the EVA system

As with many aspects of design and development, while one area was being evaluated, other areas were evolving, resulting in changes to the original concepts and methodology. Such was the case with the ATM canister, as work-stations and EVA egress locations were being discussed. A 1-g mock-up of the canister was developed to assess the access and to define the detailed design of each telescope and film camera unit. From the expected (but not final) camera configuration, the general size of the access door, the experiment locations and the reach envelope available were determined. As the programme developed, so the design of the experiments became more defined, allowing far more accurate estimates of the EVA requirements in the extraction and replacement of film cassettes. It also became easier when one of the experiments changed from a camera requiring a cassette film to only a video display on the ATM console, thereby decreasing the number of cassette changes from seven to six.

With the refinements in the position of the EVA hatches, the final work-station positions, and the location of the cassette receptacles in the experiments, it was time to consider the methods of moving the astronaut and the cassette from the OWS to

A Skylab EVA to replace ATM film cassettes.

the ATM and back without damage to the film magazine or the spacecraft, and without overtaxing him – all within the allotted four hours of EVA. It must be remembered that at this time (December 1966–mid-1969), the only American EVA experience was the Gemini EVAs (which on the whole were successful, although they raised concerns about work-load, restraint and fatigue) and a short EVA demonstration of the Apollo lunar pressure garment in Earth orbit (which was shortened due to the medical condition of one of the crew). The Soviets had completed only two short EVAs: the pioneering EVA by Leonov in March 1965 (portrayed as successful, but with difficulties that remained unclear for several years), and a brief vehicle-to-vehicle transfer in January 1969. The techniques for extended Earth-orbital EVAs in support of definite objectives at a space station were untried and, up to 1973, largely unknown. The Skylab series would have to demonstrate that EVA from a space station was possible, practical and highly desirable for keeping the vehicle flying. The fact that it had been achieved after almost a decade of planning is largely ignored in reviews of the programme and of space stations in general.

Evaluation of transfer modes
One of the most challenging hurdles in developing the ATM EVA system consisted of the contradictory requirements for each of the two work-stations, in terms of reach capability and visibility, and in relation to the other work-station. As the system developed, the 1-g and zero-g simulations played an increasingly important part in defining the system and understanding the 'domino' effect that even the smallest change would have when introduced into the system.

The whole process began with a 1-g walk through of the ATM film exchange *before* detailed drawings were produced outlining the EVA timeline. When a particular system or procedure appeared promising in 1 g, simulations were conducted in the water tank to verify the system. When a specific aspect of a procedure or an item of equipment required more definite evaluation, this was conducted onboard a KC-135 flying a parabolic curve. For the development of ATM EVA techniques, such parabolic tests included the evaluation of umbilical stowage, transferrance of packages, and studies of mass-handling dynamics and camera operation. The MSFC Neutral Buoyancy Simulator was used for Skylab EVA simulations, as it was large enough to evaluate an EVA task as a single procedure.

Using existing workshop designs, a wide range of transfer methods were proposed, prepared and demonstrated, and as the OWS EVA programme developed, a better understanding of an astronaut's capabilities on orbit (also based on the experiences from Gemini, and increasing use of the water tank) finally reduced the complexity of ATM transfer systems. By using the water tank, evaluating 1-g simulations, and adapting the KC-135 flights to verify test results, predictions of what would occur on orbit shifted the emphasis from research into potential solutions to a verification programme for a short-list of candidate concepts.

In the early studies, when the EVAs were to start from the air-lock module, a variety of telescopic booms had been envisaged, but in light of concerns over high work-loads for the crew, automated robotic arms were devised, in much the same way as the Shuttle and ISS robotic arms have evolved. One of the more advanced ideas was designated 'The Serpentine' – a segmented design that was highly automated and capable of transferring the film magazines and the astronaut to and from the work-stations. It received much early emphasis because of its potential in saving crew work-loads; but it was far too complex and too heavy, and thus created a mass penalty. Another method of transfer was 'roll up' booms, which had been successfully used as a method of deploying long antennae on unmanned satellites.

The next task was to establish how the astronauts could translate from the EVA hatch to the work-site. Two methods were favoured: by using the booms themselves, or by using a series of fixed hand-rails on the exterior of the multiple docking adapter, the LM cabin and the ATM rack. It was finally decided to proceed with hand-rails, because of the lack of stiffness in the proposed booms and potential hazards from metal straps used to form the boom elements into a longer structure.

The next significant change in the EVA programme came with the decision to use the LM forward hatch as the EVA access path. With this decision emerged the preference for simpler rail systems over the more complicated booms and robotic arms. In early studies, 'aided' (powered and manually controlled) rail systems were evaluated, together with pivoting arms that would enable packages to be swung between the hatch area and the work-stations. Interestingly, a rail-based system (the 'trolley'), designed to allow the astronaut to translate himself and the film at the same time, is now being installed on the ISS after tests on earlier Shuttle missions. More than three decades ago, Ed Gibson evaluated such concepts for the OWS. In underwater tests conducted in 1969, the system demonstrated that it could indeed move both astronaut and payload at the same time. However, in order to do so it required complex and heavy fabrication and a great deal of development time to create a reliable and easy rolling system. It was also found that off-centre loads tended to jam the rollers, especially when curves were being negotiated.

When the 'trolley' was abandonded, the work-stations could be redesigned to provide more direct access from the LM hatch. From here a pivoting arm, activated by a handle at the LM hatch work-station, could control the lowering of the arm down to the LM-end work-station to service the appropriate camera. For the film magazines at the Sun-end work-station, a dual rail system (the 'skateboard') extended all the way to the Sun end. Gibson evaluated two systems underwater. One of them could be pushed in front of, and be controlled by, the EVA astronaut, while the other used a simple hand-crank device operated from the LM hatch work-station, allowing the equipment to be moved to the work-station separately as the astronaut translated to install or remove the cassettes.

The dual rail system provided better stability during translation, especially through the solar arrays, where there was concern that the astronaut might inadvertently damage the panels by contact, especially by trailing legs. To alleviate this, a trough-like grid-work (the 'coal chute') was included, in which the astronaut could insert the toes of his EVA boots to stabilise his lower body as he translated or worked. When the configuration of the OWS changed from the spent stage (Wet) concept to the pre-fitted (Dry) concept in 1969, these were the working designs for

EVA transfer to the ATM. In both 1-g mock-ups and underwater simulations, the flip-over arm and skateboard concept both proved effective. The flip-over arm concept resurfaced two decades later, but not on am American station. Boom-arm devices called Strela (Arrow) were developed by the Soviets, and were later installed on Mir.

Development of the final Skylab ATM EVA system
The decision to eliminate the LM-supported ATM structure, and to launch the telescope system with the rest of the cluster, unmanned on a Saturn V, significantly changed the proposed workshop EVA programme. With less direct translation and transfer than the Wet Workshop configuration, the Dry Workshop saw a return to using the air-lock module for egress. Evaluation of the two work-stations on the LM revealed that the Sun-end work-station could be moved to within 45° in a direct line from the air-lock module hatch, and the LM-end work-station, while 90° from its optimum location, was still accessible for crew translation, and access and did not seriously hamper activities. It was later called the central work-station.

Located near to the EVA hatch (using a spare Gemini hatch that matched the curvature of the air-lock module, and had proven ease of operation and opening arc), a work-station was added to the fixed air-lock shroud area. Film cassettes could be stowed here, and it contained a temporary stowage hook for equipment. There were also hand-rails and foot-restraints and access to all the transfer equipment and restraints required during the EVA. During Skylab EVAs, the first crew-man out of the station occupied this area and acted as 'EVA manager', providing support by supplying film equipment and controlling the umbilical, while the second crew-man performed the physical translation to and from the ATM work-stations. The third crew-member remained inside the OWS to monitor the systems and the progress of his two colleagues outside, and to provide photodocumentation when required.

In early Dry Workshop studies, consideration was again given to the 'skateboard' device for transferring film from the fixed air-lock shroud work-station to the ATM. However, it soon became clear that due to the interfaces between the multiple docking adapter and the ATM, a smooth continuous rail system was difficult to align, and so the concept was dropped. The next concept featured extending booms, mounted at the shroud work-station and 'pre-aimed' at the two ATM work-stations, and used for the transfer of film canisters, as a separate crew translation aid, and for moving other equipment for EVA purposes. The idea of an astronaut carrying film cassettes to and from the work-station was quickly rejected on the grounds of possible damage to the magazines or the spacecraft structure, and potential harm to the astronauts. It would also have increased the translation work-load and added to the overall EVA timeline.

Once again, development of the systems and procedures was evaluated stage by stage in the 1-g and neutral buoyancy simulators. If the 1-g tests proved a feasible concept, it was fabricated and installed in the water tank to determine the best location for foot-restraints, the operation of the booms, and the transfer paths to and from the work-stations. On Skylab there were three booms – one to the Sun-end work-station, one to the central work-station, and a spare. Their positions were determined from the results of these 1-g simulations, while their accessibility for replacement and their aiming points were determined in the water tank. Using this information, an alternative transfer method – the 'clothes line' assembly – was also installed, mounted as close as possible to the proposed travel path of the booms. In case of a failure on a film transfer boom, a 'replacement work-station' was also designed. This temporary work-station consisted of a hand-rail for use as a toe-bar. and a small aluminium channel section for use as a heel restraint, mounted near the air-lock tunnel to allow direct access to all three transfer booms.

One aspect causing some concern in the development of the EVAs at Skylab was the restraint and support of the astronauts' umbilicals. Early in the evaluations, dozens of clamps were installed to route the umbilicals through, but gradually these were found unnecessary. Eventually, umbilical clamps were located only at each work-station and in two places (one for each astronaut) near the EVA hatch. Those at the hatch were primary, while the others remained available for use at the discretion of each astronaut.

Due to the design of the Sun-end work-station – which was not in line of sight of the air-lock – an intermediate work-station was devised, because no method of transfer assembly could negotiate corners. The transfer work-station was developed by solely use of the water tank, because of its unique location and relationship to other work-stations. With this work-station, together with the decision to stow the experiment S082A and S082B film magazines using a film transfer 'tree', and providing a temporary stowage area for the S082 cameras during film exchange, the development of the Skylab EVA system finally flown on the missions was complete. An EVA Critical Design Review, held in November 1970, recommended little modification prior to commitment to flight crew training.

A series of contingency EVA scenarios was also developed at the same time as the overall OWS EVA programme (1966–73), using a step-by-step method in approaching a problem, evaluating a system or procedure, and testing the best solution. The late inclusion of the solar array deployment by EVA was an excellent example of this procedure. Initially, three translation methods were short-listed, after which two were chosen. The favoured options were then tested and the most promising selected, and the three best options for deploying the arrays followed the same route. To those who developed the Skylab EVA system over many years, this step-by-step approach proved to be 'a powerful tool which need not be time-consuming or expensive.'

EVA FROM SKYLAB, 1973–74

In the pre-mission planning for the Skylab mission, there were to be six separate EVA periods (each of approximately 2.5 hours) totalling 29 man-hours, designed to retrieve and replace ATM film cassettes, retrieve particle sample collectors from experiments D024 (Thermal Control Coatings) and S230 (Magnetosphere Particle Composition), and make visual observations of the exterior of the station. There would be four complete changes of ATM film, which entailed handling a total cargo mass of approximately 680 kg, plus two retrievals of samples from experiment D024, and four from experiment S230.

However, by the end of the third mission in February 1974, a total of ten EVAs had been accomplished in 82.5 man-hours. These EVAs not only accomplished all the pre-flight mission objectives, but were also fundamental in saving the station for operational use after the loss of the micrometeoroid shield and one main solar array during the ascent phase. The astronauts also managed to complete an additional eighteen mission objectives and thirteen in-flight repairs.[10]

Skylab's planned EVA programme
Prior to the launch of Skylab 1 (the unmanned OWS) on 14 May 1973, mission planners had determined the sequence of EVA tasks and objectives for each of the three manned missions (at that stage planned for 28, 56 and 56 days).

Skylab 2 A single EVA was planned for Mission Day 26. The sole EVA of 2 hrs 30 min, by the Commander and the Science Pilot, was aimed at full retrieval of the six ATM cassettes, a partial installation of experiments S052 and S054, and sample retrieval from D024.

Skylab 3 Three EVAs were planned for this 56-day mission. The first, conducted by the Science Pilot and Pilot, would take place on Mission Day 4, last for 2 hours, and feature partial installation of S056, H-alpha 1, S082A and S082B, and collector retrieval from S230. The second EVA of the mission, on Mission Day 29, would be completed by the Commander and Science Pilot in 2 hrs 45 min, and would feature full retrieval and full installation. The third and final EVA, planned for Mission Day 55, would last 2 hrs 30 min, and would be conducted by the Commander and Pilot. The two astronauts would complete a full retrieval, a partial installation (S052 and S054), sample retrievals from D024 and collector retrieval from S230. The total mission EVA time would be 7 hrs 15 min spent on three EVAs.

Skylab 4 Two EVAs were manifested for this final visit to the Skylab station. The first, by the Science Pilot and Pilot on Mission Day 4, would last 2 hours, and would feature partial installation of S056, H-alpha 1, S082A and S082B, and collector retrieval from S230. The second EVA, carried out by the Commander and Science Pilot on Mission Day 54, would last 2 hrs 45 min. During the EVA, the two astronauts would make a full retrieval of all ATM cassettes using the clothes-line method. This would leave the ATM empty of film cassettes and end the ATM programme. The astronauts would also collect the final samples from S230 before ending their EVA. The total mission EVA time was planned at 4 hrs 45 min on two EVAs.

It is interesting to note that in this planning, the Science Pilot seems to be the primary EVA astronaut, performing the only EVA with the Commander on the first mission, two of the three on SL-3, and both on SL-4. The Commander would have performed the one EVA from SL-2, two of the three from SL-3, and the second on SL-3, while the Pilot was not assigned to an EVA on the first mission, and would assist on the first and third on SL-3 and on the first from SL-4.

A Skylab repair EVA.

Rescuing Skylab: the first EVAs

The first EVA associated with Skylab, on 25 May 1973, was a stand-up EVA (SUEVA) from the open CM hatch. The Pilot attempted to free the jammed remaining solar array from the small strip of aluminium, but was unable to exert enough force to free the strap and allow deployment of the panel. Paul Weitz, working with the upper part of his body out of the hatch, assembled a 4.5-m pole from three 1.5-m lengths passed out to him by Science Pilot Joe Kerwin. At the end of this pole was a 'shepherd hook', which Weitz used in an attempt to pry the strap from the trapped array. The CM hatch was adequate for a simple SUEVA, to retrieve sample packages (as on Apollo 9) or for photodocumentation (as conducted on Apollo 15–17), but Weitz was trying to impart a force via the pole and hook (later exchanged for a prying tool) to force the aluminium strap free. But even with Kerwin holding his legs, he could not exert sufficient force. Indeed, the only result was that the CSM was pulled towards Skylab, forcing Conrad to control the spacecraft. He commented that it was impossible to free the trapped wing with the limited tools available. Docking with the station, they transferred inside and deployed a solar shield parasol-type array from the sunward scientific air-lock in the side of the OWS, until they could perform a full EVA outside the station. The SUEVA tethers, hooks and poles, designed to deploy and restrain a SUEVA sail to the exterior of the station, would be reused during the activities designed to free the trapped wing.

During the next EVA, on 7 June, Conrad and Kerwin successfully freed the jammed solar array by using a procedure evaluated by a ground team headed by astronaut Rusty Schweickart, and approved on 4 June.[11] The astronauts received the EVA timeline and instructions via the onboard teleprinter, and fabricated the required items from materials already onboard the station or from additional items taken to the station in the CSM. The teleprinter was a new aid to an orbiting crew, and allowed better discussion with the ground controllers to further understand what they were being asked to accomplish. As an addition to the basic EVA training and the experience of procedure specialists, this ensured success for this challenging and important EVA (it was only the thirteenth time that American astronauts had performed an EVA in space). The crew confirmed their understanding of the techniques and new information conveyed to them during a televised partially suited practice session inside the OWS on 6 June. The excursion was a contingency EVA consisting of two parts: severing the strap restraining the array, and then deploying the array in its proper position.

Cutting loose the debris
For this task, four tools were used: the SUEVA cable cutter, vice grips, a pry bar, and (revealing the astronauts' ingenuity and versatility in adapting what was at hand) a bone saw taken from the onboard dental kit! With the practice of the procedures to cut the strap, and the appropriate tools available, all that the astronauts had to do was to make their way to the location where the strap was to be severed. But this added a new problem, because it was in a location that was not designed to support an EVA crew-member, with no translation aids or methods of restraint. Fortunately, Gemini had demonstrated the difficulty of performing even the simplest tasks on EVA without translation devices and restraints, and with this knowledge the planners provided a solution to a problem that had not existed before launch.

Conrad and Kerwin performed the EVA by moving from the air-lock module in the fixed shroud area through the air-lock module trusses to the long discone antenna boom at the forward end of the OWS. They assembled three of the 1.5-m SUEVA poles and two spare twin-pole sunshade poles by screwing them together, and then attached a cable cutter to the one end. A spare EVA waist tether was used to fix the device to the antenna boom. This not only allowed translation to the solar array beam, but was also the method to cut the strapping. Next, a 6-m length of rope was attached from the cable cutter back to the air-lock area for safe operation of the cutter from a distance of 9 m. The operation required one of the astronauts to attach the cutters to the debris and translate down the pole to assist in cutting, while the other astronaut moved to the fixed air-lock shroud area to activate the cutters.

Deployment of the solar panel
Prior to physically cutting the debris, the astronauts had to attach a beam erection tether. It was believed that the electronic damper that was designed to slow the deployment of the array was frozen, and so the development team, led by Schweickart, devised the beam erection tether to help force the wing open. It was constructed from a 9.8-m section of SUEVA sail tethers that was no longer required, plus two EVA waist tethers. Hooks were attached to holes in the wing array, while the other end was also attached by a hook near the discone antenna boom. With no firm footholds available to pull on the rope, the astronauts had to translate to the

centre of the beam erection tether and hold the rope over their shoulders to restrain themselves against the hull of the OWS, stand up, and then strain and pull on the beam erection tether to deploy it. It was also supposed to prevent the astronauts from flying off into space, but once again problems became evident when they tried to execute the idea.

Kerwin found that gaining a firm foothold at the discone area, in order to position the pole/cutter device, was more difficult than in the water tank at Huntsville, because for some reason its design differed from that on Skylab in orbit. After a frustrating 30 minutes of trying to position the pole with one hand while holding on with the other, he doubled up the tether to reduce its length 0.9 m, to hold him more firmly to the station while still allowing partial use of his other hand. Finally, the aluminium strap was snared in the jaws of the cutter. Meanwhile, Conrad began to attach the beam erection tether hooks to the discone antenna, and then the two small hooks to both holes on the first solar wing. Yet again, however, he found that the hardware in the ground mock-ups was different from the flight model, with the second, smaller hook not fitting the respective hole. Tightening the strap as much as possible, they cut the cable, and Conrad placed the beam erection tether over his shoulder, after which he was joined by Kerwin to pull everything free. With both astronauts straining against the beam erection tether, the damper suddenly gave way and the array sprang out, and they were flung off the hull of the station. Seconds later, when they recovered their composure and position, the wing was fully deployed, and the electricity meters on the ground were recording the increasing power supply. The EVA was an outstanding success, but before the astronauts returned to the inside of the ATM, the films had to be changed, and other tasks had to be performed.

SKYLAB EVA TASKS

With the space station saved, the science programme could continue. The three missions were flown as planned, except that the second mission was extended to 59 days and the third to 84 days. All crews worked on a variety of EVA tasks at the station – some of them planned, and others that were added as the need arose.

The Apollo Telescope Mount

The primary purpose of EVA on Skylab was the replacement of used film cassettes. This was accomplished on the third EVA on SL-2, all three SL-3 EVAs, and three of the four EVAs on SL-4. Problems with some of the ATM equipment also allowed the astronauts to demonstrate their ability to service and repair faults during scheduled EVAs.

On Skylab 2, following the deployment of the solar array, the astronauts fixed a faulty film feed system on S082A by replacing the cartridge with a new unit, and after the S054 aperture door behaved erratically, they secured the door in open. During their last EVA, after an internal relay inside a battery charger regulator module had become stuck, the stuck array was freed with a well-placed hammer blow to a particular screw head. The candidate screw was revealed in a sketch teleprinted up from the ground. On the same EVA, the occulting disc of S052 was cleaned after glare was observed on one of the discs. The problem was resolved by using a lens brush to clean debris from the disc.

During the final EVA on SL-2, Conrad had difficulty maintaining his position when trying to install a sample piece of the JSC sail material around an ATM strut, until Weitz came to his aid. Conrad recalled the difficulty which his Gemini 11 colleague Dick Gordon had experienced on his EVA almost seven years earlier: 'It was a real Dick Gordon operation. I was really floating away out there trying to hold on until Paul came down and put his feet on my shoulders. Then I could hold on with my legs, and I finally got that sail cloth on.' The effort caused Conrad to overheat, and caused an increase in heat loads in his suit.

On the first EVA from SL-3, a ramp latch on the ATM S055 telescope aperture door (on the Sun end) became stuck. The problem was solved by a crew-man removing two 1.11-cm bolts that secured the latch. The problem of attempting this while wearing bulky EVA gloves and holding a tool designed for use inside the OWS, was solved by modifying a box-end wrench from the OWS tool-kit, building up the handle with layers of tape, and attaching a tether for improved grip. On their next EVA, the astronauts removed and replaced the ATM rate gyro processor package. This had shown signs of deterioration after the first mission, so the second crew took with them a new set of gyros. These had to be installed inside the vehicle, but connected to the vehicle control system by EVA. Had the gyro system failed completely, Skylab would have to have been immediately abandoned, as there was no replacement hardware or alternative system. To remove and replace electrical connections that were not originally planned to be replaced, a 7.3-m-long cable had to be looped to the computer interface on the ATM. To complete this task, specially designed connector removal pliers were included in the tool-kit taken to Skylab in the CSM. The astronauts also again cleaned the S052 occulting disc on their third and final EVA.

On the first EVA of SL-4 at the ATM, the crew secured the aperture door of the Hydrogen-alpha 2 telescope in the open position, and on their second EVA they repaired the S054 X-ray spectrographic telescope filter wheel, which had stopped between filters. After removing the film magazine, the astronauts pressed two small retaining devices on the instrument, after which they opened the shutter mechanism to gain access to the stuck filter wheel. Viewing the operation was only possible by using a mirror to look inside, and the repair was carried out by manipulating a screwdriver through the open shutter to select the permanent filter-wheel position. It operated on only one filter wheel for the remainder of the mission. Although trained in the operation of the experiment, the astronauts were not familiar with its inner workings, as they would normally not have had to access the inside of the instrument. Detailed messages were therefore teleprinted up to the crew, who clearly understood them. The flight controllers followed the EVA operation by using a flight-configured S054 camera in MCC. This allowed real-time understanding of how the astronauts were proceeding, and offered additional help and support through the CapCom. During the second EVA, the astronauts also secured open the aperture door of S082A.

Experiments

The S149 particle collection experiment was installed during the first EVA on SL-3. This experiment was supposed to have been deployed from the Sun-side workshop scientific air-lock, but the deployment of the parasol prevented this, forcing a redesign of support equipment, which was then launched on SL-3, to mount the experiment on the ATM solar shield disk. Two foil detector cuffs from S230 Magnetic Particle Composition were retrieved, and two calibration clips were installed on the first EVA on SL-3. During the third SL-3 EVA the S149 experiment samples were retrieved, together with the final set of D024 disc and strip samples and one S230 collector. Also installed was a sample of parasol thermal shield, attached to a clipboard and to the S-10 hand-rail in direct view of the Sun. This provided further information on the performance and deterioration of such material after prolonged exposure to solar light and heat.

During the first EVA from SL-4, the crew installed the T025 Coronagraph Contamination Measurement experiment.This was another experiment originally designed to operate through the scientific air-lock now occupied by the parasol shield, but instead, a specially designed mounting attached the experiment to the ATM truss close to the air-lock shroud EVA work-station. It was operated on each

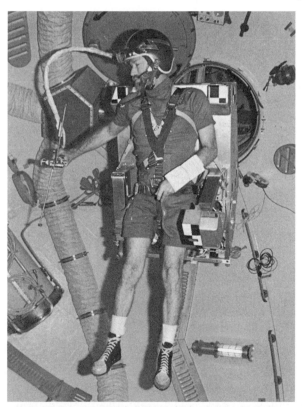

Test-flying the M509 MMU device inside Skylab.

of the four SL-4 EVAs for contamination measurement and photodocumentation of comet Kohoutek.

On their first EVA the crew also inspected and repaired the S193 Earth Resources experiment. They analysed and repaired the instrument pointing system by inspecting and cleaning the pitch and roll axis potentiometers, disconnected three electronic cables, and installed a 'jumper box' to isolate the pitch and roll functions, determining that the unit had failed. Using a standard ratchet assembly (removing two Allen head screws), the crew removed the pitch-axis launch lock, replaced it with a gimbal lock, and secured it at 0° pitch. An electrical power disable plug was then installed on the electronic assembly to remove power from the pitch-axis servo. The cover over the jumper box was also replaced.

For the S228 Transuranic Cosmic Ray experiment, a passive detector was installed on an external clipboard to hand-rail S-10 during the second EVA on SL-3. The crew also retrieved and replaced the S149 Particle Collection experiment samples, a new set of D024 disc and strip samples on the experiment panel, and two thermal shield material samples (identical to the material used to fabricate the twin-pole sunshade), using double-sided adhesive tape for attachment. They then installed a new magnetospheric particle collector (for the S230 experiment), removed a pair of collaboration clips, and replaced one clip over the new collector.

The second SL-4 EVA saw the installation of the S020 UV X-Ray Solar Photography experiment. This was another instrument planned to be deployed through the SAL, but instead the crew mounted it on an ATM truss near the EVA work-station at the fixed air-lock shroud location, using a special mounting that was launched on SL-3 but not installed. They also retrieved the S149 experiment for the final time

On the third SL-4 EVA, the stand-up EVA sail material, which had been installed during the final SL-2 EVA, was retrieved from the +Y ATM truss. The crew also retrieved a sample of air-lock module micrometeoroid cover material from the air-lock module EVA hatch hinge area, used a digital thermometer to take a temperature reading of the S020 experiment housing, and installed, operated and retrieved the S020, T025 and S201 photographic experiments.

The twin pole assembly
The installation of this sunshade was the first EVA task for the SL-3 astronauts. The deployment began with installation of the equipment near the fixed air-lock shroud. After moving to the Sun-end work-station, Lousma received the foot-restraints, sunshade base plate and sunshade bag assembly via the transfer boom. Garriott, stationed in the fixed air-lock shroud, constructed the twin poles and transferred them to Lousma, who installed them in the base plate. Then, using halyards on the pole assemblies, it was pulled out to deploy. Lousma then tied off the reefing lines against the ATM truss, and secured the second shield in place for the duration of the station's orbital life.

Visual observations
All the EVAs conducted included verbal and visual observations of the exterior of the OWS, ATM, CSM and other elements of the station. In particular, Jack Lousma

conducted a visual inspection from the ATM Sun-end work-station of the CSM, which had previously identified leakage problems from CSM thruster quads A and B. Lousma did not identify the source, although there was an obvious leak; but the leaks later stopped, allowing the mission to run to its full duration of 59 days. On this EVA the crew also inspected a leak in the air-lock module/OWS coolant loop.

Photography of comet Kohoutek

During the second SL-4 EVA, an opportunity was taken to visually observe and photograph comet Kohoutek. The crew mounted experiment S201 Far Ultraviolet Camera on the ATM truss previously used for the T025 and S020 experiments. The space station was then manoeuvred immediately before experiment operation so that the photocathode was shadowed under the ATM solar array. The experiment was then activated by a crew-man and completed three sequences of ten photographs. The T025 camera was also used for cometary photography.

Skylab's last EVA: close-out activities

On 3 February 1974, the final EVA from Skylab was completed. It was the thirty-eighth American EVA and the fortieth world EVA, but it was not expected that it would be the last American EVA for more than nine years, until the first EVA on the sixth Shuttle mission in April 1984. Originally intended only to remove all six ATM camera cassettes and retrieve one of the S230 collectors on Mission Day 54, it took place on Mission Day 80, four days before landing, and included twenty separate tasks.

The crew used the clothes-line back-up retrieval system for transportation of the ATM film magazines, installed two lines to the main ATM EVA work-stations, and manually activated them for the transfer of the film. After the flight, the crew indicated that the clothes-line system was easier to use in zero g than in the underwater simulations (probably due to the viscosity of the water), and that even if the clothes-line and hook systems had failed, it would not have been difficult to hand-carry the magazines.

A zero-g fixture cover plate, approximately 14 cm in diameter, was removed from the Sun-end solar disc with a ratchet wrench, with a straight screwdriver blade employed to remove several flat-headed screws. Adhesive tape was used to retain the cover and screws during removal and transportation back to the OWS.

Photographs of the OWS cluster were taken with a Nikon camera and a data acquisition camera. The data acquisition camera was taken to the Sun-end of the ATM to take documentary images from the higher position, and was also used for photodocumentation of selected targets from the fixed air-lock shroud and of the activity of the astronauts during the EVA. The Nikon was used for specific targets of importance.

The astronauts also remounted and operated the S020 and T025 experiments and the S149 particle collection experiment. They took temperature readings of the S020 housing, and retrieved a variety of samples for return to Earth for post-flight evaluation of the components after nine months in space.

The retrieved and retuned items included an air-lock module micrometeoroid cover sample, the S230 particle collector, the S228 panel and parasol materials samples that were attached to the clipboard, the D024 discs and strip samples, the twin pole sunshade sample, EVA communication cue cards, and an extension from the ATM central boom for thermal evaluation.

Summary
Following the return of the third flight crew in February 1974, a series of five Lesson Learnt documents was compiled by Skylab teams at NASA HQ, JSC, KSC, and the Skylab and Saturn Program Offices at MSFC. The reports from the Marshall Skylab Program office and from JSC in Houston provide some insight into the lessons learned from the Skylab EVA programme.[12]

The Marshall report emphasised that the lack of translation and stability aids for EVA hampered attempts to free the jammed solar array, and added: 'The EVA access area should not be limited by the lack of EVA hand-holds or stability aids, but should encompass the entire vehicle. These aids should be either integral to the exterior design or allow for the simple attachment of portable devices.' This lesson was certainly learned, because at the times of these reports, thought was being given to developing the EVA systems for the Space Shuttle. In the mid- to late 1970s it was already clear that the Shuttle would serve as a valuable tool in the service and repair of large space structures, and so the provision of EVA aids on those spacecraft *prior to launch* would assist in any future EVA-based servicing and repair. The most notable of these designs was the array of hand-holds across the exterior of the HST.

The Marshall document also indicated that a high-fidelity mock-up should be available (ideally at the contractor's facility) for the development of EVA work-stations and operations. On Skylab it was found that even the model in the water tank at Marshall differed from the one flown, and any EVA problem frustrated both the space crew and the ground crew (as with the attachment of the solar shields). But were lessons leaned here? It appears not, because in 1984, when the astronauts of STS-51A were attempting to berth the captured Westar and Palapa communication satellites in their specially designed cargo bay housings, it was found that the ground mock-ups and designs differed from those in orbit. A few centimetres may not seem much, but the satellites had to be manhandled, and recovery techniques had to be adapted.

Marshall also highlighted another problem on Skylab, in that venting from the suits was more propulsive than expected, and imparted motion on the spacecraft, thus activating thruster usage during periods of EVA. This could have affected safety, and Marshall therefore recommended that future spacesuit designs should include non-propulsive vents.

The JSC report recognised the need for adequate restraint during EVA. A universal foot-restraint was certainly an EVA requirement, and experience from IVA revealed that foot-restraints should be attached to the spacecraft, not to the crew-member. This also applied to EVA restraints. With regard to on-orbit repair and maintenance, the JSC report considered the following:

1 EVA should be considered as a normal means of repair.
2 Proper procedures, tools and equipment for crew use should be developed.
3 In designing flight equipment, provision should be made for in-flight maintenance. (On the SL-2 EVA, the astronauts realised that by taping items to their suit they could reach them while wearing the pressurised glove. This idea was adapted for the Shuttle, to provide tool tethers and carriers so that the equipment was at hand when required.)
4 In the design requirements of a programme, consideration should be given to EVA inspection and repair issues (as later incorporated for the HST).
5 Tool and retainer boxes and bungee cords should be provided for the effective containment of loose nuts, bolts, washers and tools (adopted for Shuttle EVAs).
6 An adequate work-site, repair bench or other suitable location should be provided, and be equipped with restraints for tools and equipment (and presumably the astronaut, although this was not mentioned in the report).
7 A suite of spares should be carried for hardware items, (probably) for servicing and/or replacement (again adapted (eventually) for Shuttle and ISS operations).
8 The sizes of screws, bolts, nuts and similar items should be standardised in future spacecraft design. (Skylab astronauts were involved in developing EVA techniques for the ISS, and were instrumental in this and other issues arising from Skylab.)
9 A high-fidelity maintenance training simulator should be provided. (This was followed up for the Shuttle and the ISS.)
10 The fluid and gas systems should be capable of being serviced from the interior of the spacecraft (IVA). Fluid and gas connectors (B-nuts, welds and solder joints) should be located and configured so that they can be inspected by the crew for leaks. (This technique was demonstrated on one EVA during STS-41G in October 1984, and performed operationally on a series of EVAs from Salyut 7 by cosmonauts the same year.)
11 Exterior protective covers for experiments and equipment should be designed for easy manual operation on EVA as well as by automatic opening. EVA manual override might be necessary if the automatic opening fails. (This option was trained for on many Shuttle contingency EVA procedures, and was performed on STS-37 during the deployment of appendages on the Compton Gamma Ray Observatory in April 1991.)

The JSC report also stated: 'The Skylab crew[s] demonstrated conclusively the capability to perform a wide variety of on-orbit maintenance and servicing tasks.' This capability was also demonstrated on Russian stations and several Shuttle missions, and continues on the ISS. The report also emphasised: 'The crew-man's manual dexterity was not noticeably impaired by prolonged exposure to zero g. However, EVA tasks continue to be hampered by the dexterity limitations of the suit and gloves.' This is a problem that is improving with the ISS, but is not yet totally resolved. Under the category 'Capability for EVA Access, EVA Paths and Hand-holds', the JSC report stated: 'EVA paths should be established and hand-holds provided so that the crew can traverse to any point on the exterior of the spacecraft.

If fixed restraints are not feasible, alternative design concepts should be considered.'
Some direct paths were planned for Skylab (for example, to and from the ATM
work-stations), but all were limited by the 18-m EVA umbilicals, and in some cases
by the lack of hand-holds. Universal foot-restraints were designed for the later
Skylab missions.

A further summary of Skylab EVA results was included in the post-programme
summary held in California during August 1974. The conclusions were as follows:

- EVAs were essential in repairing the crippled Skylab cluster and in keeping it
  running throughout all the Skylab missions.
- The inherent flexibilities of manned spacecraft are realised most when EVA
  capability is provided.
- Given adequate provisions (restraints, tools, lighting and procedures), an
  EVA crew-man can accomplish most of the tasks he could accomplish on the
  ground. These tasks included the following activities: a) thermal shield
  deployment; b) solar array erection; c) pointing $0°.25$ (T025 experiment); d)
  diagnostic analysis and repair (S193 experiment).
- Underwater simulation provides a good duplication of the EVA environment,
  and is very effective for procedures development and crew training. Tasks are
  a little easier to perform in a zero-g environment than in an underwater
  simulation.
- The thorough EVA training of flight crews and ground crews was a key
  element in the successful accomplishment of the many unplanned Skylab
  EVA operations.
- The uplink and downlink capabilities provided by the teleprinter, television
  and air-to-ground communications were key factors in the planning and
  execution of the unplanned Skylab EVA operation.

Moreover, the Skylab EVA operations were more closely linked to the
pioneering excursions from Gemini, even though they came after the highly
successful Apollo programme had been completed. During Apollo, nineteen
separate EVA periods were conducted, but only four of these took place in space,
while the rest were on the lunar surface. When it is considered that prior to Apollo
11 and the first Moonwalk, the previous difficulties and experiences on Gemini
were followed only by a short (and abbreviated) Earth-orbital EVA on Apollo 9,
the success of Skylab becomes clearer. The ten highly successful Skylab EVAs
clearly showed that the difficulties encountered on Gemini EVAs had been
successfully addressed and overcome.

The conclusions highlighted that with the provision of proper restraints and tools,
good lighting conditions and clear procedures, a wider range of tasks could be
accomplished, far exceeding the earlier tasks assigned to Gemini. It had already
become clear, towards the end of Gemini, that underwater training was a key to
flight and back-up crew preparation, and it remains the same four decades later with
the ISS. Indeed, it is not only the flight crews that undergo EVA training. At NASA
at least, many of the EVA flight controllers have donned suits and entered various
water facilities to simulate what they expect the crew to do in space.[13]

During the Skylab 2 technical crew debriefing at JSC, the first Skylab crew added their own comments on why they considered that their stand-up EVA did not work and the solar array deployment EVA was successful.[14] Pete Conrad reflected on the question of why they could not put the cutter on the aluminium strap and cut it from their position in the CSM. The crew had discussed this the day after the stand-up EVA, and agreed that if Weitz had been able to see that the strap was sticking out sufficiently, then they could have severed the strap on the stand-up EVA. Conrad could see a side view of it through the CM windows, but Weitz had a more direct view. He was surprised that when he used the hook on the wing and heaved, he pulled the two spacecraft closer together. His efforts had imparted motion on the workshop and activated the thruster attitude control sub-system, which fired directly at the astronauts to move the station away, with the hook caught under the wing and beginning to pull Weitz out of the CM hatch. Conrad had to control the CSM to compensate for the firing of the thruster attitude control sub-system, and the crew almost cast the pole and hook loose. Weitz recalled his feeling that it seemed to be a crude procedure to try to use brute force as he leaned out of the hatch. When Conrad and Kerwin conducted their EVA on 7 June, they could see exactly what the problem had been. They discovered that a single screw had driven through the aluminium skin, while others had merely scratched the surface. The screw had penetrated the skin of the SAS panel, and the whole strap had become entangled around it.

Good communications between the team on the ground and those in space was also cited as essential. The final Skylab mission is often (erroneously) remembered for a breakdown in communications between the flight controllers and the astronauts, but it is perfectly clear from reading the flight records that in many cases this is simply not true. The repair to the S193 experiment and the S054 filter wheel required in-depth communications with the ground, as the astronauts were unfamiliar with the inner workings of the experiments, which were not intended for EVA maintenance and repair.

Such factors have continued to be important to today's EVA programme, and have been enhanced by further in-flight experience, advances in technology and telecommunications, and benefits from other programmes through an international exchange of ideas, results and resources.

Skylab was a quantum leap for developing the techniques of EVA in space, let alone at a space station. The standards set during those few months in 1973 and 1974 have been a benchmark for all space station EVAs that have followed. Sadly, there would be no further Skylab stations or missions, and the development of EVA techniques at a space station became a Russian domain for the next twenty years, with their Salyut series and the highly successful Mir space station. Ironically, the next time Americans performed EVA at a space station was at Mir in 1996.

SALYUT TAKES OVER, 1977–86

The Skylab astronauts successfully demonstrated that EVAs were essential in supporting a space station, but unfortunately for the Americans it was not possible to capitalise on this experience in the short term. There was no commitment to a space station to follow Skylab, and after the joint American–Soviet Apollo–Soyuz docking mission in July 1975, no American would even venture into space for almost six years, until the advent of the Shuttle. The Shuttle evolved from a space infrastructure plan in the late 1960s – a reusable ferry system to large space stations in support of regular flights to a lunar orbital space station and surface research base, as well as a starting point for the first manned planetary missions to Mars. By 1981, all that was left of this grand plan was the Shuttle, and although it would respond well to its new roles of deployment, retrieval and repair missions, and short scientific missions using the European Spacelab module, it would not be until the mid-1990s that the Shuttle would finally fulfil its original function and visit a space station (albeit a Russian station). In 1971 – two years before Skylab was launched – the Soviets launched Salyut, the first of a series designed with a view to the creation of a permanently manned orbital space station. During the Salyut (and subsequently Mir) programme, the Soviets demonstrated the true value of long-duration space habitation, supported, among other things, by a regular EVA programme. It would take several years to establish it, but by the end of the Mir programme in 2001, space station EVA operations at the ISS could draw confidence from the cosmonauts' experience of activities outside their space stations over more than two decades.

The early Salyuts
The first Salyut space station was launched in April 1971. The first mission to fly to it – Soyuz 10 – failed to hard dock, and crew transfer was prevented. The second mission – Soyuz 11 – set a new endurance record of 23 days in space onboard the station, but the prime crew was killed during the descent phase of the mission in June 1971. The programme was further set back with the loss of the next space station (undesignated) in a launch mishap in July 1972, and the on-orbit failures of two stations – Salyut 2 and Cosmos 557 – in April and May 1973. The next successful space station, Salyut 3, was launched in July 1974, but was occupied by a crew for only fourteen days, as a second crew failed to dock, adding to the frustrations. In 1975 the next station – Salyut 4 – fared better with a 30-day and a 63-day residence, and in 1976 this was followed by a new station – Salyut 5 – which hosted a 49-day and an 18-day residence, with another failed docking attempt. In these six years, seven space stations were launched and ten crews were sent to them, with only six actually living on board. In all of these operations, no Soviet cosmonaut performed EVA. This was a great surprise for Western observers, in light of Skylab's success, the fact that the Soviets had performed the first EVA in 1965, and that they had seemingly abandoned EVA operations since the Soyuz 5/4 transfer in 1969.

By 1977, speculation in the West suggested that the Soviets would embark on their first EVAs for more than eight years from the next station, Salyut 6. In May 1976, while attending a space congress meeting in Philadelphia, Soyuz 18 Commander Pyotr Klimuk, who had spent 63 days onboard Salyut 4, indicated that EVA had been planned from Salyut 4, but that rescheduling and heavy training programmes had led to the removal of exits from that station.[15] Western interpretations of events in 1976 and 1977 then suggested that EVAs were also

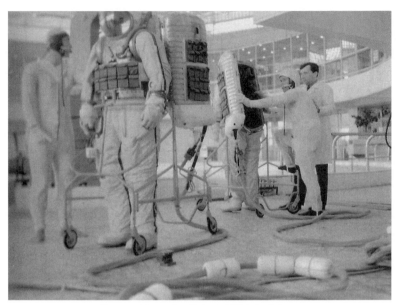

Preparations for Salyut EVA training.

planned for Salyut 5, but none were forthcoming. The Commander of the Soyuz 21 mission to Salyut 5 was been Boris Volynov, who had also commanded Soyuz 5, the last mission during which a cosmonaut left a vehicle in space. This established the supposed link to EVA preparations on Salyut 5.

The next mission – Soyuz 22 – was a solo Earth observation/resources mission, and was not a direct part of the Salyut programme; but the background of the two cosmonauts again presented Western observers with grounds to suggest that EVA was imminent. Commander Bykovsky had been involved in the original Soyuz 2/1 mission for which EVA transfer was planned in April 1967, before the difficulties onboard Soyuz 1 forced abandonment of the Soyuz 2 mission as originally envisaged. Bykovsky was then involved with EVA training for the Soyuz 5/4 mission, while his Soyuz 22 Flight Engineer, Vladimir Akyonov, had performed Soyuz-to-Soyuz EVA simulations a decade earlier. Western interpretation – which happened to be misinterpretation – suggested that all this was in preparation for EVA on the next mission to Salyut 5. When that mission – Soyuz 23 – was launched in October 1976, Soviet reports seemed to support this assertion.

The Flight Engineer of that crew, Valeri Rozhdestvensky, was a former Soviet Navy diver in charge of a team of deep-sea divers, and had completed a course of 'special training involving zero-g simulations and parachute jumps'. Did this mean that EVA was assigned to their mission? Reports that Rozhdestvensky had commented that he expected his former career as a diver to prove useful during the mission seemed to indicate this, but in the event the crew failed to dock with the station, and after two days the spacecraft re-entered and ended its mission – in a lake! Four months later, Soyuz 24 managed to dock with Salyut 5, and the crew was able to complete an 18-day residence; but once again they did not step outside, even though Western observers were convinced that EVA was included in the initial planning.

The Soyuz 24 crew also appeared to have an EVA background, and had been the back-up crew to Soyuz 23. Commander Gorbatko had served as a back-up to Leonov on Voskhod 2 and Khrunov on Soyuz 5, and Flight Engineer Yuri Glazkov had written a PhD thesis on EVA. Western interpretation of crew reports indicating that they had reduced the levels of the atmosphere, again gave rise to speculation that at the very least, EVA simulations were being conducted. Earlier reports in the West indicated that in 1972 the Soviets had unsuccessfully tried to purchase an Apollo pressure garment to analyse its make-up, but that the sale had been blocked by the US Government. Then, in June 1973, as part of the opening of Soviet space facilities to the West under the Apollo–Soyuz Test Project, Vladimir Shatalov guided a group of Western journalists around the Yuri Gagarin Cosmonaut Training Facility. On a mock-up Salyut that they were shown, the three solar arrays were located further back than the forward pair of arrays on the first Salyut, and revealed an aperture on one side of the forward transfer compartment, near the front docking port, that had the appearance of a large circular EVA hatch.

Salyut 6, launched in September 1977, would prove to be a milestone in cosmonautics, setting several records and firmly establishing space station operations as the main Soviet activity in space. When the first Soviet EVA since 1969 was completed in December 1977, the Western interpretation of events over the previous few years seemed to be vindicated. However, the story had begun much earlier than the mid-1970s.[16]

Evolution of Soviet space station EVA hardware
As with many aspects of the Soviet space programme, the techniques of working in space – including EVA from orbital platforms – were first proposed by the 'Father of Cosmonautics', Konstantin Tsiolkovsky, in the early part of the twentieth century. By the mid-1960s, the first cosmonauts had orbited the Earth, and Leonov had performed the first EVA. At Korolyov's OKB-1, plans for a new spacecraft, Soyuz, were underway, and there were dreams of sending cosmonauts to the Moon in response to the call by President Kennedy. But there was also an alternative programme from a second design bureau, which was laying the groundwork for what was to follow in the 1970s and 1980s. The Central Design Bureau of Machine Engineering (TsKBM), under the direction of V.N. Chelomei, was looking not to the Moon, but to the creation of an orbital space station called Almaz (Diamond), with military objectives.

Under the Almaz programme, cosmonauts were required to transfer from a transportation ferry vehicle to the orbital station by EVA. Based on their experiences in other programmes, suit manufacturer Zvezda was tasked with supplying the EVA suit for the Almaz programme. The Yastreb suit was far more advanced than the Orlan semi-rigid suit (selected for the L3 mission Commander) and the Oriol soft suit, and it was therefore logical to select this suit for the military station programme. Work progressed in designing the documentation, completing fit-and-function checks in a mock-up of the air-lock, and planning the manufacture of training and flight suits.

There was some debate over the difficulty of providing one suit to cover all requirements in both IVA and EVA scenarios; but it was soon realised that it is impractical to use a semi-rigid EVA suit in a rescue mode, as the requirements are so different, and equally, it is impractical to use a smaller, lighter and more mobile rescue suit for an EVA during which the suit must retain a positive pressure and incorporate numerous devices not required for a rescue. By 1969 – after studies had determined that EVA would become a regular activity from space stations – Zvezda proposed the development of two types of suit. The first would be a lightweight and individually tailored rescue suit, and the second would be a development of the Orlan L3 semi-rigid design that could be adjusted to fit the wearer and periodically serviced and repaired. It also made sense to launch such suits with the space station, so they could remain on orbit as crews changed, and could be used by more than one crew as required, thus saving launch and landing weight penalties and costly refurbishment on the ground.

A government directive of 3 July 1968 had authorised feasibility studies to be conducted with Yastreb-derived suits and the Almaz station. However, after further evaluation of the Orlan-type suit in a mock-up of the proposed Almaz station, a new 'decision document', approved in November 1969, stated that a 'new and more advanced spacesuit of the Orlan type' would be used on future orbiting stations.

Under terms of the new agreement, the Orlan suit was designed to:[17]

- Support EVA of one or two crew-members, performed from an air-lock with a 78.5-cm diameter hatch.
- Maintain onboard equipment on the outer surface of the space vehicle.
- Carry out operations with 'departure from the vehicle, and manoeuvring with the use of an individual propulsion system' (to be developed under a separate agreement).
- Accommodate the development of a separate self-contained back-pack with radio communications, telemetry and power supply sub-systems.
- Support a five-hour EVA from the Almaz station and have the capability to perform from two to four such EVAs over a ten-week period.
- (Pending the results of a feasibility study) have facilities for crew recharging of the life support system with both oxygen and water and replacement of the contaminant control cartridges onboard the space station.

By 1970, delays in preparing the Almaz station for launch, and the impending launch of the American Skylab station (planned for late 1972) led to a Soviet government resolution being issued on 9 February 1970. This resolution was based, in part, upon the recommendations of a group of specialists from TsKBM at the end of 1969. These specialists stated that in the interests of 'scientific and national economy purposes', a basic civilian station could be prepared and launched sooner than the more advanced Almaz military station. The new station would be a hybrid of the hull of the Almaz design and flight proven components from Soyuz. Designated DOS (*Russian*, long-term orbiting station), work progressed towards an initial launch in 1971. This first manned space station was originally to be called Zarya (Dawn), but the name was changed to Salyut (a salute to Gagarin's historic

In-flight servicing of Orlan EVA suits.

mission a decade earlier) shortly before launch on 19 April 1971, when it was realised that Zarya had been used as the code name for ground control.

EVA from Salyut

With the decision to fly a 'civilian' Salyut before the 'military' Almaz, work on an EVA suit to support DOS activities began in the early months of 1970. Modifications to the basic Orlan/Almaz suit were required for use on DOS, mainly for extended and multiple use, long-term storage and orbital maintenance. To identify the variant, the 'D' from DOS was used, to create the Orlan D configuration. The modifications were approved in April 1970, and changes in the documentation revealed that on DOS, Orlan D suits would:

- Remain on orbit for up to three months.
- Have an operational life of at least ten hours, providing they were adequately maintained and serviced on orbit.
- Support three or four separate EVAs, each of 2–4 hours duration.
- Accommodate crew-supported recharging of back-pack storage bottles with oxygen, refilling of cooling water, and replacement of contaminant control cartridges.

The overall mass of two suits, filled and charged for use, and including all onboard components, was not to exceed 216 kg.

Over the next year, studies were conducted on the upgraded Orlan suit and its back-pack, and exactly how they would be stored on the Salyut over a long period, with a number of test suits and mock-ups fabricated to evaluate systems, procedures and components. Test included models designed to evaluate suit-to-station connections, and the method of ventilation and on-orbit drying after and between EVA periods. At this point, the traverse distance was determined to be 15–20 m from the forward EVA hatch, and the electrical/communication umbilicals were therefore fabricated to support EVAs of up to three hours at that distance and as a safety tether for each cosmonauts. A set of replacement and spare parts was evolved for on-orbit change and replacement of parts. A system of orbital checkout and test equipment and a network of EVA hand-rails was planned for the exterior of the Salyut, around the hatch and in the places where an EVA might probably take place, such as towards the docked Soyuz, across the surface to the solar arrays, or to the rear of the station.

Planning the first Orlan/DOS EVA
With the decision to fly DOS before Almaz, and with the design of the EVA suit and support equipment defined and its operating enveloped agreed, the next task was to assign the suits to a station and formulate a plan for using them operationally.

A programme of tests and evaluations was planned, with cosmonauts in the Hydro Laboratory at TsPK and in altitude chambers at the Air Force Scientific Research Institute (GKNII). With delays to the Almaz programme and a decision not to fly Orlan suits on the first Salyut (probably as a result of the hybrid design and the lead time required to incorporate all the EVA support equipment and facilities), no EVAs were apparently assigned to Salyuts launched in 1972 or 1973. By May 1973, with the loss of three Salyuts and the creation of the Apollo–Soyuz Test Project, Zvezda had placed a request to TsKBM to try to implement an Orlan suit system on a Salyut as soon as possible, in order to flight-test the equipment and procedures and to gain some operational experience in support of the space station and other (lunar) programmes under development.

This was authorised in January 1974,[18] with documentation and specification issues being resolved over the next nine months. On 18 September 1974, the final decision was made to fly Orlan D suits on DOS 5 (which became Salyut 6). Ahead of the mission planners lay two more years of suit tests and integrated tests, and the development of a training programme for the cosmonauts. It was the middle of 1977 before Orlan flight units no.33 and no.34 were ready for installation on the Salyut, with flight units no.35, no.36 and no.38 also prepared as back-up systems. While work on Orlan D for Salyut progressed, work on proving Orlan for Almaz apparently did not cease until after the end of Salyut 6 operations and the termination of the Almaz programme in the early 1980s.

It appears that EVAs were not planned before Salyut 6 in 1977. It is conceivable that elements of the EVA hardware, systems and procedures were flight-evaluated on earlier stations, or that EVAs were part of early planning, but the hardware was simply not ready in time for launch on the unmanned stations. The difficulties with docking, the maintenance of a reasonable environment onboard the station, the work-load, the objectives of extending the duration, the safety of the crew, and flight scheduling, as well as other factors and influences – all of these removed any chance of EVA from Salyut 4 and probably Salyut 5.

SALYUT EVA OPERATIONS, 1977–86

The first EVAs from a Russian space station took place under the Salyut programme between 1977 and 1986.[19] Three EVAs were accomplished from Salyut 6 (1977–79), and thirteen excursions were made from Salyut 7 (1982–86). Considering the lack of previous Russian EVA experience of *any type*, the achievements of these sixteen EVAs were even more outstanding. By 1977, Soviet cosmonauts had logged only two EVAs and less than two man-hours outside their spacecraft – in sharp contrast to the twenty-four Americans who had logged more than 250 man-hours during thirty-seven EVAs on EVA on Gemini, Apollo and Skylab.

The first Salyut EVA

Salyut 6 was launched on 29 September 1977, and on 9 October was followed by a two-man crew on Soyuz 25, as part of the sixtieth anniversary celebrations of the October Revolution, and around the twentieth anniversary of Sputnik. Western observers expected an EVA to be conducted as a celebration of events; but unfortunately, it was not to be. The Soyuz docking system failed to hard dock with the forward port on the Salyut (at the time, it was not disclosed that there was a second, rear port) and the crew, flying a Soyuz reliant on chemical batteries rather than solar arrays, had only limited power available for independent flight. After only two days in orbit, the cosmonauts were back on Earth. As a result, the manning

Preparations for EVA in the forward transfer compartment of Salyut 6.

schedule for future operations was changed, and a stand-up EVA inspection was added to the objectives of the next crew. The first use of the Orlan D was supposed to have been a demonstration of IVA in the depressurised forward transfer compartment, but the failure of the Soyuz 25 docking necessitated a change to include a docking at the rear port and a SUEVA visual inspection of the forward docking port on Soyuz 26, to ensure that it was available for future use.

After a brief training programme to include the new EVA objective, on 10 December 1977 Commander Yuri Romanenko and Flight Engineer Georgi Grechko launched into space and docked successfully with the second port of the Salyut. Ten days into the mission they began the EVA to inspect the forward port. This first Soviet EVA from a space station took place through the front docking port. Grechko opened the docking hatch and floated out about halfway to inspect the docking cone and capture latches to ensure that they were in working order. It was a simple inspection, and after only a few minutes he reported that the docking facility had not suffered any obvious damage that would prevent its future use. The total EVA depressurisation time was less than 90 minutes, and the SUEVA was completed in just 20 minutes. Grechko was soon back inside, his objective accomplished; and that was the end of the first Salyut EVA. However, an exaggerated story, related by Grechko, was maintained for many years after the mission. Grechko frequently suggested that he 'saved' Romanenko from floating out into space after he could not resist looking out of the hatch. Romanenko repeatedly dismissed this as a bad joke, and Grechko eventually admitted that the story was misinterpreted, and that there was never a real danger, as his electrical/communication umbilical was attached to the inside of the station.[20]

Salyut EVA operations
Over the following nine years (1977–86), EVA operations from Salyut stations expanded from a simple SUEVA inspection to major repair and construction activities. The learning curve of these EVAs was valuable in the planning of far more ambitious EVA operations on Mir. Salyut EVA activities are summarised below.

Experiments
The second EVA from Salyut, in July 1978, was the first operational EVA, designed to evaluate the Orlan suits in open space and to retrieve a number of experiment sample cassettes mounted on the exterior of the station. It also presented the Soviets with an opportunity to evaluate the EVA hatch for the first time. News reports claimed that the crew was 'reaping the scientific harvest' almost as soon as they began the EVA. Flight Engineer Alexander Ivanchenko made a full exit, and Commander Vladimir Kovolyonok remained close to the hatch area, to assist his colleague. Three panels of material samples, attached to the station prior to launch, were retrieved and replaced after ten months exposure. These panels consisted of samples of rubber, plastics, steel, glass, ceramics and duraluminium, and were designed to evaluate their durability in the environment of space and their potential as components of new space vehicles.

Particles of dust, micrometeoroids and space debris litter the orbit of any

An EVA from Salyut, using the Orlan D suit.

spacecraft, and constantly bombard the surfaces of the spacecraft, instruments and the crew outside. Large items of space debris – such as spent rocket stages, defunct satellites and the remains of spacecraft – can be tracked from the ground, but there are millions of smaller items of space junk and microscopic particles that can cause serious problems in a collision with long-term space platforms at orbital velocities. Carefully monitoring of the condition of the station is therefore a major task. When a small micrometeoroid panel was removed from the exterior of Salyut 6, hundreds of small craters were revealed in an area of only 0.63 m.

Medusa was an experiment designed to study biopolymers. It consisted of three parts. Two of these parts – one open, and one closed – were outside the station, while the third remained inside the station as a control. The exposure of enclosed nucleic acids to space would provide researchers with information on whether life might have originated there. Post-flight analysis revealed that the exposed samples contained material similar to nucleotides, and that direct sunlight had apparently stimulated them. Medusa experiments were conducted throughout the life of Salyut 6 and Salyut 7. Some of the exposed materials on Salyut 6 were recovered during a contingency EVA in 1979, while others remained on the space station for the rest of its orbital life, as there were no further EVAs planned to retrieve them. Four years later, in July 1982, the first EVA from Salyut 7 repeated the Medusa experiment by retrieving sample cassettes installed before launch and replacing them with new samples. During EVA on 25 July 1984, the Medusa samples and several other cassettes were exchanged by Dzhanibekov and Savitskaya, and again a year later by the next resident crew (Dzhanibekov and Savinykh).

On Salyut 6 in 1978, Kovolyonok's crew had deployed an instrument to record background cosmic radiation, and it was reported that in 1982 the first EVA crew from Salyut 7 (Berezovoi and Lebedev) deployed a similar instrument – a 10-m long synthetic-aperture radar antenna along the side of the Salyut, to be retrieved a couple of years later.[21]

In August 1985, Dzhanibekov and Savinykh installed an sample of solar-cell material, which would be retrieved by a later crew in 1986. After return to Earth it would be used to evaluate the effects of vacuum and radiation, and to determine its suitability for use in constructing solar arrays. They also installed a French experiment called Comet, designed to capture dust particles from comets. For this, a detector block consisting of four rectangular boxes was mounted on a central spine. When installed, the doors were closed, and would later be opened to collect material comets Halley and Giacobini–Zinner.

The samples from the exposure experiments, samples cassettes and the Comet experiment were collected by Kizim and Solovyov during the first EVA of the final expedition to the Salyut 7 station in May 1986. During the same EVA they installed, on the exterior of one of Salyut's windows, an optical device called Boss, designed to test optical wavelength transmitters for telemetry broadcasting. During the second EVA three days later, they also installed the Fon (Background) instrument on top of a retracted girder. This was designed to measure the density of the 'atmosphere' around the station produced by outgassing, which affected precision measurements. The recorded data were transmitted via the Boss device to receiving stations on the ground.

The same crew also installed a micro-test unit containing samples of aluminium and magnesium alloy, in an experiment designed to 'strain' the metals to determine their response and performance under stress and loads. This information would be used for determining the most suitable material for constructing other large structures in space. The final EVA task from a Salyut was the retrieval of the solar array material deployed in August 1985. It was intended that these samples would have been retrieved on a later mission to the station – which according to some reports included the use of the Buran shuttle to visit the Salyut – but in the event, no further missions visited the station, and the remaining samples were lost with Salyut 7 in February 1997.

Evaluation, inspection and observations
EVAs from Salyut 6 and Salyut 7 presented the first opportunity for the Soviets to evaluate their support equipment and chosen method of exit in space. During the second EVA, conducted in July 1978, the crew evaluated the strength and security of the Yakor (Anchor) restraint device, evaluated the station's response to their movements outside the vehicle, and used TV to record their activities, which Kovolyonok described as 'very agitated ... difficult, but interesting'. During the

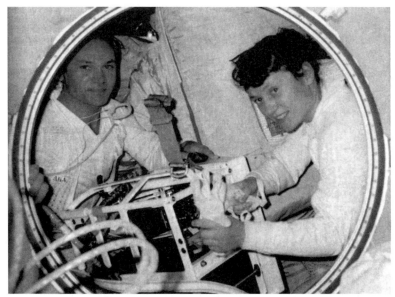

Wearing cooling undergarments, the crew of Salyut 7 prepare EVA equipment.

EVA they deployed portable lamps to allow them to continue working during the night-time pass of each orbit. Fascinated by the view, at one point they turned off the lamps to marvel at the stars and at the lights on the surface of the Earth. They also watched bright meteors entering the Earth's atmosphere below them, and marvelled at the sunrise and sunset from their vantage point outside the station. In little more than two hours the cosmonauts had successfully demonstrated the Orlan suit and Salyut EVA system, thereby boosting confidence in plans for more complicated and demanding EVAs. With their work completed earlier than planned, the cosmonauts were told to return to the hatch; but they replied that they preferred to take their time, as it was the first time in 45 days that they had been allowed 'outside for a walk'.

The first EVA from Salyut 7, in July 1982, was a similar event. Valentin Lebedev conducted the main EVA tasks supported by his Commander Anatoli Berezovoi from near the hatch area. Further evaluation of the restraint devices was completed, together with an evaluation of the use of tools outside the station. Using an improved EVA spanner, Lebedev performed a range of demonstrations under the Istok experiment. The spanner featured small spheres around a central rod, and was designed to grip bolts more firmly to prevent the possibility of slivers of metal penetrating the spacesuit. He also worked on the joining of metal parts, including the assembly of lengths of pipe and the evaluation of a 'memory metal' that shrunk to a predetermined form at room temperature to create an hermetic seal. These were then subjected to an internal pressure of 150 atmospheres to check their integrity and properties. Lebedev also evaluated the use of a variety of threaded components. This was a valuable exercise in determining which materials, procedures and tools would suit future EVA operations involving the connection and sealing of electrical, mechanical and liquid/gas supply lines.

Svetlana Savitskaya during her historic EVA.

During an EVA on July 1984, the crew tested a multipurpose EVA tool called URI (a Soviet acronym). This tool could be used for welding, soldering, spraying and cutting, and thus obviated the need for several implements previously tested in missions since 1969 (under the automatic Vulkan and Ispaatel programmes). The cosmonaut assigned to carry out the test was Svetlana Savitskaya, who in 1982 had been the second woman to fly into space – nineteen years after Valentina Tereshkova flew. Now, only weeks before astronaut Kathryn Sullivan was due to walk in space during STS-41G, Savitskaya would perform another Soviet space 'spectacular' by becoming the first woman to walk in space – nineteen years after Alexei Leonov ventured out onto the first ever EVA.

There were six cosmonauts onboard the Salyut at the time of this EVA, and so emergency procedures had to be conducted in case a problem should prevent the EVA crew from returning to the station. As the Soyuz could carry only three cosmonauts, and there were four still inside the Salyut, Solovyov transferred to the Soyuz T-11 for the duration of the EVA. The transfer compartment was depressurised for the excursion, and if a problem had occurred then Dzhanibekov and Savitskaya would have returned via Soyuz T-11. Kizim and research cosmonaut Yuri Atkov from the resident crew, and Igor Volk form the visiting crew, would have had to use Soyuz T-12, docked at the other port. However, the EVA was successfully completed.

Savitskaya used the URI to cut a sample of titanium, and in only one minute produced a 10-cm cut in the 0.5-mm-thick sample. Her next task was to weld together two pre-cut 1-mm samples, and despite being a little uneven the three tack welds and seam appeared to be good. During these tests, Savitskaya completed six welding experiments, using four samples of stainless steel and two of titanium mounted on four sample boards. Six soldering experiments included samples of tin and lead and the use of an electron emitter to melt solder in a special crucible, which was sufficiently heated in 45 seconds. She also completed a programme of six cutting experiments – three of titanium and three of stainless steel. The operation was supposed to be completed during orbital daylight, but Savitskaya complained that the Sun was shining directly into her eyes and was hindering her progress. However, she eventually completed her tasks, after which the two EVA cosmonauts swapped places and Dzhanibekov operated the URI. He likened it to spraying paint on a wall, and predicted that it would be widely used on future missions (although it has not actually been used very much). At the end of the EVA, the equipment was disconnected, packed away, and passed back into the hatch area.

During the final EVAs from Salyut 7 in May 1986, cosmonauts Leonid Kizim and Vladimir Solovyov completed an EVA programme originally intended for the previous resident crew – Vladimir Vasyutin, Viktor Savinykh and Alexander Volkov – whose mission had been curtailed due to the illness of the Commander, Vasyutin. During their first EVA, on 28 May 1986, the crew installed a beam erection device designated Ferma Postroitel. It was stored in a cylinder – 1 m in diameter, 1 m tall, and weighing 150 kg – and inside this cylinder was a folded lattice girder designated URS. It was installed on a base platform, had three methods of deployment – manual, semi-automatic and fully automatic – and could extend to 15 m above the Salyut to reveal a 40 × 40-cm cross-section lattice weighing 20 kg. It was tested for stability, deployment and retraction in space, and was subjected to further tests during the final Salyut EVA on 31 May 1986. These tests were stated as being carried out in conjunction with the development of similar devices designed to move cosmonauts and equipment from one point on the station to another. Tests were conducted on the vibration of the girder as it was extended, and, under the Mayak (Beacon) experiment, a TV camera focused on a light at the end of the girder to record its oscillations for post-flight analysis. A small seismometer also recorded low-frequency vibrations on the structure. Towards the end of the EVA, Kizim traversed up the girder, and stopped only a few metres short of its furthest end as vibrations spread up the whole structure. The cosmonauts also tested an improved version of the URI welding gun designed for easier operation than its precursor. These tests were allegedly a step on the road to the welding of lattice and pin structures to form larger structures in space, assembled by 'space construction workers'.

Contingencies

In the planning for Salyut 6 EVAs, officials had indicated that any excursion would be conducted in connection with a variety of 'hypothetical emergencies'. Cosmonauts assigned to the series of missions were therefore trained in a range of contingency situations, including the repair of solar arrays and the closing or opening of aperture covers for experiments and telescopes.

Exiting the Salyut EVA hatch.

The third EVA from Salyut 6, in August 1979, fell under this category. During the final month of a long six-month mission, cosmonauts Vladimir Lyakhov and Valeri Ryumin were conducting experiments with a 10-m diameter radio telescope dish, designated KRT-10. It was supposed to deploy from the rear docking port upon the separation of Progress 7, when the cargo ship undocked from the station, but it apparently did not deploy correctly. After a month of observations, the cosmonauts activated a separation device consisting of pyrotechnical bolts and springs to jettison the unwanted telescope and free the aft port for further use. They expected to see the antenna float by, but were alarmed to discover that it was still attached to the station, and was blocking the rear port and seriously restricting future operations.

The crew could have left the antenna attached, and departed using their Soyuz at the other docking port, but the orientation system would then not control the station and, as the refuelling connection system for Progress craft on this station was only at the rear, resupply would have been blocked, effectively ending the life of the station. The only option was to go outside and attempt to free the telescope from the rear docking port. But this posed its own problems (as Ryumin later outlined in his mission diary). Firstly, the two cosmonauts had not trained for conducting unplanned EVAs; and secondly, after six months in orbit they were long out of practice with the Orlan D suits and procedures. These suits were the same as those used since 1977, and, given where the telescope was entangled, had not been previously used to traverse such a distance. It was to be a demanding and challenging operation, but one that both cosmonauts were sure they could achieve, although neither had performed EVA before and this would be only the fifth Soviet EVA operation. Tension both in space and on the ground was high as the cosmonauts began their task at orbital sunrise on 15 August 1979. Ryumin would make the translation to the aft of the station, while Lyakhov would remain in the hatch area, feeding the umbilicals out and ready to heave on the safety line attached to his colleague. When Ryumin finally reached the rear of Salyut, he noted that the four cables from the antenna had pierced the thermal blankets and had become entangled. He therefore decided to cut the steel cables to free the unit, but, aware that the whole antenna could surround him like a fishing net, he had to be careful,.

As the Sun shone directly into his eyes, Ryumin lowered his visor. However, his rapid breathing caused fogging of the inner faceplate (a problem encountered on Gemini more than a decade earlier), and so he raised it again. After he cut the first cables the antenna swung toward him, but after Lyakhov warned of this he avoided the oscillations of the dish and cut the rest of the cables. He then rested before using a 1.5-m 'long forked stick' to prod the structure away from the station. Surprised at how easy the operation had been, he took the opportunity to examine the condition of the station from his vantage point near the aft docking port. 'It looked well worn, with the skin-plating fading and ragged in some spots,' he later wrote. Expecting that designers of space stations (of which he was one) would appreciate samples of the dust collecting on the exterior of the station, he wiped a porthole with a cloth and then stuffed the cloth into an exterior pocket in his suit. On the way back to the hatch he reported his success to the ground, and in his earphones was greeted with the sound of huge applause from Mission Control. Those on the ground were surprised that the operation had progressed so smoothly and quickly, but the cosmonauts' actions had restored the station to operational use – and it continued in use for a further two years.

Construction

One of the limiting factors of scientific and operational research onboard a station is the power required to supply the onboard systems and the growing array of scientific hardware. Incorporated into the design of the Salyut 7 station, therefore, was provision for cosmonauts, during EVA, to attach additional smaller solar arrays to the three large main panels launched on the station. Salyut 7 was launched in April 1982, but this task was not planned until the second resident crew – Vladimir Titov, Gennedi Strekalov and Aleksandr Serebrov – arrived at the station in spring 1983. The training programme for the crew included additional EVA training (for Titov and Strekalov) to attach the first two sets of additional panels to one of the main arrays, monitored by Serebrov from inside the Salyut's working compartment. Unfortunately, the failure of Soyuz T-8 to dock with the Salyut in April 1983 seriously changed these plans.

Lyakhov and Alexandrov were cycled to fly a replacement mission by June, but had not trained extensively in the operation of attaching the additional solar panels. Therefore, Titov and Strekalov were prepared to visit Salyut 7 in September, to conduct the solar-array EVA. Another problem arose on 9 September, when the station suffered an oxidiser leak in its onboard propellant system, which almost forced the abandonment of the station. However, analysis showed that the problem could be worked around until the extent of the damage was understood. Lyakhov and Alexander Alexandrov would remain on the station, while Titov and Strekalov would complete the EVA. Once again, however, fate intervened. On 26 September 1983 a faulty launch vehicle exploded on the pad just 90 seconds before launch. Fortunately, the capsule rescue system aborted the launch and catapulted the crew to safety before the launch vehicle exploded and engulfed the pad in a ball of flame and debris. As a result it was decided that Lyakhov and Alexandrov would conduct the solar-array EVAs, but over two excursions instead of one.

The extra solar panels were already onboard the station, in the cargo of the Cosmos 1443 module that had automatically docked to Salyut 7 in March 1983. When deployed, each panel extended to 5 m long and 1.5 m wide, and was designed to provide up to 50% of the power of the main array when installed. The installation involved as many as forty-eight individual operations to attach the structure to the main array and then winch the additional panels to their full length – if everything proceeded according to plan. In a contingency operation this task increased to 189 operations. Designers and mission planners established a set of guidelines and rules for deployment of the panels, and training for the task was demanding – which is why Titov and Strekalov were recycled to a new launch to try to board the station a second time. Lyakhov and Alexandrov had undergone very little training for such an operation, as it was not part of their planned objectives, and although Lyakhov had performed EVA from Salyut 6 four years earlier he had remained at the hatch area and did not have many tasks to perform. This excursion would be a team effort, with both men working to complete the deployment.

The two EVAs – described as complicated and laborious work – were completed on 1 and 3 November 1983; but the cosmonauts' inexperience in EVA training was evident, as on both exits it took them more than forty minutes to translate from the hatch area to the base of the larger solar array. They carried a selection of hand-tools and a multipurpose hand-held manipulator that allowed them to reach out across the whole length of the main array, and the folded arrays (DSB; a Russian acronym) were stored in a metal container. At the site, the cosmonauts installed two foot-restraints (Yakor) to help retain their position, attached safety tethers, and then attached a connection to hoist the additional panel to the main array, after which an electrical connection was plugged into the main array. The whole process was repeated on the second EVA by having mission controllers turn the array 180° to face the cosmonauts' work-station so that they did not need to move to the rear of the array to work.

Throughout the EVAs a full-scale replica of both the main and auxiliary array structure was constructed in front of the flight control consoles at Mission Control, to allow them to follow the progress of the two cosmonauts in orbit. A team of support cosmonauts – identified as Leonid Kizim and Vladimir Solovyov – also conducted a duplicate operation in the Hydro Laboratory at TsPK. Despite their lack of training and EVA experience, the cosmonauts worked well to complete the task; but it had required two EVAs to fit out one panel, which was costly in terms of time and consumables and the increased work-load on the crew. Alexandrov was at one point amazed by the whole experience of EVA, and his fascination led him to casually discard a small unwanted item to see what would happen. But Mission Control rebuked him for this action, as the station's stellar positioning system could have been confused due to reflection of light off the object.

Six months later, in May 1984, Kizim and Solovyov were themselves in space, conducting their fifth EVA. Their previous four EVAs had prepared them for the repair of the failed propellant line on the Salyut engine system (see below), and their experience in training for these tasks and in performing outside the station became evident when in little more than three hours they completed the attachment of the two sets of additional arrays to a second main array on the Salyut. They encountered problems when attempting to tie knots in a bundle of attachment wires – which they likened to 'trying to thread a needle while wearing boxing gloves' – and one of the hoisting handles broke; but during a single EVA lasting 3 hrs 5 min they completed a task which had taken the previous and less experienced team nearly six hours on two EVAs. The new solar cells provided Salyut with an additional 6 amps over the existing cells on the main arrays. The third pair of additional arrays was finally installed by Dzhanibekov and Savinykh during their EVA in August 1985.

A major repair

The leak of an oxidiser pipe in the Salyut during September 1983 was a potentially serious setback to future operations with the station, which at that time had been in orbit for only seventeen months. An investigation into the incident pinpointed the leak in the aft engine compartment around the rear transfer compartment, and it was also determined that, with adequate training, the next primary crew could perform a series of EVAs to bypass the faulty component and seal the leaking pipe. To accomplish this they would have to work in an area of the station not designed for prolonged EVA support, and cut through the hull of the station. This was an impressive and risky venture, and was, according to the official commentary, designed to prove the 'capability and technological experience of conducting operations in outer space.' Kizim and Solovyov increased their training programme, and logged more than thirty sessions in the Hydro Laboratory, the IL-76 flying laboratory, and 1-g simulations and dress rehearsals. Their programme also included the installation of the solar array and more than two hundred contingency procedures. In order to accomplish the task, a set of thirty different tools (some of them specially developed for the task) was made available for them to take into orbit. The crew was launched in February 1984, and the whole operation – including the installation of the solar array – involved six EVAs, totalling 22 hrs 45 min, over a period of five months.[22]

The first four EVAs involved preparation of the work-site and accessing the hull of the station. Progress 20, docked to the rear port, supported the EVAs, and featured a pre-fitted special platform attached to the forward end to allow the cosmonauts to 'stand' on it. The two cosmonauts prepared the work-site on the initial EVA by installing a 5-m ladder and transferring a 40-kg tool-kit to the work-station. They then pierced and cut an area for a plastic hatch cover in the unpressurised section that contained the hydraulic connections to the station's propulsion system. On the next EVA they punctured the outer skin over the reservoir conduit of the propulsion system, where they successfully installed a valve which they checked for air-tightness by passing nitrogen through the system. On the third EVA, two filler tubes and a metal bypass line were connected, and this new conduit was then added to the main reservoir and tested for air-tightness. The fourth EVA was used to install and check a second conduit; and when this task was completed, a sequence of pressure checks on the whole system pinpointed the leak on the onboard control panels monitored by Atkov inside the station. These EVAs had served only to identify the leak and install a bypass system, and not actually seal the leak. This was the purpose of the fifth EVA.

Once the problem had been identified, new tools and procedures were developed and tested in simulations on the ground. The tests were carried out by Vladimir Dzhanibekov, who would command the Soyuz T-12 visiting mission to the station and instruct the resident crew on the operation to seal the leak. In the original plan, Dzhanibekov would have gone outside to put his training into practice and complete the repair, but after all their hard work, Kizim and Solovyov quite rightly objected to this plan, and stated that they wanted to finish the job themselves. So, Dzhanibekov would instruct them by using models of the clamping device and training videos. This training included simulations inside the Salyut, with the crew fully suited and practising on a full-size model which they had previously attached to the inner wall of the station. The breached pipe would be sealed using a pair of moveable clamps attached to a piston-type actuator fed by compressed air. When placed over the pipe this would crimp it closed with 250-atmosphere pressure. This time, however, the cosmonauts did not have the assistance of a docked Progress or a work platform, and needed to secure everything to the outer rim of the Salyut's working compartment. After opening the thermal cover and installing the clamp, the cosmonauts checked the seal, activated the clamp, and replaced the thermal cover for the last time. On the way back to the hatch they used a special cutting tool to retrieve four sections of the solar-array elements for post-flight analysis on Earth, to assess the deterioration of silicon cells in orbit. The cutting tool enabled them to cut and remove the samples without touching the other parts of the array with their gloved hands.

The two cosmonauts spent almost a full day outside the spacecraft and established a new Soviet cumulative EVA record of 22 hrs 45 min on one mission (only Gene Cernan – Gemini 9 and Apollo 17 – had spent more time outside a spacecraft), and their work enabled further visits to the station. Having restored Salyut to an operational level, it came as a bitter blow, a few months after the cosmonauts returned home, to find that the unmanned station had lost onboard power and radio contact with Mission Control. In summer 1985, therefore, it was therefore decided to launch another rescue mission to attempt to dock to the 'dead station' and restore it to working order so that it could continue its manned programme. The station would eventually support two more resident crews and a further three EVAs before Mir took over in 1986.

Summary

By the end of the final mission to Salyut 7 in the summer of 1986, Soviet cosmonauts had increased Soviet EVA duration from less than one hour to more than a hundred hours in less than nine years. Kizim and Solovyov had both logged an impressive eight EVAs in two missions, and logged 31 hrs 36 min each. The experience gained on Salyut led to more detailed planning of regular EVAs from Mir, using improved Orlan suits and a wider range of support equipment. Although not evident in 1986, Mir itself would become a test-bed for EVAs from a permanent manned space station, and would provide important lessons for the International Space Station.

The Salyut series also provided useful lessons:

- EVA could be used to deploy and retrieve small experiment packages that could be be exposed outside the pressurised compartment without the need for complicated air-locks or recovery techniques, and be flown over the lifetime of the station.
- The observational skills of the cosmonauts allowed them to report, photodocument and inspect areas of the station for deterioration and damage, and install and test sample materials, over various time periods, for use in future spacecraft construction techniques.
- A generic training programme for EVA operations could – with additional training by instruction from the ground, visiting crews, models and videos – be used to improve techniques, and be flexible enough to upgrade the skills of a crew that had not trained for certain EVA procedures or had experienced a long gap between training and performing the task in orbit.
- Several construction and assembly tasks were demonstrated that could be expanded upon for larger space structures.
- Contingency EVA operations could be conducted, providing there were hand-holds and foot-restraints, and an array of suitable tools (also determined by the Skylab astronauts).
- On-orbit repairs and modifications to the station could be performed in an area not designed or intended for EVA repair or on-orbit maintenance (again underlining what the Skylab astronauts had clearly demonstrated).
- EVA provided a means of upgrading station systems to improve the capabilities and scope of research tasks planned for the station.

EARLY SPACE STATION EVAS: CONCLUSION

In a review of both Skylab and Salyut EVAs, several factors become clear. Firstly, given adequate time, training and materials, the life of the station could be saved or rescued by a combined team effort of ground crew and flight crews when threatened by system failures or certain structural failures, which would not be possible in a totally automated mission. This was demonstrated in the deployment of solar arrays and shields on Skylab, and during the repair of Salyut 7's propulsion system.

On both Skylab and Salyut, small additional experiments were taken outside the pressurised hull, operated (and in some cases repaired), and then retrieved for return to Earth. Large instruments (such as the Skylab ATM) could be periodically serviced or, in the event of an operational problem (Salyut 6's KRT-10 antenna), maintained by the crew.

Limited on-orbit servicing was possible on multiple EVAs in the same suits, provided that sufficient support, maintenance, cleaning, spares and operational systems were available.

Man-handling large items to and from hatches across the hull of the station increased the crew's work-load and highlighted the need for transportation devices, suitable work-stations, tool support systems, and manipulation hooks and girders to reach inaccessible places and save the crew's energy. It was also important to provide clear routes across the surface of the station to save time when accessing the work-station, as well as proper lighting to work on the dark side of each orbit.

The use of water tanks to train for station EVAs clearly proved important for both Skylab and Salyut. With a crew in space, the ability to put support teams in tanks to work through a problem, before the orbital crew attempted to resolve it, presented every chance of finding a workable solution. In Earth orbit, of course, from where communication between Earth and the spacecraft takes only seconds, this is more practical than it would be for EVAs beyond the Earth–Moon system.

Moreover, the ability of astronauts and cosmonauts to utilise EVA when encountering difficulty, and to work around and overcome problems and continue the mission, was reward in itself. Confidence is built upon both success and failure, and with these early EVAs the ability to overcome failures provided a sound base for work on larger and more permanent structures in space: Mir, and the International Space Station.

REFERENCES

1 *Skylab: A Chronology*, NASA SP-4011, 1977, p. 16, entry for 18–21 May 1961.
2 Peebles, Curtis, 'The Manned Orbiting Laboratory', *Spaceflight*, **22**, No. 6, June 1980, 252; Preliminary Technical Development Plan for MOL Status as of 30 June 1964, SSM-50A, pp. 6-22–6-23, Space Systems Division, USAF Systems Command.
3 Preliminary Technical Development Plan for MOL Status as of 30 June 1964, SSM-50A, pp. 6-23–6-24, Space Systems Division, USAF Systems Command.
4 Shayler, David J., *Apollo: The Lost and Forgotten Missions*, Springer–Praxis, 2002, pp. 14–22.
5 *Ibid.*, pp. 39–41.
6 The Boeing Company Report, D2-84010-1, Multipurpose Mission Module, 0-{?} Utilization Study, November 1965.
7 Shayler, David J., *Apollo: The Lost and Forgotten Missions*, Springer–Praxis, 2002, pp. 61–62.
8 Heckman, Richard T., 'Skylab EVA System Development, George C. Marshall Space Flight Center', paper (AAS 74-121) presented to the American Astronautical Society Twentieth Annual Meeting, Los Angeles, California, 20–22 August 1974.
9 Memo from R. Bryan Erb, Assistant Chief, Structures and Mechanics Division, MSC, to Major William R. Pogue, CB. Subject: Access to ATM experiments from LM ascent stage hatch opening, dated 22 December 1966. Curt Michel Collection (ATM files/CB Memos), Fondren Library Archives, Rice University, Houston, Texas.

10 Schultz, David C., Kain, Robert R. and Millican, R. Scott, 'Skylab EVA', in *The Skylab Results*, Proceedings, Volume 1, Twentieth AAS Meeting, 20–22 August 1974, NASA JSC, AAS74-120; Shayler, David J., *Skylab: America's Space Station*, Springer–Praxis, 2001.

11 Portree, David S. and Treviño, Robert C., *Walking to Olympus: An EVA Chronology*, NASA Monographs in Aerospace History, No. 7, October 1997, pp. 30–32; Shayler, David J., *Skylab: America's Space Station*, Springer–Praxis, 2001, pp. 180–184.

12 *Lessons Learnt on the Skylab Program*, MSFC Skylab Program Office, 22 February 1974, and JSC, Houston, Texas, 6 March 1974.

13 Astro Info Service interview with EVA/Shuttle Flight Director Glenda Laws, Houston, Texas, 8 May 2002.

14 Skylab 1/2 Technical Crew Debrief, 30 June 1973, JSC-08053, Training Office, Crew Training and Simulator Division, NASA JSC, Houston, Texas.

15 Shayler, David J., 'Resumption of Soviet EVA?', *Spaceflight*, **19**, September 1977, 317.

16 Abramov, Isaak P. and Skoog, Å. Ingemaar, *Russian Spacesuits*, Springer–Praxis, 2003.

17 *Ibid.*, p. 48.

18 *Ibid.*, p. 150.

19 Kidger, Neville, 'Above the Planet: Salyut EVA Operations', *Spaceflight*, **31**, No. 2, February 1989, 45–49, No. 3, March 1989, 102–105, No. 4, April 1989, 139–141, and No. 5, May 1989, 154–155; *Walking to Olympus*, NASA Monograph No. 7.

20 Astro Info Service interviews with Yuri Romanenko, July 1989, and Georgy Grechko, April 1994.

21 Kidger, Neville, 'Above the Planet: Salyut EVA Operations', *Spaceflight*, **31**, No. 3, March 1989, 104.

22 *Ibid.*, No. 4, April 1989, 139–140, No. 5, and May 1989, 154; Ovchinnikov V.S., 'Experience of the Salyut 7 Propulsion System Repair Operations', Moscow Aviation Institute, USSR, IAF Paper, Thirtieth-Eighth Congress, Brighton, England, October 1987.

Service calls

The American Space Shuttle system evolved over a period of at least twenty-five years, from the theoretical studies after the end of the Second World War through the X-series of rocket research planes and lifting bodies of the 1950s and 1960s. In 1972 – after many years of budget confrontations and debates concerning designs – it was decided to build the Shuttle system in the form which it has retained to the present day: a reusable manned orbiter, an expendable external tank, and recoverable twin solid rocket boosters. At the time of this decision, the Apollo 16 astronauts were exploring the surface of the Moon, and demonstrating the standard of EVA technology and techniques that had been built up over the previous decade.

It is worth recalling the limited scope of EVA planning in the summer of 1972. There was just one remaining Moon landing (Apollo 17) on the launch manifest, with three excursions planned on the surface and a deep-space EVA to retrieve the SIM-bay film cassettes. Then, in 1973 America launched Skylab using former hardware from the Moon programme to accommodate three teams of astronauts. As well as the onboard operations, each crew would complete several EVAs outside the station – primarily to retrieve and replace solar telescope film cassettes and to deploy and retrieve scientific experiment packages.

After Skylab, all that the American astronauts could be certain of was the short rendezvous and docking mission with a Soviet Soyuz spacecraft, and no EVAs were planned for that mission. Indeed, the Soviets were coming to terms with the loss of a Soyuz crew in a tragic accident – the second in little more than four years – and they were concentrating on developing a safe and reliable Soyuz ferry system to Salyut space stations before embarking on their own EVA programme. Although the capacity for Salyut EVAs was in development, five more years were to elapse before another cosmonaut would exit a spacecraft in orbit.

After demonstrating to the world the ability to don a spacesuit and 'step outside' a spacecraft travelling at twenty-five times the speed of sound around the Earth, or walk and even drive across the dusty lunar surface, the prospect of continuing EVA activities beyond 1973 seemed a little bleak. Images of astronauts walking in space or on the Moon were familiar in advertisements and on the covers of space books, but there was a continuing debate on the future of EVA. There was, of course, the new

Space Shuttle programme, and here, dreams and desires began to run wild once again. As in the 1960s, with the projections for lunar exploration after Apollo, the Space Shuttle was soon portrayed as the answer to all of America's questions on how to reach space quickly, repeatedly and cheaply – or so it seemed.

THE SPACE SHUTTLE ERA, 1981–2012

Over the following three decades, a very different picture of the Shuttle system developed. It proved costly, complicated and, as we have seen on two occasions, tragically flawed in some of its design features and changing management policies. The Shuttle hardware can theoretically fly up to about 2020/21, but is to be retired around 2012. Certainly for the foreseeable future, there is nothing else that has the launch capability, flexibility and rescue capability of the three remaining orbiters. Unlike the reviews and changes after the *Challenger* disaster, there is no post-*Columbia* call for a replacement orbiter; and, indeed, there are no major spares available to build one. There are some who agree that the Shuttle should be grounded; but what would replace it, and how can the ISS be completed without it? Once the main construction of the ISS is completed, the role of the Shuttle will be phased out, although one of the vehicles might be kept flightworthy for 'mission unique' roles such as the urgent delivery of replacement elements for the space station. Only time will tell.

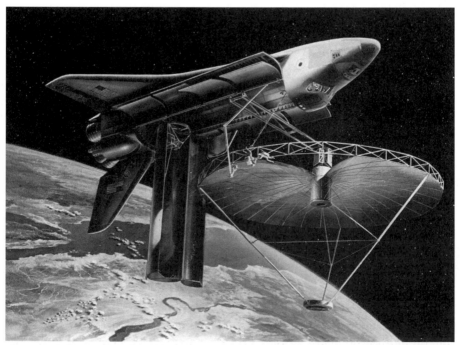

An artist's impression from the 1970s, of Shuttle-supported space-construction EVA.

It is clear that the fleet's performance to date, on well over a hundred missions into space, has demonstrated the Shuttle's versatility. As a manned heavy-lift launch vehicle system it has repeatedly provided access to space by supporting short scientific research flights and satellite deployment and retrieval missions. More recently it has demonstrated the role for which it was originally planned – as a manned ferry vehicle to support the construction of and supply logistics to a large manned space station in Earth orbit.

By 1998, commercial satellite deployments from the Shuttle had ceased, as had the scientific research flights onboard the European Spacelab modules, the 'secret' military missions, and the deployment of spacecraft to be sent to other parts of the Solar System. What remains for the Shuttle is the resumption and completion of space station construction tasks, and the supply and reboost capabilities that it offers to sustain the station in its early years. One or two visits to the HST have yet to be decided upon. However, there is one capability that links all operations on the Shuttle, from the very first flight in 1981 to the very last flight (currently planned for 2012): the capacity for *every* Shuttle crew to perform a spacewalk, whether or not EVA is planned. All Shuttle crews have team members who train for and support at least a contingency EVA operation, and the ability of the Shuttle to support extensive EVA operations has been essential for the servicing, repair and retrieval of stranded satellites, the maintenance of the HST, and the creation and expansion of the ISS.

Shuttle EVA opportunities

Following the decision to proceed with the Shuttle, NASA immediately promoted its ability to support EVA in addition to all its other proposed mission roles.[1] Because of the Shuttle's ability to deliver a varied payload, EVA became a baseline capability, and in its brief, NASA stated: 'EVA can provide sensible, reliable and cost-efficient servicing operations for these payloads because EVA gives the payload designer the option of orbital equipment maintenance, repair and replacement without the need to return the payload to Earth or, in the worse case, to abandon it as useless space junk. Having EVA capability can maximise the scientific return of each mission.'

Thirty years after these words were written, the statement remains true of Shuttle EVAs. For Shuttle missions, three types were defined:

- *Planned* An EVA that is included in the mission plan prior to launch so that the crew can complete a primary mission objective.
- *Unscheduled* Not planned before the mission began, but required to achieve operational success from a payload or to advance the overall accomplishments of the mission.
- *Contingency* An EVA that is required to safely return the crew. These could fall into a number of scenarios, some of which could be trained for prior to launch.

It was decided that on every Shuttle mission there would be provision and equipment to support at least three two-person EVAs, each up to six hours duration. Two of the EVAs would be assigned for payload operations, but the third would always be a capability to perform a contingency EVA (such as manually closing the payload doors) up to the end of the orbital phase of the mission. As the programme developed, additional EVAs were added, depending on the objectives of the mission, the capabilities of the crew and orbiter, and the duration of the flight. In the early years of the programme, and as part of the commercialisation promotion of Shuttle capabilities, any supplementary costs attributed to specialist EVA payload operations would be discussed under the terms of agreement between NASA and payload customers.

Orbiter provisions for EVA
The two-deck design of the Shuttle allowed for the use of the upper flight-deck as an observation platform using the various windows. It also contained the controls, displays and monitoring devices for the orbiter and the payload, and controls for the Remote Manipulator System. The mid-deck included stowage for the EVA support equipment, a preparation area for the EVAs, and access to and from the air-lock.

The air-lock is the means to exit the vehicle into the vacuum of space without the need to depressurise the entire crew compartment. It is also a useful storage area for

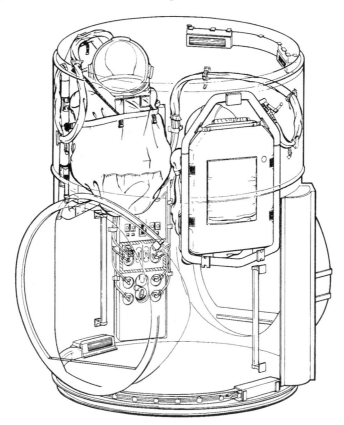

The Shuttle air-lock.

the EMUs, and includes a location for the interfaces and displays required to support EMU operations and servicing. It is removable, and can be located in one of three places, depending on the type of payload carried. Its main location is inside the mid-deck compartment, which provides for maximum use of the payload-bay volume. It can also be rotated 180° and positioned in the payload bay, still connected to the aft cabin bulkhead, or can be placed on top of the pressurised tunnel system for access to habitable payload structures (Spacelab/Spacehab) or into a space station through the orbiter docking adapter, while retaining the Shuttle's capacity to support EVAs. The air-lock is cylindrical, 160 cm in diameter, and 211 cm long, and has a pair of 1-m D-shaped openings with pressure-sealed hatches.

The unpressurised payload bay of the Shuttle has a usable payload envelope, 4.6 m in diameter and 18.3 m long. Provisions for EVA include hand-rails for crew translation on the fore and aft bulkheads and along each of the payload-bay door hinge lines. These are designed to withstand crew-induced loads of up to 900 N in any direction. Retractable tethers and their slide-wire management systems on each side of the payload bay support safety requirements, but do not interfere with payload ground handling or orbital operation envelopes. Safety tether attachment points of up to 2,550 N are also included in the design. Seven cargo bay floodlights and up to four cargo bay CCTV cameras are also provided in the payload bay, together with an EVA tool stowage box that can be located in several places, depending on the mission flow. There are also a number of payload accommodation points to retain the various types of cargo on the keel or walls of the payload bay.

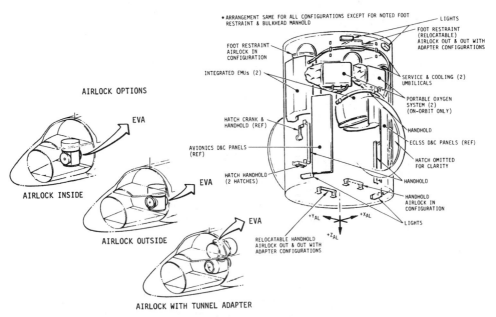

The location of the Shuttle air-lock.

Additional stowage space is also available underneath the payload bay liner in the first 1.22 m of the payload bay near the hatch area. During three missions in 1984, the MMUs were stowed on the port and starboard longeron at the fore end of the payload bay, while the RMS is also located and stowed along the port longeron. The orbiter EVA communication mode of operation provides automatic relay to the EVA astronauts on the S-band or Ku-band voice transmission, so that the EVA crew, the orbiter crew and ground flight controllers can all talk to each other. The global Tracking and Data Relay Satellite System (TDRSS), introduced on Shuttle deployment missions, gradually replaced many of the ground hand-over stations used during early programmes, to offer much better communications with the crew.

Shuttle EVA capabilities

According to early NASA statements, Shuttle EVA operations were aimed at demechanising the operational tasks and reducing the amount of automation, as well as simplifying testing and quality assurance programmes, reducing manufacturing costs, and increasing the probability of mission success. The original capabilities of Shuttle EVA operations were as follows:

The Shuttle EVA hatch area on the crew compartment aft bulkhead.

- Inspection, photography and possible manual override of vehicle and payload systems, mechanisms and components.
- Installation, removal or transfer of film cassettes, material samples, protective covers, instrumentation and launch or entry tie-downs.
- Operation of equipment, including tools, cameras and cleaning devices.
- Cleaning optical surfaces.
- Connection, disconnection and storage of fluid and electrical umbilicals.
- Repair, replacement, calibration and inspection of modular equipment and instrumentation on the spacecraft or payloads.
- Deployment, retraction and repositioning of antennae, booms and solar panels.
- Attachment and release of crew and equipment restraints.
- Performance of experiments.
- Cargo transfer.

Typical Shuttle EVA timeline
According to the early guidelines, a typical Shuttle EVA would occupy a complete working flight-day. Three hours would be required to purge nitrogen from the blood of the astronauts to prevent the bends, and at least 1.5 hours would be required for EVA preparation (suit donning and air-lock decompression). The EVA would occupy up to six hours, followed by a further 1.5 hours of post-EVA operations, but in reality the crews managed to complete much of the preparation in the days leading up to the EVA, and a change in operating procedures resulted in partial reduction of the cabin pressure to reduce the time spent in pre-breathing prior to exit. However, even now – twenty years after the first Shuttle EVA – because of the sheer work-load, most of the flight crews spend a full working day in completing and supporting one period of EVA of between three and eight hours.

For each Shuttle-based EVA, NASA stated that it would take responsibility for two one-person or two-person EVAs on each flight of up to six hours, plus the provision and training of the EVA crew, use of the RMS in supporting an EVA, and the provision of EVA support equipment (tools, restraints, lights, and so on). Payload options on early flights included the availability of additional EVA consumables, flight or payload-specific EVA crew training, special RMS end effectors, special support equipment, and the provision and use of the MMU.

During the 1970s, NASA's analysis of cost-effectiveness suggested that the inclusion of EVA provisions in the design of payload systems would reduce costs compared to those same systems if developed without EVA provisions. According to NASA this could lead to potential savings of millions of dollars in eliminating automated servicing equipment, or the inability to support EVA servicing and repair over a longer payload lifetime. These factors led to an important decision in providing EVA support capability in some multi-mission modular spacecraft (Solar Max), the large space observatories (such as the HST) and retained payload-bay cargo (Spacelab), although the potential to include EVA servicing and repair was not always taken up by STS payload customers in their spacecraft designs (as demonstrated in several satellite configurations).

Orbital flight-test EVA plans

In the initial documentation it was stated that Mission Specialists would fill the primary EVA role, with the Pilot as the second crew-member; but this was for operational missions with four or more crew-members. On the first four orbital flight-tests of the Shuttle (April 1981–July 1982) there would be only two crew-members (Commander and Pilot), and therefore only *contingency* one-person EVAs were possible. The mission Commanders – John Young on STS-1, Joe Engle on STS-2, Jack Lousma on STS-3, and Ken Mattingly on STS-4 – trained for these EVAs, and had the primary pressure garment on board to support them. The Pilots – Bob Crippen, Dick Truly, Gordon Fullerton and Hank Hartsfield – were each back-up to the Commander in case they should be required to perform the EVA, and, according to some reports, had EMUs on board in case they were needed. The Shuttle suit had sizing capabilities, but a second suit as a back-up was a sensible option. The Pilots would have remained inside to 'mind the store', and would go outside only if

STS-1 Commander John Young prepares for underwater orbital flight-test contingency EVA training.

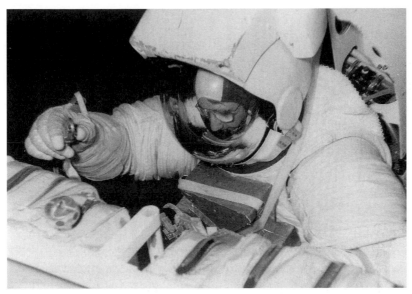

Orbital flight-test contingency EVA training in the water tank at JSC.

necessary. This was a particularly wise decision, as it was still so early in the operational programme, and these four flights were the only means of learning how the Shuttle operated in space. Unlike earlier programmes, there were no unmanned test flights prior to STS-1.

In order to maintain the 'buddy' EVA safety rule, one-man EVAs would not be routine, but the 'emergency only' nature of orbital flight-test contingency EVAs would have necessitated relaxing that rule when the crew was short on manpower. There were precedents – such as one-man surface EVAs on limited-duration Apollo landing missions[2] – that accepted the extra risk of one-man EVAs in improbable (contingency) situations during development. Each of the orbital flight-test Shuttle Commanders had previous EVA training experience, and all except Engle had accomplished EVA on previous missions, although Engle had trained as back-up LM Pilot for Apollo 14. The four Pilots – all former MOL astronauts –had accrued limited EVA simulation time, but like Engle, no in-flight EVA experience. Because Jack Lousma had the most orbital EVA experience (on Skylab), he was assigned (with Story Musgrave) to assist with contingency EVA techniques, tools and equipment, including 1-g and parabolic training and about 40–50 hours in the water tank.[3]

Early contingency EVAs

The orbital flight-test crews prepared for the following contingencies:

1   Failure of payload bay door latches, either the bulkhead or centre-line latches, or both. The crews carried about four of each type that could be manually fastened over the latching mechanism when the doors were closed but not latched. There were tethers and a portable foot-restraint with which to steady themselves while installing the latches.

2 Failure of radiator panels to return from their extended position away from the payload-bay doors back to their stowed position flush with the doors. A tool was provided to override the locks in the radiator panel hinge line and to rotate the panels to the stowed position.
3 Failure of the payload-bay doors to close. An override on the door hinge locks released them, and then a block and tackle system was installed between the edge of each door and the end of the bulkhead to individually winch each door closed.
4 Loss of capacity to return the RMS to its stowed location. They would first try the block-and-tackle system, but a back-up jettison capability was also available if the RMS could not be stowed.
5 Loss of capability to rotate the RMS stowage assembly inboard to clear the payload-bay doors when closed. This utilised the same methods as in item 4.

After the completion of STS-4 and the commencement of operational flights, the inclusion of other equipment extending outside the payload-bay envelope has, over the years, required additional contingency EVAs to ensure that they have been cleared before closure of the payload-bay doors. This has included stowing the Ku-band antenna and the rendezvous radar, jettisoning the Spacelab Scientific Air-lock and Experiment Tables, lowering the inertial upper stage air-frame tilting mechanism, and clearing deployed docking interfaces.

EVA crew roles and responsibilities
The change from operational EVAs with Pilot/Mission Specialist to Mission Specialist/Mission Specialist combination appears to have occurred around the time that the orbital flight-test crews were training (1978–82). It became clear that the Pilot and the Commander would be fully occupied in flying into and out of orbit and maintaining the vehicle in space, and simply would not have the time to train for EVA. It was a more sensible distribution of the work-load to assign primary EVA operations (and later the RMS operations and the role of the centre-seat Flight Engineer) to the Mission Specialist crew-members.[4]

Training for these orbital flight-test missions focused mainly on manual closing of payload-bay doors. For more operational EVAs there was established a training protocol with three categories. *Advanced training* would be used only once, to develop the initial experience in using the Shuttle EMU and orbiter EVA equipment and procedures, and would become part of the astronaut candidate's advanced training programme prior to assignment to a flight crew; *flight-specific training* would be employed to achieve a specific payload EVA once a crew-member had been assigned to a flight crew; and *recurring training* was to maintain the skills of an EVA-experienced crew-member between flight assignments.

As the programme developed it became clear that EVA support roles would be required. These included operating the RMS during EVA, photodocumentation, an IVA crew-member to assist and support EVA preparations, suit donning, post-EVA activities, and, more recently, an EVA choreographer to coordinate activities – outside on the EVA (by following the timeline), and inside by working with the RMS operator, with the Commander handling the operation of the Shuttle. The expansion of EVA operations with the HST and the ISS has seen 'teams' of astronauts perform EVAs on one flight day and then support EVA operations by the second 'team' the next day. On some missions there have also been assignments as a back-up EVA astronaut, providing redundancy for primary EVA astronauts if they are unable to complete an assigned EVA. This 'back-up responsibility' is reflected in the majority of crew training for both systems and payloads, and cross-training offers additional flexibility during real-time mission operations. In addition, on some missions in which EVA is an important objective (such as at the HST), a spare EMU is flown as an additional contingency. On HST missions it has also been the practice to reassign members of a previous service mission to the next one (including several astronauts with EVA experience), to help smooth translation to a new crew, to transfer flight experience to achieve mission success, and to reduce some of the dedicated training time. On such flights, these crew-members are designated as EV1, EV2, EV3 and EVA4, and they usually perform the EVA tasks in the following sequence:

EVA	Prime	Back-up/IVA
1	EV1/EV2	EV3/EV4
2	EV3/EV4	EV1/EV2
3	EV1/EV2	EV3/EV4
4	EV3/EV4	EV1/EV2

Several astronauts experienced in EVA also support mission operations as CapCom during EVAs, perform ground simulations, or are on standby at the WETF in case underwater simulations are required during a mission in flight. The experienced EVA astronauts also work in the Astronaut Office as crew representatives during the development of new EVA tests and procedures, so that they are available to offer advice.

Developing the skill

No EVAs were planned on the orbital flight-test missions. The objective was to evaluate the Shuttle vehicle, systems and procedures, the RMS, and payload-carrying capabilities during pre-launch preparation, countdown, ascent, orbital operations, entry, landing and turnaround. A demonstration EVA was first assigned to the fifth Shuttle mission – the first 'operational' mission at the end of 1982.

Following STS-1 in April 1981, it was decided to test one of the EMUs from the flight to determine whether it would have worked correctly had it been used, and whether the rigours of the mission had damaged it during its storage in the mid-deck air-lock. The test revealed, however, that there were no anomalies in the suit storage system during a nominal mission profile.

Prior to the mission, an extended programme of Shuttle EMU verification tests had been completed by prime contractor Hamilton Standard before delivery to NASA. Astronaut George Nelson was then assigned to test the suit in the vacuum chamber at JSC, in cooperation with orbital flight-test prime and back-up crew-members, during one week in October 1980. Tests were run at normal sea-level pressure (with the chamber door open), followed by a sealed test-run inside the closed chamber, which was reduced to 5 psi for the transfer from external power (chamber facility systems) to internal suit power and oxygen supply. To evaluate the suit's cooling capabilities and mobility in use, Nelson also walked on a treadmill, while holding hand-rails inside the chamber, while the suit was supported by a weight relief system to allow mobility at 1 g.

A demonstration of EMU suiting was planned for Flight Day 3 during STS-2 in November 1981. Commander Joe Engle would don an EMU and follow EVA procedures, but stop just short of depressurising the air-lock.[5] However, this procedure was cancelled in-flight when the minimum mission profile (54 hours) had to be flown due to the loss of a fuel cell. The test was transferred to STS-4 (July 1982), and Commander Ken Mattingly performed a successful suited demonstration of EVA preparation to the point of air-lock depressurisation on Flight Day 4.

STS-5: a frustrating start

As a repair EVA had already been tentatively assigned to STS-13 (Solar Max), and plans were in hand to test-fly the MMU from STS-11, the completion of a 3.5-hour EVA to demonstrate the operational capability of the EMU in a vacuum was of high priority. STS-4 had tested operations in zero gravity, and although EMUs had been carried on all four orbital flight-tests, none had been used outside the vehicle. The plan was to allow STS-5 astronauts Bill Lenoir and Joe Allen to evaluate the air-lock, suits and restraint devices, and test prototype tools and repair devices, some of which were in development for the Solar Max repair mission. It was an ambitious task for the first excursion by Americans outside a spacecraft since February 1974. It was also frustrated. During preparations on Flight Day 5, Lenoir reported a problem with his EMU regulator and pressurisation levels, after which, Joe Allen's suit developed a problem with a suit fan. All in-flight attempts to solve the problems were unsuccessful, and the EVA was therefore cancelled, although some useful suiting and air-lock activities were achieved.[6]

Success with STS-6

After the failure of the STS-5 EVA, Paul Weitz, the Commander of STS-6, lobbied to have the demonstration EVA assigned to his flight. Corrective action was implemented to ensure that similar problems did not recur, and due to the importance of the success of this EVA it was indeed assigned to STS-6 (April 1983). The original flight plan for that mission was to deploy the first TDRSS (TDRSS-A) by the inertial upper stage in order to improve communications with the Shuttle, and then stay in orbit for a few days to accrue some orbital experience with the new orbiter *Challenger* (OV-099) and to complete some small experiments.[7]

The first American EVA in nine years took place 7 April 1983 (Flight Day 4) in the payload bay of *Challenger*. It was performed by Story Musgrave and Don Peterson, and lasted for 4 hrs 17 min. During this excursion, the two astronauts completed a number of tasks to evaluate the Shuttle EVA baseline system. They translated from the air-lock along the length of the cargo bay to work at the aft bulkhead, and evaluated translation rates, hand-holds, slide wires and tethers. They also evaluated the dynamics of the safety tethers by using port and starboard tethers and translating across the payload bay from one side to the other. There was an

The first Shuttle EVA demonstration.

evaluation of the mobility of the EMU in zero g, and the use of portable foot-restraints, and the two men completed operations at the cargo bay storage assembly (tool-box) and evaluated foot-restraints, cargo bay storage assembly door operations, and the removal, handling and storage of tools. They also demonstrated the lowering of the inertial upper stage tilt table (the inertial upper stage/TDRSS had been successfully deployed earlier in the mission) with the winch not connected to the table, conducted a forward bulkhead winch simulation, demonstrated the payload bay retention (strapping) device, translated with a large mass (a bag of latch tools) down the payload bay door slide wire, and finally stored the equipment and re-entered the air-lock.[8]

Generally, the astronauts moved easily around the cargo bay and tested a variety of contingency procedures for possible future use, as well as using several tools specially designed for impending EVAs. They evaluated equipment handing techniques and the dynamics of safety tethers, the reliability of foot-restraints, hand-rails and the end-to-end slide-wire system, and qualified the suits, air-lock and baseline equipment for operational use. The mission planners were thus able to implement much more ambitious EVAs from the Shuttle. After almost a decade, American astronauts were finally capable of working outside their new spacecraft; and over the next three years this capability would be dramatically expanded.

SHUTTLE EVA SATELLITE SERVICING OPERATIONS, 1984–97

During the era of Shuttle commercial satellite deployments (1982–92), the opportunity arose to retrieve, service, repair and redeploy satellites that had encountered operational problems. In one case, two stranded satellites were brought back to Earth for later redeployment, and on another mission the Shuttle astronauts evaluated the future potential of refuelling a satellite to extend its operational life in orbit. This series of missions and demanding EVAs broadened the experience of EVA operations not only for the Shuttle but for future servicing operations from larger space platforms.

Multimission Modular Spacecraft

During the development of the Shuttle system in the 1970s, its capability to rendezvous with a previously launched unmanned satellite indicated that, in certain cases, the design of the satellite would be affected by the ability of the astronauts to service and repair the onboard systems; and, of course, the development of Shuttle-based hardware, equipment and procedures would have to be evolved in line with these satellites.

The development of Multimission Modular Spacecraft (MMS) represented a significant step toward an integrated manned and unmanned programme in which a reusable space platform could carry out a wide range of scientific research programmes from Earth orbit, and could be serviced or repaired in the cargo bay of the Shuttle before being redeployed to continue its mission. The planning of these (essentially man-tended) periodic visits by a Shuttle crew would be dependent on the missions of the satellite and the real-time operational status of the hardware. The MMS was a standard reusable space platform which incorporated modularised sub-systems, handling power, communications, data handling, attitude control, and other scientific instruments peculiar to the mission. The satellite could be launched by unmanned expendable launch vehicles as well as from the Shuttle, to operate in an orbit capable of being reached by the Shuttle for servicing, repair or, if required, retrieval for return to Earth for refitting and later relaunch, possibly with a completely different set of experiments and research objectives. The design was ideal for the Shuttle, and the theory that the Shuttle could rendezvous with and repair this type of satellite was to be put to the test very early in the programme.

The first MMS – Solar Max – was launched on an unmanned vehicle early in 1980. Its objective was to conduct a variety of observations, using its seven onboard instruments, during the period of maximum activity in the (approximately) 11-year solar cycle, to supplement the information obtained from the Skylab programme a decade earlier, and the unmanned Orbiting Solar Observatory (OSO) series. Unfortunately, only months into the mission the satellite suffered a serious power failure when several fuses blew as a result of a power spike. As most of the instruments required pointing-system power to aim at the Sun, the mission was threatened with failure; but Solar Max was an MMS-based design, it was therefore decided that a visiting Shuttle crew should attempt to repair it.

Before the mission (originally designated STS-13) could be launched, a few items needed to be addressed. In order to retain some science from Solar Max, the satellite was placed in a spin. Its roll rate was still 1 rpm – beyond the range of the RMS, which was unable to grapple it and bring it into the bay. A special trunnion pin attachment was therefore developed. This device reproduced the RMS end effector,

Mission module kits being used during Solar Max repairs.

and was attached to the front of an astronaut flying the MMU, to match the roll rates of the satellite. It was then possible to manually dock with the satellite, use the MMU thrusters to slow it down, and attach an RMS grapple fixture to it to allow the arm to grapple it and lower it into the bay. The problem was that with only the STS-6 demonstration EVA, the Trunnion Pin Attachment Device (TPAD), and more importantly, the MMU, had never been tested in space. Moreover, little work had been undertaken with the RMS, and so several Solar Max-related evaluations were assigned to imminent flights.

By the time the missions flew in early 1984, NASA had changed the Shuttle numbering system. Following STS-9, the missions were assigned a three-digit code in preparation for the planned (but never achieved) rate of one flight per week – up to twenty-six from KSC in Florida, and up to twenty-six from Vandenberg Air Force Base in California. The code system was based on the sequence of scheduled launches during a fiscal year (from September). For example: STS-, followed by 4 (1984), 1 (KSC launches) or 2 (Vandenberg launches), and A (first flight in that fiscal year) or B (second flight in that fiscal year). STS-9 was also STS-41A, although the latter designation was hardly ever used. Because of flight delays, however, some Shuttle flights fell out of sequence or were cancelled. This led to some confusion.

STS-11 became STS-41B (STS-10 was already cancelled), and STS-13 became STS-41C and flew before what was STS-12 (STS-41D). This practise ceased after the *Challenger* accident in January 1986. It was designed to track mission-assigned payload, and it is unclear why the next Shuttle mission number was not assigned on successful launch (which was the policy during the earlier programmes of unmanned planetary probes such as Mariner).

Paving the way
The practise of completing EVA tasks in preparation for later missions was developed during the Solar Max repair mission. The in-cabin tests and evaluations of the EMU on STS-4, and the cancelled STS-5 mission, led to the demonstration EVA on STS-6. On the next mission, STS-7, the crew demonstrated a reduction in the cabin atmosphere from 14.7 psi to 10.2 psi, although the 20/80% oxygen/nitrogen mix was retained. In addition, they also demonstrated the ability of the RMS to capture a previously deployed satellite. The Shuttle Pallet Satellite (SPAS) was a free-flying satellite capable of supporting a range of scientific experiments, and the mission demonstrated the manoeuvrability and control ability of the Shuttle in rendezvousing and grappling the satellite later in the mission. This was the first item that a Shuttle reboosted with another object, and the operation clearly demonstrated an abiltiy to rendezvous with larger satellite such as Solar Max.

On STS-41B, the astronauts successfully demonstrated the MMU (see below). Bruce McCandless first flew the unit out to 50 m and then to 100 m from the orbiter, and Bob Stewart later also test-flew the unit. With the test-flights completed, the astronauts also carried out tests and evaluations pertaining to the Solar Max mission. After attaching a Manipulator Foot-Restraint to the end of the RMS, Stewart placed his feet in the device and was moved around the payload bay by RMS operator Ron McNair. This evaluated the abiltiy of the RMS to support an astronaut during EVA in the payload bay. Foot-restraints had been evaluated on previous EVAs, having been identified, during Gemini, as one of the requirements to ensure stability when working in a zero-g environment. The difference in using the RMS-mounted foot-restraint was that from the aft flight deck the RMS operator could manoeuvre the EVA astronaut to any location in the payload bay, thus offering a far greater scope for EVA operations than from a static foot-restraint. The EVA crew also evaluated tools intended for the Solar Max mission so that they were qualified to operate in both vacuum and in zero g. During the second EVA the astronauts were to test the TPAD by flying the MMU to dock with the slowly rotating SPAS attached to the end of the RMS; but this was prevented by an electrical fault in the RMS, and so the demonstration had to be confined to the SPAS still mounted statically in the Shuttle payload bay and not turning. The process of attaching the TPAD to the trunnion pin was accomplished several times.

The evaluation of a Shuttle-to-satellite refuelling system was completed on the mission – not that Solar Max needed refuelling, but there was another satellite in trouble. Landsat 4's communications system was experiencing problems, and part of it had failed. It was in too high an orbit for the Shuttle to reach it directly but it would be manoeuvred to a lower orbit, at the expense of its onboard propellant

The first untethered demonstration flights of the Shuttle MMU were completed during STS-41B in February 1984.

levels. If a repair was manifested, then a refuelling operation would also have to be accomplished. In evaluating the orbital refuelling system, the astronauts successfully used the valve and the pump to transfer freon red dye. On a later flight, a transfer of hydrazine propellant would be attempted before manifesting the Landsat 4 repair and refuelling mission. One last demonstration on the mission, reflecting the usefulness of astronauts on EVA, was the recovery of a portable foot-restraint that had become loose and was floating off into space. McCandless attempted to translate down the payload bay sill to reach the device, but it was beyond his reach. Thinking quickly, he called to his Commander, Vance Brand, to move *Challenger* so that he could reach the restraint, and this proved effective. NASA related this to the potential of rescuing a stranded astronaut with a failed MMU.

A job well done
Just two months later, STS-41C was in orbit chasing Solar Max at an orbital altitude of 560 km – the highest orbit to date. *Challenger* was stationed about 100 m away from the satellite. Goddard Spaceflight Center had powered down most of its instruments and eliminated most of its spin-rate, but it was still slowly tuning. George ('Pinky') Nelson, flying the MMU, approached the satellite, and, ensuring

that the solar arrays did not strike his helmet, he slowly approached the TPAD over the trunnion pin. But he could not engage it with the capsule, and after a failed second attempt he withdrew and instinctively grabbed the tip of one of the solar arrays with the hope that he could stop the rotation. However, the result was that the axial spin changed to a pitching roll, which could have disabled the satellite. The power being drawn from the storage batteries would soon be depleted due to the loss of solar lock, and Nelson's only option was to return to the orbiter whilst the ground tried to determine how to capture the satellite. With his nitrogen running low, he returned to the Shuttle, frustrated in his attempt at locking on to the satellite. This was adverse publicity, and the media was quick to pick up on the astronaut's inability to deliver one of the objectives of the Shuttle EVA programme: to retrieve and refurbish payloads. But this was only the fourth Shuttle EVA (after one on STS-6 and two on STS-41B), and nothing like this had been tried before. It was also only the second mission during which the MMUs had been used; and expecting success on every mission was highly optimistic. Fortunately, overnight, as the crew slept, Goddard Spaceflight Center managed to reorientate with Solar Max. Two days later, with a more stable satellite, the crew tried again, but this time with the RMS. Care had to be taken in case the residual motion should damp the RMS, but Terry Hart found the operation relatively easy as he snatched the satellite out of orbit and lowered it into the payload bay and into the U-shaped Flight Support Station. This station was mounted on a tilt table which could turn and tilt to provide more access to the astronauts serving the spacecraft. Drawing power from an umbilical connected to Shuttle systems, the satellite was now ready for the crew to complete the repair on their second EVA.

The design of the MMS featured a support bus of three 1-m-square boxes 120° apart, and the ability of the tilt table to turn the vehicle to face the astronaut reduced

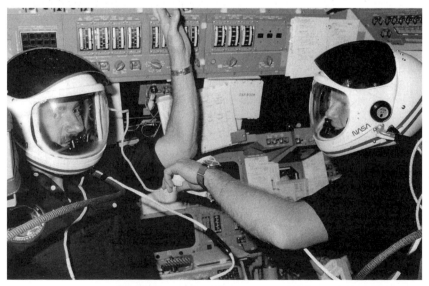

The EVA pre-breathe process, *c.*1984.

the amount of work required to complete the task. With the release of a set of bolts, the boxes could be completely removed and exchanged with a new unit. The astronauts rode on top of the RMS, and replaced the 227-kg attitude control and electronics box in an hour less than the time allocated for the task. With the extra time available, and with renewed confidence, the crew then attempted to repair the coronagraph, which had not operated correctly since the beginning of the mission. This unit was not originally intended for on-orbit replacement, and required repair and not exchange, and so access to it was far more awkward,.

It was this operation that set the precedent for abnormal procedures that would highlight future Shuttle EVA servicing techniques. The first problem was to gain access to the experiment. Using adapted scissors to cut into the layers of thermal blankets, the astronauts used ducting tape to move the cut blankets out of the way. The removal of six small screws presented another problem, especially as each astronaut was wearing a pressurised spacesuit glove whilst holding a screwdriver; and each screw had to be retrieved so that it would not float back inside the area and potentially cause a problem. Also required was the removal of a further twenty-two screws around the instrument controller, and the re-establishing of the connections

On STS-41C, George Nelson attempts to slow the rotation of Solar Max by hand to enable him to capture it.

of eleven circuits. The return of the screws was impractical, and so ducting tape was used to hinge and seal the working area. The insulation was then replaced and taped in place to cover the work area. As this area was not in a location that faced the Sun when the satellite was in operating mode, it was not expected that the operation of the instrument would be degraded due to the deterioration of the tape or a change in thermal properties due to not using the screws to seal the unit.

The returned components were to be analysed on Earth for their value in use as cost-effective components in satellite manufacturing. Logically, if certain elements of a spacecraft composition exhibited signs of degradation, then by using the Shuttle these components could be replaced. Occasional servicing or replacement and upgrades could be planned, and the extension of the life of some satellites, rather than the launch of full replacements, would be cost-effective. On the other hand, should it be found that components were degrading, then the options in extending the life of the satellite, either by upgrades or full replacement, could be more accurately planned. Solar Max had been in orbit for four years, and as part of this programme the crew also recovered thermal blankets and parts of its aluminium structure for return to Earth. During their work in the payload bay, the crew praised the design and operation of their servicing tools, as not a single problem arose during their use.

But was the operation successful? The following day – 12 April – Solar Max was redeployed by the crew, and was left to carry on its observational programme. It was again a fully operating satellite on station, it was able to orientate itself, and the coronagraph produced results better than expected. Two weeks later the satellite recorded the largest solar flare since 1978; but ironically, this increased solar activity contributed to the atmospheric degrading of the orbits of satellites, including that of Solar Max. There were talks of a possible refuelling mission to boost the satellite to a higher orbit during the 1990 STS-32 mission; but this was not to be, and Solar Max finally ended its mission with a destructive re-entry in December 1989.

As with all EVAs, the crew evaluated their equipment and procedures and suggested improvements to the system. As with STS-41B, the tension of the safety tethers was approximately twice the level required on the take up reel. For the astronauts this was distracting and fatiguing, and indicated that a reduction in tension was required. Equipment designed for EVA should be constructed with sufficient tolerance to ensure that its use in conjunction with gloves becomes much easier. Payload-bay lights for night-time operations were 'adequate', but needed to be supplemented for work-site operations. EMU-mounted lights worked very well. The power tool 'performed superbly, and should be made part of the standard Shuttle tool-kit [for] both EVA and IVA', the crew reported. On the other hand, the grey tape carried for orbiter contingency operations simply did not work, and it was suggested it should be replaced by the three-layer Kapton tape used on Solar Max. Generic restraints were suggested for future use. The wrist tether, used to secure the moveable tripod foot-restraint, had to be removed every time the restraint was adjusted, which proved to be time-consuming and tedious. The Manipulator Foot-Restraint worked well, and greatly assisted the replacement tasks on Solar Max. Moving a large mass (226.8 kg) did not present any problems provided that the movement rate was kept low. The crew reported that 'handling larger masses would be equally feasible as long as there were adequate hand-holds.' The EVA scissors, designed with extended blunt ends, were not the ideal tool for splitting tape and cutting. They had to be pushed well into the tie-wrapped bundles before the cutting surfaces worked, and extreme care was need to prevent inadvertent damage.[9]

An orbital service station?

The STS-41G mission in October 1984 was to demonstrate the orbital refuelling system for the transfer of toxic hydrazine fuel, and evaluate the provisional stowage and assembly tool-box located in the payload bay. Because of the scientific cargo – the 10.7 × 2.1-m Shuttle Imaging Radar B (SIR-B) antenna – the equipment had to be located on a Mission Peculiar Equipment Support Structure (MPESS) at the aft of the payload bay. The EVA led to further experience in potentially refuelling satellites in orbit and in evaluating operations in a full payload bay. NASA indicated a 2,268-kg capacity tanker system which would be fitted in the payload bay and, using standard hook-up facilities, be used to refuel satellites such as Landsat and the Compton Gamma Ray Observatory. For reasons of crew safety, the transfer experiment was successfully demonstrated by command from the aft flight deck after the EVA, and led to the approval of the Landsat 4 satellite. However, because that satellite was in a polar orbit, the mission, manifested for 1987, had to be launched from Vandenberg Air Force Base in California; but as a result of the 1986 *Challenger* accident, launches from Vandenberg were abandoned, and due to the subsequent delay in the return to the flight programme it was not possible to attempt the refuelling of Landsat 4.

This EVA also presented the opportunity to evaluate EVA procedures and equipment prior to embarking on more ambitious activities on later missions. A 35-mm camera, flashlight and LCD evaluation box were strapped to the wrist of Dave Leestma in the air-lock, who suggested that it would be more practical to have all tools located outside at the EVA work-site or at a provisional stowage assembly area, rather than having to carry them out through the air-lock. Due to the payload configuration, both tethers were routed down the port slide-wire, and while this did not cause undue problems it requires extra care by the two astronauts (Leestma and Kathryn Sullivan) to avoid crossed tethers and entanglement. They also discovered that the tether reel tension was too high. This exerted a force that was fatiguing and distracting, and they therefore requested a reduction of the tension for future EVAs.

The LCD evaluation box was found to be readable in bright sunlight and in shadow, but it required back-lighting when used in dark shadows and during night-time activities. On extended EVAs this would assist in providing the astronaut with visual display data on consumables. Evaluation of the provisional stowage assembly revealed that the cover latch retention strap was easy to unsnap but difficult to snap – and it was left unsnapped. The astronauts suggested that it be replaced with a locking tab similar to that on the reel tether box – not only to secure the cover, but also to initiate standardisation of securing devices rather than a variety of retention devices for the various stowage facilities. The tools inside the box were found to be easily restrained.

A set of seven unique tools was supplied for access to the manual fill valve of the orbiting refuelling system, to provide redundant seals for crew safety and to control fluid flow to the 'satellite'. Once engaged, the fuel transfer unit and valve became a permanent feature, and provided a standard interface for refuelling. The astronauts had only to perform the fuel line connection task, and all would proceed according to plan – even down to the loosening of the foot-restraint, as during WETF training. It was recommended that a means should be found to secure the foot-restraint to ensure that it remained in the required position and did not work loose during an EVA operation. It was found that the safety-tether tension force remained very noticeable during orbital refuelling system tasks. On one occasion, Leestma lost his grip when the tether pulled him to the slide-wire rather quickly, and this emphased the need to reduce the tension.

Prior to completion of the EVA the astronaut manually stowed the Ku-band antenna that had experienced problems earlier in the mission, and reported that the operation (a contingency EVA task) was easily achieved manually. The SIR-B antenna also had to be pushed shut by the end of the RMS. The astronauts also visually inspected the antenna, and found that the insulation layers between the outer and inner leaf were pressed together, and might have been the cause of the problem with the latch. The crew also used an Intersan hydrazine vapour detector to ensure that they had not been exposed, and then carried it on their EMUs back to the crew compartment.[10]

Rescue and repair
In February 1984, during the STS-41B mission, two satellites were successfully deployed; but each of the payload assist module solid rocket motors failed to work correctly, and burbed for ten seconds instead of the planned 83 seconds. They ejected as planned, but as a result of the short burn-time the satellites became stranded in a useless, but accessible, orbit. As a result, a recovery mission – STS-51A – was assigned to retrieve the stranded satellites from their low orbit. As these were identical satellites (HS-376), one set of tools and capture operations was devised for both. Since they were not expected to be seen again after leaving the Shuttle, recovery and capture aids were not incorporated in the design, and new devices had to be developed. Using the MMUs for the translation across to the satellites, an apogee kick motor capture device (called the 'stinger', because of its resemblance to an insect sting), mounted on the front of the MMU, would capture the satellites to bring them close to the RMS. An A-frame would be used to lock them into the payload bay, and they would then be returned to Earth for refurbishment and possible relaunch.

In April 1985, a US Navy Leasat/Syncom communication satellite was taken to orbit, but it failed to activate correctly after deployment from the Shuttle. This time the first American unscheduled EVA was performed in an attempt to activate the satellite, but unfortunately the efforts of the astronauts and the controllers proved ineffective. A repair mission was therefore planned for later in the year, and this was successfully accomplished on the STS-51I mission in August 1985. After a spate of satellite deployment difficulties, the success of the astronauts in repairing the Leasat

During STS-51A, the MMU and stinger is used to capture a rogue communication satellite.

3 satellite again demonstrated the flexibility and potential of Shuttle EVA operations.

In 1991, STS-37 astronauts helped to configure the Compton Gamma Ray Observatory for deployment, and a year later the first three-person EVA assisted in the capture and deployment of a stranded Intelsat satellite. These were the last EVAs which involved satellite retrieval and repairs before the era of the ISS, and they represented a further step in the development of EVA servicing techniques The additional skills and lessons learned from these missions included the following recommendations for EVA operations.

STS-51A

The stinger worked as designed, and the rogue satellites were captured by the insertion of the device into the apogee motor of the satellite, and by closing the grapple ring and 'docking' the astronaut to the satellite. The MMU cancelled rotational rates, and the astronaut then moved close to the RMS for capture by means of a grapple fixture on the side of the stinger. Whilst still attached to the satellite, the other astronauts secured the bracket to the top of it, after which the MMU Pilot pitched over to allow relocation of the RMS on the new grapple fixture on the frame previously installed. This was followed by final separation by the MMU Pilot, and lowering into the payload bay by the RMS. However, the bracket failed to operate as designed, because the dimensions of the common clamp of the bracket were slightly different from those of the retention ring on the satellite. Once again, the detailed design specifications on file were not as exactly as thought. It was found that a waveguide extension had blocked the bracket retention system and prevented it from locking on to the top of the satellite. A back-up plan, devised only a few weeks earlier, was therefore brought into play. Instead of using the bracket grapple, the RMS continued to use the stinger grapple, whilst the MMU Pilot disengaged and stowed the unit, made his way to the portable foot-restraint, and grabbed hold of the still-folded main antenna. The RMS was then disengaged, and the second astronaut installed a nozzle cover and an A-frame stowage mount. Both astronauts pitched the satellite over into the bay, and manually attached the A-frame to the support structure floor of the payload bay. For the first operation, Joe Allen flew the MMU to capture Palapa, whilst Dale Gardner remained in the payload bay; and on the second EVA the astronauts reversed the order of tasks and adopted the capture scenario, without the attempt to attach the bracket, so that installation into the bay was a much smoother process. Both omni-antennae were severed on top of the satellites to allow closure of the payload-bay doors. Both satellite were returned to Earth, refurbished, sold and successfully relaunched, with new identification names. Recommendations from this experience were as follows:

- A suitable method of securing equipment in the air-lock should be devised. Due to insufficient provision for dedicated stowage, the crew has to resort to the use of tape, tethers, and straps.
- LED displays, for day-time read-out in direct sunlight, should be improved. During both EVAs, Gardner encountered considerable wearing of his EVA glove, due to chaffing by the knurled handle of the torque wrench. It was recommended that such knurling should not be incorporated in future tools.
- A one-handed tether hook should be developed so that a hand would be free whilst tethering. On earlier EVAs, the astronauts experienced excessive tether forces, and this was not remedied until flights after STS-51A. The crew also recommended that the safety tether reel should be improved.
- Extra tethers should be incorporated on planned EVAs missions, and the stinger and A-frame should be improved (although they were never flown again).
- Handle extensions should be incorporated in all EVA zipper mechanisms.

The crew concluded that it was relatively easy for a suited EVA crew-man to hold steady or move slowly, and to retain good control with a 900-kg mass with dimensions up to 3 m, provided that there is adequate restraint and guidance by a second crew-member. In zero g they found it easier to handle the larger items than the much smaller items. Even the HS-376 satellite, with no proper hand-hold, was easier to move due to the numerous places where it could be gently gripped; and the firing of the small vernier thrusters on the Shuttle, or movement of the RMS, did not affect the grip. Rather than have a single crew-men guide the satellite, it was easier for both crew-men to move and berth the satellite with one hand, with the other hand on the support structure.

The entire STS-51A operation proved physically and mentally exhausting for all five crew-members. The Commander and the Flight Engineer (MS2) were involved with rendezvous, supporting the EVAs by station-keeping, operating the RMS, and taking photographs, while the Pilot assisted the EVA crew by directing and monitoring their activities. The EVA required three hours of preparation followed by ten hours of continuous work, with no time to rest or eat. The EVA crew-members were exhausted and tired, and the onboard crew-members, who had had to continually look into the bright payload bay, suffered eye-strain and long-lasting headaches. The crew suggested that these type of EVA should not be undertaken prior to Flight Day 4, to allow for full adaptation to spaceflight, and for rest, and recommended that EVA capability in relationship to flight days be planned as follows:

Day 1 or 2	Contingency EVA only
Day 3	Earliest day for simple short-duration planned EVA
Day 4	Earliest day for complex planned EVA
Day 5	Best day for complex planned EVA

They also recommended that one day of rest between EVAs was highly desirable and probably mandatory, and that for complex EVAs a crew of six was advantageous, as the extra hands would be useful in helping with any orbiter problems that arose, assisting in documentation, and supporting intravehicular operations when necessary. The role of intravehicular crew-member was highly desirable, and as, on STS-51A, the IVA crew-member – Dave Walker – was also the Pilot, there arose a conflict of Pilot and EVA training. The integration of the IVA crew-member with the EVA crew, their equipment and tasks, would probably result in more complex EVA operations. It was planned that the Pilot on STS-51A would be back-up to the Commander/Flight Engineer (MS2) for rendezvous and operations, but this was not possible, due to his involvement in the EVA tasks. Cross-training was also beneficial, with the Flight Engineer (MS2)/RMS primary operator – Anna Fisher – and Pilot receiving previous EVA and MMU training. During the EVA, communications between the EVA crew and Mission Control via the IVA crew-member prevented simultaneous communication by the various crew-members.[11]

STS-51D

By the time that the STS-51D crew rendezvoused with the Syncom after its failure to operate the programme start-up sequence, they thought that they were about to undertake some type of EVA, the manual dexterity of a human being preferable to that of the arm. But they were soon informed that the controllers wanted them to fabricate a 'fly-swatter' device from a plastic flexible document cover, with slits to allow the RMS to snag the switch that was thought to be in error on the satellite. A device rather like a lacrosse stick was also prepared as an alternative method of activating the satellite. The crew reported that the instructions were excellent, and that the fabrication of the device – using a bone saw and knife, and sticky tape – presented no real difficulties. As this was the first unplanned EVA in the programme, communications to and from the ground were vital in reviewing each step of the EVA operation, in understanding what was required, and in assessing whether the

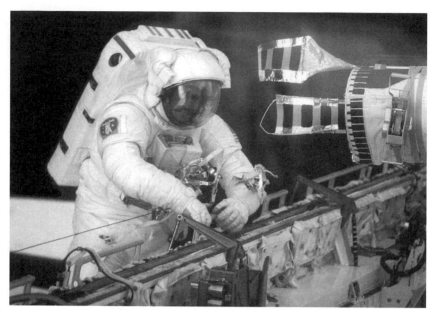

The 'fly-swatter' device is attached to the RMS during STS-51D.

final result was as planned. The crew also commented on their ability to rely on their hours of contingency EVA training in their preparation for an EVA, even though they had not trained for one.

Altitude chamber runs and contingency EVA training help to boost confidence levels in EVA crew-men yet to perform an EVA. And despite some 'flailing around' – probably due to the difference in performing simulated EVAs in water tank as opposed to the real thing – the crew felt very comfortable once outside. They did, however, suggest the inclusion of additional restraint mechanisms for when retrieving tools from the storage areas. During the attachment of the 'fly-swatter' and 'lacrosse stick', the EVA crew – Jeff Hoffman and Dave Griggs – reported the value of real-time communications with support astronaut Jerry Ross at Mission Control, who had evaluated the straps attached to the RMS to secure the devices. Hoffman noted that his EVA helmet hit the fly-swatter a couple of times, and considered that had they first performed the operation in the water-tank facility they would have been more aware of the clearance during the EVA. His visibility was restricted by his helmet, and the call came from CapCom Jerry Ross, who was monitoring the EVA from the ground. This teamwork in having a EVA crew observed by both the IVA crew and Mission Control offered additional safety and support options that proved valuable during real-time execution of the EVA. However, the crew again commented on the design of snap-locks on some of the equipment. This had been noted during previous missions, but no modifications had yet been introduced.[12]

STS-51I
During the early stages of the Syncom IV/Leasat repair mission, the RMS suffered

blown fuses. This restricted arm movement to one joint at a time, without the aid of computer assistance, and so to carry out the repairs, two EVAs, rather than one, were scheduled. Past experience in adapting EVA plans, the assignment of experienced EVA astronauts as CapCom during EVA operations (STS-41C astronaut George Nelson was CapCom for this EVA), and confidence in the ability of astronauts to complete EVAs by using Shuttle support equipment and procedures, allowed van Hoften – who had also worked on the Solar Max repairs – to attach, on his first attempt, a specifically designed capture bar to bridge the gap between two ground-processing sockets, and to then grab, by hand, the slowly spinning satellite on its next revolution. A second handling bar was then attached, whilst the first was replaced with an RMS grapple fixture to allow the robotic arm to hold the satellite, thus freeing the astronauts to complete the actual repair. During certain parts of the EVA the crew reported that they felt cold, as did several astronauts on earlier mission. Because of the length of time spent 'standing' still on the foot-restraints and holding items for several minutes, there was no movement of the arms and leg movement, and they therefore cooled more quickly. Plans were evaluated to investigate the thermal properties of the gloves, boots and suit limbs on later EVAs. The operation of safing the satellite, using specialist tools and shorting plugs, proceeded smoothly as they installed a bypass electrical harness around the faulty switch. They reported that the onboard batteries had not frozen as feared, and that the omnidirectional antenna had deployed, indicating life in the satellite, which afterwards was left overnight on the RMS in a safed mode. The following day the crew inspected the engine bell of the satellite, which 'looked very clean'. The spun bypass unit was powered up, and its lights came on as expected, indicating the correct functioning of the bypass system. The grapple bar was then removed, and the satellite was redeployed in orbit by van Hoften, on the RMS, who held one of the capture bars and manually spun it up to 3 rpm. The crew reported that the solar cells appeared to be undamaged, and that they had removed a piece of sticky tape from the 'fly-swatter'. Their efforts to control the 4.3-m-diameter satellite were somewhat hindered due to difficulty in seeing each other; but they had successfully captured, repaired and deployed the satellite, which slowly progressed to geosynchronous orbit over several months. In less than fourteen months, Shuttle crews had rendezvoused with four errant satellites, and performed EVA programmes to restore their operational status. Confidence in the success of future EVAs could not have been greater.

STS-37 and STS-49
In 1986 the loss of *Challenger* had a devastating effect on the Shuttle programme and on the American manned spaceflight programme in general. The Shuttles were grounded, crews were disbanded, and missions were cancelled, as the full investigation into the accident progressed. Additional safety features required that future EVAs would be carried out tethered inside the payload bay, and the MMUs were grounded; and there were to be no daring last-minute rescue EVAs until the fleet had been restored to full flight operations. It would be a further six years before another astronaut left the air-lock of the Shuttle in space.

During the deployment of the Compton Gamma Ray Observatory from STS-37 in April 1991, its antenna determinedly remained in its stowed position. The astronauts had already been assigned an EVA to evaluate tools and procedures for the space station, and it was therefore decided to take the opportunity to conduct an unscheduled EVA to try to manually deploy the antenna. This did not require specialised tools or equipment; just a well-placed gloved hand to shake it loose. Jerry Ross grabbed hold of the observatory's flight support structure with one hand, and used his other hand to shake the boom. Both astronauts then set up a foot-restraint, and removed a pin, which allowed them to pull out the antenna. This they had practised during their contingency EVA training in the WETF, and this training proved useful in this operation, although they both reported difficulty in locating hand-holds on the satellite in the darkness. At the end of the EVA the two astronauts remained suited in the air-lock in case of further problems in deploying the satellite. But their services were not required, and as the satellite drifted away from the Shuttle they could not resist one last look out of the hatch to see it depart.

The final commercial satellite retrieval and repair mission of the pre-ISS era – to a satellite that had been stranded in low Earth orbit since March 1990 – was undertaken in May 1992. During launch, the payload deployment system failed and did not disengage the satellite. In order to save the satellite, controllers ordered that it be separated from the perigee-kick motor, and that it use its own small thrusters to place it in a low orbit offering the opportunity for a Shuttle crew to go to the rescue. There had been sufficient time to plan for the recovery of Intelsat VI, but it took two years to evolve the recovery and EVA plan. The manifest dictated that to attach a new kick motor to the Intelsat, to allow it to be boosted to its operational orbit, a Shuttle payload bay had to have adequate space to accommodate and work on the 5.2 × 3.2-m satellite. The crew also had to have time to train for such an operation. The first flight of the new orbiter was STS-49, which was already assigned to demonstration EVAs in support of ISS EVA development. The evolved idea was that an astronaut on the end of the RMS would hold a capture bar that could be attached to the lower support ring of the satellite. This also included an RMS grapple device that would allow movement over a new kick motor located in the payload bay, and attached to the satellite by an adapter on the motor and capture bar. It appeared to be a straightforward procedure, but, as previously discovered, these types of EVA are not so simple.

Intelsat was slowly rotating, but at 0.5 rpm it was within the parameters that simulations in the WETF had determined were acceptable when using the hand-held capture bar. On the first attempt, the capture bar did not capture, due to a lack of force to activate the capture latches. A second attempt was more successful, but when Pierre Thuot, on the end of the arm, attempted to hold the bar to slow the rotation, it slipped off and added a coning motion to the rotating satellite. For three hours Thuot tried repeatedly to attach the capture bar to the satellite, but he did not succeed. Once again, training on the ground could not exactly replicate the events on orbit. In addition, the RMS ceased movement several times, as it was placed in a configuration which its drive joints could not support. It was clear that the onboard liquid propellant was affecting the stabilisation of the satellite, and wobbling it every time the astronaut released his hold. Had he been able to secure the satellite, Thuot would have had to hold it very still until the liquid ceased to move, and then be extremely careful that he did not impart new movement to begin the cycle all over again. Although the WETF proved valuable in simulating the basic capture and EVA timeline, it could not reproduce the dynamics of the liquid inside the actual flight vehicle. The following day, Thuot and Richard Hieb ventured outside to try again. The satellite had been dampened by Intelsat controllers, but even with much greater care in positioning, Thuot on the RMS, and more gentle forces on the capture bar, the bar simply refused to connect as designed. Five attempts were made before the task was abandoned for the second time. It was time to rethink the operation.

Ground support teams began to examine the problem. It was clear that the capture bar would not work, and so a new idea emerged, both on the ground and in orbit. By using struts from the planned Assembly of Station by EVA Methods (ASEM) experiment, a triangular structure could be assembled on top of the MPESS in the centre of the payload bay, allowing not two but three astronauts to perform EVA using foot-restraints attached to the MPESS, the RMS and the starboard sill. *Endeavour* would then move underneath and allow the three pairs of hands to reach

The first three-person EVA is enacted to capture the Intelsat during STS-49.

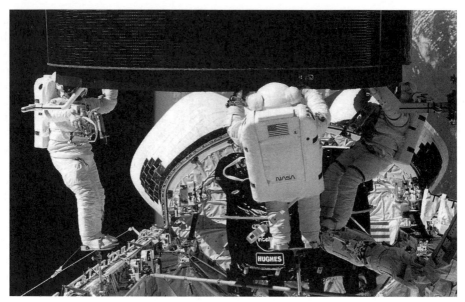

The three-person EVA technique.

up and grab the satellite by hand. The questions that arose were as follows. How can the capture bar that includes the RMS grapple, be replaced to allow the lowering of the satellite into the bay? Can two astronauts man-handle the satellite, secure it, and still work on it, or is a third pair of hands required? Are there provisions on board for a third person to perform EVA, and can the air-lock support such an operation?

It was the crew who put forward a suggestion for a three-person EVA, as they had a spare EMU on board; and as Tom Akers and Kathy Thornton were scheduled to perform the ASEM EVA, it was decided to allow Akers to accompany Thuot and Hieb outside for a final attempt at capturing the Intelsat. *Endeavour* had sufficient propellant on board for one more attempt, but before authorisation, the technique was to be demonstrated in the WETF at Houston. This evaluation was performed by Story Musgrave, Richard Clifford and James Voss, who simulated fitting three astronauts in the Shuttle air-lock, and determined exactly where to locate each crew-member in order to capture the satellite. They determined that although it was a tight fit in the air-lock, a three-person EVA was possible. The plan was approved.

The world's first three-person EVA – the hundredth EVA – began when Akers, Hieb and Thuot left the air-lock to position themselves to capture the satellite. Thuot rode the RMS, Hieb positioned himself near the starboard payload-bay wall, and Akers was on top of the ASEM strut on top of the MPESS across the centre of the payload bay. All three of them grabbed the satellite, and then Hieb attached the capture bar whilst his colleagues held on. It had already been determined that their handling of the satellite would not exceed the glove-touch temperature limit of 160° C. With the RMS lowering the satellite over the kick motor, the astronauts successfully attached the unit to the satellite and then returned to the air-lock. Thornton managed to eject the satellite back into orbit on his third attempt – which caused a few hearts to palpitate – but two days later the kick motor was ignited to send the satellite on its long-awaited transition to geostationary orbit. Once again the Shuttle astronauts and the NASA team had demonstrated their ability to overcome difficulties on EVA and achieve success. It was a boost in confidence in EVA operations expected at the ISS later in the decade. It was also the last commercial satellite EVA of the Shuttle programme to date. Between 1984 and 1992 the Shuttle astronauts and their ground support teams had, in fourteen EVAs, demonstrated their ability to rendezvous, capture, repair, retrieve, service and redeploy a variety of satellites, and to adapt procedures, tools and equipment to overcome a variety of in-flight difficulties not encountered during ground simulations. These skills were put to operational effect during the series of HST servicing missions between 1993 and 2002.

MMU activities

On Gemini 9, Gene Cernan had been unable to evaluate the AMU. During Skylab, the evaluation of the manoeuvring devices inside the pressurised OWS revealed that the back-pack device was much more effective than the foot-controlled device, and provided an adequate zero-g demonstration of the unit by several astronauts, one of whom (Garriott) had not previously trained on the unit. The use of the Shuttle MMU during three flights in 1984 clearly demonstrated its usefulness in EVAs away from the payload bay of the Shuttle. Although it was not used in the final capture of Solar Max, it was the Trunnion Pin Attachment Device, and not the MMU, that was faulty. The use of the MMUs during STS-51A was critical in capturing the two comsats and returning them to the payload bay. Evaluation of the MMU was accomplished by six astronauts on STS-41C, during which the units operated flawlessly, with propellant consumption estimates falling within 7% of flight usage. This was in marked contrast to STS-41B, during which propellant usage was approximately 50% greater than predicted. This was found to be due to actual versus predicted flight activities, with in-flight tests being more extended than originally planned. The amount of available EVA propellant was increased as a result of more accurate flight estimates, and was proven during STS-41C. By the time of the STS-51A mission, understanding of MMU operations led to improved training procedures and piloting techniques. Fuel could be save either by pushing off from the payload bay structure in order to translate across the payload bay, or by awaiting orbital sunrise before activating the thrusters on the unit. On this flight, MMU propellant was underestimated by 3–19% for both Pilots, due to differences in piloting techniques and attempts to keep the Sun out of the Pilots' eyes. The MMU was, according to all six Pilots, a joy to fly, and on STS-51A it was much quicker and more accurate to change the attitude of a satellite with the MMU than to position or reposition the RMS. The problem with the MMU was its size. In confined places – such as those expected to be encountered on the ISS – its bulk would restrict operations close to the station, and following the *Challenger* accident in 1986 its expected use on other retrieval and repair missions and its use as a rescue system or inspection device was unrealised. Like the Soviet MMU on Mir, the American MMU, after years of development and evaluation, had become obsolete.

What was required was a smaller and less bulky device to accommodate the rescue of a stranded astronaut whilst on EVA, whether at the Shuttle or the ISS. This requirement resulted in the development of the SAFER unit.

EVA incidents

On several EVAs, astronauts had complained about the thermal qualities of the gloves and boots, due to their feeling cold in their extremities as they remained in one position for a length of time. On STS-41C, Nelson's urine collection device failed during the second EVA, and a considerable quantity of urine was released into his EMU. However, the liquid-cooling and ventilation garment soaked up the liquid, and none of it reached the vent loop. The EMU oxygen flow and the LiOH/charcoal cartridge removed most of the odour from the suit, but there followed a rather unpleasant post-flight clean-up operation. Nelson's helmet became fogged because he had reduced the flow of coolant liquid after he felt cold, and this in turn reduced the ventilation, leading to condensation. During STS-41G, Dave Leestma experienced some difficulty in removing his right boot from the foot-restraint. This

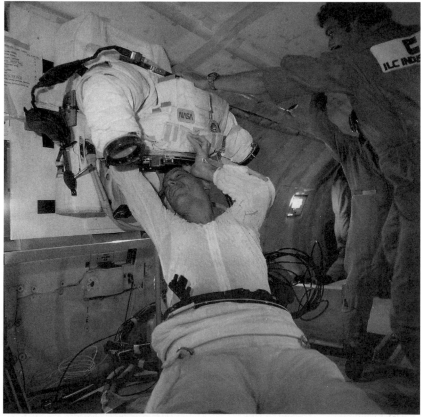

Repetitive training for EVAs from the Shuttle – such as the tight squeeze into the EMU during reduced-gravity aircraft flights – ensured familiarity with procedures and equipment once in orbit.

was caused by a thermal micrometeoroid cover overlap at the heel of the boot, which caught in the heel clip of the foot-restraint. Improvements to the boot heel were evaluated on STS-51A. Dale Gardner's EVA glove palms were abraised by the knurling on some of the tools during STS-51A, but their integrity was not affected.

Summary
Between 1983 and 1992, fourteen out of nineteen Shuttle EVAs were involved with the servicing of satellites. The role of space servicing was about to take on a new role with the HST, but these fourteen EVAs had clearly set a precedent for EVA operations from the Shuttle. By the mid-1990s, the operational constraints of scheduled, unscheduled and contingency EVAs from the Shuttle were amended; and regardless of the type of EVA, a series of procedures, using a detailed checklist, were clearly defined:

1 *Quick-don mask pre-breathe* Wearing of a face mask for pre-breathe and cabin depressurisation.
2 *Cabin depressurisation to 10.2 psi* Reduction from 14.7 psi.
3 *EMU checkout* Preliminary checkout of EMU systems and components prior to donning.
4 *Mid-deck preparation and preparation for donning* Configuration of the EMU, its ancillary components, and EVA equipment for crew-member donning.
5 *Suit donning* Assistance by the IVA crew-member during donning of each EVA crew-member into the Shuttle EMU and ancillary components, approximately forty minutes in duration.
6 *EMU check* Configuration and checkout of the EMU prior to EMU purge.
7 *EMU purge* Nitrogen purge of the EMU prior to pre-breathe.

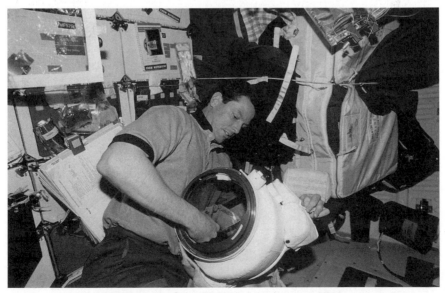

Maintenance of the EMU helmet between EVAs.

8 *Pre-breathe* Crew-member acclimatisation to lower chamber pressure (40–70 minutes).
9 *Preparation for depressurisation* Configuration of the air-lock for exit, and closing of the inner hatch prior to air-lock depressurisation.
10 *Air-lock depressurisation* Configuration and checkout of the EMU, air-lock depressurisation, and opening of the outer air-lock hatch.
11 *EVA* Air-lock exit, completion of the planned, unplanned or contingency EVA by assigned crew-members as required by the flight plan or in flight developments (up to 8.5 hours), air-lock re-entry, and closing of the outer hatch.
12 *Air-lock repressurisation* Air-lock repressurisation, and opening of the inner air-lock hatch.
13 *Post-EVA* Shut-down of the EMU systems, and doffing of the EMU and ancillary components.
14 *EMU maintenance/recharge* Change-out or recharge of the EMU battery and lithium hydroxide cartridge, general cleaning of the EMU for subsequent EVAs, and recharge of the EMU water system.
15 *Post-EVA entry preparation* Reconfiguration and restowage of the EMU and air-lock equipment.

General mission constraints were applied to Shuttle EVA activities, although they did not define specific activities peculiar to a given mission. These guidelines were more generic:

• The maximum schedules duration for a planned EVA is six hours.
• All scheduled and unscheduled EVAs require two crew-members.

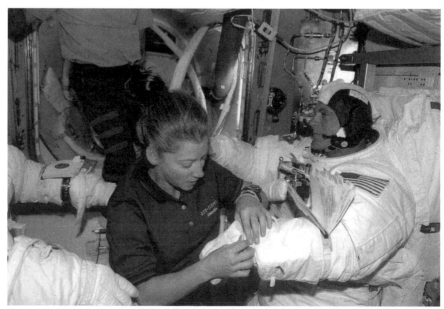

The IVA astronaut, assisting the EVA crew, is an integral part of each EVA, and often acts as the alternative or back-up EVA astronaut and EVA choreographer.

- No scheduled or unscheduled EVA shall take place on Flight Day 1 (MET up to 24 hours).
- No scheduled EVA shall be planned to take place prior to Flight Day 4 (MET 72 hours), unless: a specific flight requirement is dedicated to such EVA; or the payload customer had specifically negotiated with NASA for the early scheduled EVA capability and an exception has been duly processed.
- The latest that an EVA may be scheduled in pre-flight planning is two days prior to end of the mission (landing minus 48 hours).
- No unscheduled EVA shall be planned to take place prior to Flight Day 3 (MET plus 48 hours), unless: a specific payload has no alternative but to use EVA as its third level of redundancy for the purpose of critical back-up to deploy/operations that would prevent the loss of payload; payload customers have been made aware of the inherent risk that EVA may not be able to be performed due to crew status, and the customer has thus agreed to accept this risk; the payload customer has previously negotiated with NASA for the early unscheduled EVA capability and an exception is duly processed.
- The response times required prior to an unscheduled payload EVA were: upon the discovery of a failure leading to an EVA with approximately 24 hours allocated for EVA preparatory prior to starting EVA maintenance on the failed component; if the above case occurs on launch day a 44-hour EVA preparation time is allotted; if a payload requires a shorter EVA response time then this must be negotiated with NASA and an exception must be duly processed.
- Payload activities that may require an unscheduled EVA towards the end of a mission can only be approved if there are sufficient consumables and landing opportunities to extend the mission to perform the EVA and still preserve a minimum of two wave-off days.
- Whenever possible a minimum of one flight day must separate two scheduled EVAs for any given EVA crew-member.
- A contingency EVA will be scheduled in real time whenever it is necessary to restore the orbiter to configuration for safe return. Contingency EVAs are recognised for failures pertaining to the following orbiter systems: radiator actuators; payload-bay doors, bulkhead latches, centre-line latches, air-lock hatch, RMS, bulkhead cameras, Ku-band antennae, and ET doors.

There were also several guidelines for all EVA operations:

- Always use the 'make before break' tether protocol.
- Never use the EVA glove as a hammer.
- The EVA crew-member and equipment must remain tethered at all times.
- Slow and deliberate motion provides much greater stability than quick and jerky motions.
- Body positioning is 90% of the task.
- Each EVA crew-member should personally check his/her own EMU and EVA equipment.

The experiences on Shuttle EVAs also underlined the importance of providing flight-accurate mock-ups for training and simulations; the use of the WETF for complete EVA simulations; adequate and secure restraints for the feet, body and hands; the availability of orbiter IVA crew-members to support EVA tasks and assist in pre- and post-EVA operations and monitor EVA timelines, acting as a communication link from the EVA crew to orbiter and ground teams; the value of the combined IVA and ground visual facilities to assist the EVA crew; adequate EVA time management and task planning to ensure avoidance of fatigue and exhaustion by both IVA and EVA crew-members; adequate payload-bay and payload/work-station lighting; the availability of the RMS to support EVA tasks; flexibility in the design and execution of EVA timelines; real-time use of support astronauts in 1-g and WETF simulations; the reassignment of EVA-experienced astronauts to other EVA tasks on later missions or to support roles (WETF/ CapCom) on the ground; the cross training of IVA and RMS astronauts with EVA activities, hardware and procedures; and continued in-flight and post-flight debriefing and evaluation of procedures equipment and hardware to ensure continued development of new methods of achieving mission success. Not only were these valuable for STS EVAs – especially during the HST servicing missions – but they also refined the planning for Shuttle-based ISS construction EVAs planned from the late 1990s.

THE HUBBLE SPACE TELESCOPE SERVICING CAPABILITY

As a measure of the ability of Shuttle-based astronauts to service, repair and upgrade a vehicle on orbit, one spacecraft predominates: the Hubble Space Telescope. From the beginning of the development of flight hardware for the telescope, the option of EVA servicing and repair was evaluated. Up to 90% of the HST's onboard equipment and instrumentation was provided with a back-up or identical unit, and most of the scientific instruments provided redundancy because their functions slightly overlapped. It had been recognised early in the design that the development of the STS system allowed for periodic repair and/or replacement of equipment. The delay in the launch of the HST allowed the Shuttle system to be flight-proven and refined, and maintenance missions could be scheduled to replace equipment or instrumentation in order to extend the orbital lifetime of the facility. In the design of the telescope, as well as in the scientific and operational constraints and requirements, there was a requirement to include provision for orbital deployment, on-orbit servicing, and orbital retrieval by the Shuttle.[13]

EVA planning

Details of proposed EVA servicing of the HST began to emerge during the late 1970s. The first mission would deploy the HST in orbit, and should anything go wrong there existed the capacity to perform a range of contingency EVAs to resolve problem or return the HST to Earth. A series of planned servicing missions would then be included in the launch manifests over the orbital lifetime of the HST (at least

During most EVAs, the other IVA crew-member is the RMS operator. Here Nancy Currie is shown in the RMS operator's position at the aft flight deck. Together with the EVA crew and IVA crew-member, the RMS operator trains and works as an integral part of Shuttle-based EVAs.

fifteen years). These missions would be specifically aimed at the maintenance, servicing, repair, replacement and removal of selected items, using dedicated transfer devices and by means of servicing doors in the telescope's structure. Capture, EVA support and deployment would be accomplished by means of the RMS, and specific flight-support equipment would be provided in the payload bay to allow orbiter servicing over several days. By 1979 the original plans called for the telescope to be returned to Earth, refurbished and relaunched every five years, with an on-orbit servicing mission every 2.5 years and with hardware lifetime and system reliability requirements based on this 2.5-year interval. Throughout the HST's fifteen-year lifetime, therefore, in addition to the initial deployment and final retrieval missions there could be two retrieval missions, two redeployment missions, and at least three on-orbit servicing missions – at least nine dedicated Shuttle missions, involving a maximum of five flight seats (thirty crew-members) for the deployment and retrieval missions, and up to seven flight seats (twenty-one crew-members) available on three on-orbit servicing missions. Although several of the astronauts would probably serve on more than one mission, there remained more than fifty crew assignments for which training was required. This placed a major strain on the limited crew-training facilities, involved considerably more flight preparations, and committed at least one Shuttle to support these specialist missions.

By 1985, however, concerns had arisen over contamination and the structural loads involved in bringing the telescope back to Earth and relaunching it – with the risk of potential loss in a launch accident, and with no back-up telescope available. NASA therefore decided to eliminate the ground-servicing concept from the programme, and instead opted for on-orbit servicing based on a three-year cycle of missions. The *Challenger* accident contributed to delays, and following the launch of the telescope in 1990 the original plan would have the first servicing mission in 1993, the second in 1996, the third in 1999, the fourth in 2003, and retrieval and the end of the programme around 2005/06. This development of HST EVA tasks continued up to and beyond the deployment of the telescope in orbit during mission STS-32 in April 1990. Changes to the flight manifest due to on-orbit operations of the HST, difficulties in maintaining the Shuttle fleet, various other adverse factors (such as the problems with the wiring and fuel line in 1999, and the *Columbia* accident in 2003), and debates on the future of the HST, have, over the years, led to amendments to this plan.

During maintenance missions, the HST in held in place by a flight support structure consisting of a horseshoe-shaped cradle with a supporting latch beam, a pivot arm, and a rotating and tilting platform. There is also an orbital replacement unit carrier – a modified Spacelab pallet structure with fitted shelving and containers to hold orbital replacement units for installation into the telescope or return to Earth. The facility also holds several closed-door compartments, a system of tethers, and crew aids for use during the maintenance flights.

Early in the design of the HST, the selection of orbital replacement units was critical to the planning of flights and servicing missions within the STS programme. By selecting modular components for critical sub-systems, it could be determined which might need to be replaced, due to degradation, during the lifetime of the HST. These components were designed as complete units, and featured a self-contained box with simple connections to allow relatively straightforward replacement. In all there were seventy orbital replacement units consisting of about twenty-six components ranging from small fuse plugs to the telephone booth-sized Faint Object Camera. Several orbital replacement unit configurations were available, depending on mission requirements and degradation of equipment.

In addition to the RMS, EMU and STS EVA tool-kit, HST servicing generated its own unique tools and equipment for moving around and working on the telescope. Common items were also evaluated for use not only on the HST but also for use on other Shuttle missions, the planned Orbital Manoeuvring Vehicle (which was later cancelled), and the ISS. The HST was fitted with more than 68.5 m of EVA hand-rails (painted bright yellow), as well as guide-rails, trunnion bars and scuff-

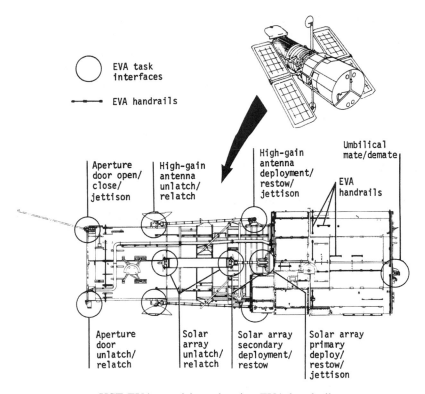

○ EVA task
  interfaces

•—■—• EVA handrails

Aperture
door open/
close/
jettison

High-gain
antenna
unlatch/
relatch

High-gain
antenna
deployment/
restow/
jettison

Umbilical
mate/demate

EVA
handrails

Aperture
door
unlatch/
relatch

Solar
array
unlatch/
relatch

Solar array
secondary
deployment/
restow

Solar array
primary
deploy/
restow/
jettison

HST EVA provision, showing EVA handrails.

plates at the fore and aft of the telescope. Thirty-one foot-restraint sockets were located across the structure, and several portable hand-holds and foot-restraints were provided for temporary use. The antennae and solar arrays could be cranked open and closed either manually or with a power wrench, and a jettison handle was attached to the socket on the aperture door and solar arrays. Portable lights and other items of equipment were also set up around the structure. All of this equipment was provided to ensure efficient servicing and maintenance.

No replacement units were carried on the deployment mission, and none were planned for the retrieval mission. The other HST EVA tasks included the demating and remating of umbilicals, the deployment of the solar arrays (latch, primary and secondary), the stowage of the solar arrays (secondary, primary and latch), the capacity to jettison the solar arrays, the unlatching, latching or jettisoning of the aperture door, and the latching, unlatching, deployment, stowage or jettisoning of the two high-gain antennae.

A significant amount of training was accomplished in the WETF throughout the ten years leading up to the deployment of the HST. Long before crews were assigned to the flights, simulations of deployment, retrieval, servicing and contingency tasks were performed by test divers, former astronauts and active astronauts. This programme of training created an essential data resource for mission-specific training once flight crews had been assigned. It became clear that assignment from a previous HST mission to another servicing mission would be beneficial to the training and performance of the flight crew, and this procedure was adopted during the programme of HST servicing missions.

The HST Shuttle missions
At the time of writing (2003), five Shuttle missions have been directly associated with HST operations, with one or two planned. In addition, several support missions have evaluated procedures and equipment to be utilised on HST-related flights.

HST deployment mission (STS-31)
The long-delayed HST deployment mission was flown in April 1990. EVA astronauts Bruce McCandless and Kathy Sullivan had trained for a number of contingency EVAs, and in addition to nominal Shuttle contingency EVAs their training included manual deployment of HST appendages (solar arrays, high gain antenne and aperture door), unscheduled maintenance and the restowing or jettisoning of deployable appendages should the HST need to be returned to Earth on the same mission; unscheduled berthing and maintenance of a retrieved HST; disconnection of the RMS end effector from the HST grapple fixture; manual HST umbilical disconnect/reconnect; and thermal blanket Velcro reattachment.

During deployment on the RMS the spacecraft became disconnected from the orbiter's power supply, and so full deployment had to be completed before the onboard batteries drained (within eight hours). As long as the solar arrays deployed, then the spacecraft could power itself; but if not, McCandless and Sullivan would have to proceed with EVA to attach an umbilical to provide electrical power. McCandless had been involved in the development of EVA procedures for the HST since the 1970s, and was entirely familiar with the requirements. As the deployment sequence proceeded, the two astronauts prepared for EVA in the mid-deck and air-lock, ready to step outside at short notice. However, the solar arrays were deployed after the overriding of a safety feature that monitored tension in the control wires. After successful deployment by the RMS, the Shuttle remained on station in case other problems should occur during the deployment and orbital set-up sequence. Upon confirmation from Goddard Spaceflight Center, the telescope was operating normally on internal power, and was receiving signals. The Shuttle then backed away to complete its mission, and the EVA crew doffed their equipment. The EVAs – for which there had been so much training and preparation – were not required.

HST preparation EVAs
The discovery that the primary mirror of the HST was at fault, and required corrective optics, is well documented. The delivery of these corrective optics was assigned to the first serving mission, and as part of the preparations for this mission (STS-61) two Shuttle flights were assigned to test planned EVA procedures at the HST under the Detailed Test Objectives programme. STS-57 (June 1993) was assigned both Detailed Test Objective (DTO) 1210 and DTO 671 (EVA Hardware for Future Scheduled EVA Missions). These encompassed testing

of equipment and procedures for both HST Service Mission 01 (SM-01) and the ISS. The astronauts took turns to ride the RMS around the payload bay, judged their ability to move large masses, and evaluated tools, foot-restraints and tethers. They also found that their hands became very cold when they moved away from the payload bay and out into space at the end of the RMS, and in their post-flight reports they stated that their hands had become numb and painful, and that they had shivered.

A second HST simulation was performed on STS-51 during September 1993. The astronauts evaluated tools stored in the provisional stowage assembly, conducted glove-warming experiments by using the lights in the payload bay, tested a portable foot-restraint developed for HST SM-01, evaluated their previous WETF training and work in the altitude chamber during in-flight activities, and evaluated high- and low-torque tether restraint. This continued work with DTO 1210 and 671 produced further confidence in planning the forthcoming HST servicing mission three months later. Further evaluations of the HST were performed on the STS-95 mission in 1998. No EVAs were planned for this mission, which included the Hubble Space Telescope Orbital Systems Test (HOST) package to support the third service mission. HOST carried the Near-Infrared Camera and Multi-Object Spectrometer (NICMOS) cooling system, the HST 486 computer, optical fibre samples, and a semiconductor data recorder to test and verify new components and technology to be used on future HST servicing missions.

Service mission 1 (STS-61)

In December 1993 this mission restored the optics on the HST by installing the Corrective Optics Space Telescope Axial Replacement (COSTAR) device. Five EVAs were completed over a period of five days. The HST was grappled by the RMS on 3 December, and on 4 December, during the first EVA, the astronauts prepared the work-site and changed the gyroscopes, the electronic control unit and the fuse plugs. During the following day the second EVA focused on the replacement of the solar arrays. The third EVA involved the replacement of the Wide Field Planetary Camera with an improved unit, the installation of more fuse plugs, and the advance completion of some tasks originally planned for the next EVA. During the fourth EVA, the Goddard High Speed Photometer was replaced with the COSTAR; and during the final EVA the Solar Array Drive Electronics (SADE) assembly was replaced, and equipment was stowed in the payload bay for the return to Earth. The telescope was redeployed into orbit on 9 December.

Service mission 2 (STS-82)

Just over three years later, in February 1997, the HST was back in the payload bay of the Shuttle for another series of five service EVAs, to be carried out by a different crew. During the first EVA, the Faint Object Spectrometer was replaced by the Near-Infrared Camera and Multi-Object Spectrometer (NICMOS), and the God-dard High-Resolution Spectrometer was replaced by the Space Telescope Imaging Spectrograph (STIS). During the second EVA, the crew replaced the worn-out fine guidance sensor with a modified version, installed an optical control electronics

HST EVA operations during one of the service missions.

enhancement kit that upgraded the original 1970s unit with a state-of-the-art 1990s version that could hold ten times the amount of data, and replaced a failed engineering and science data tape-recorder. The third EVA was used to replace a second tape-recorder unit, a data interface unit, and a reaction wheel assembly unit; the fourth EVA featured the replacement of the SADE and magnetometer covers and the installation of new thermal blankets; and during the final EVA, more patches of thermal blankets were added to the telescope structure.

Service mission 3A (STS-103)

The third servicing mission was planned for June 2000, but after three of the six gyros onboard the telescope failed the mission was split into two flights. Mission 3A was advanced to 14 October 1999, and the second mission, 3B, was scheduled for 2001. However, because of the commitment to wiring repairs on the Shuttle fleet, the mission slipped to 28 October and then 19 November. The HST requires at least three gyros to be functional and to enable precision pointing at targets, and after the failure of a fourth gyro on 13 November it was placed in 'safe mode', which constantly points the solar arrays at the Sun. After nine delays and launch scrubs, the mission was finally launched on 19 December, and a few days later the telescope was back in a Shuttle payload bay for the fourth time. This mission also featured only the third American manned mission to fly over the Christmas period since

The development of skills and EVA techniques, and years of advanced planning for HST servicing, provided sufficient confidence to allow the EVA astronaut to carefully enter the inside of the telescope during replacement of key instruments, and so expand the involvement in performing delicate and critical tasks to ensure mission success.

Skylab 4 in 1973 (which also remained in space over New Year 1974) and Apollo 8 in December 1968. Astronaut John Blaha (1996) and Dave Wolf (1997) had also been onboard Mir during Christmas and New Year. Three EVAs were planned. The first included the replacement of three rate sensor units, each containing two gyroscopes, and the installation of voltage/temperature improvement kits between the solar panels and the ten-year-old batteries. These kits were to prevent the overheating or overcharging of the batteries, and each was the size of a mobile telephone. On the second EVA the astronauts installed a new advanced computer, twenty times faster than the original unit, and a 250-kg fine guidance sensor; and the final EVA focused upon the installion of a transmitter for sending scientific data from the HST to ground receiving stations, to replace the transmitter that had failed in 1998, and a solid-state digital recorder, to replace the older, mechanical reel-to-reel machine. The HST was released on Christmas Day 1999. Due to the inability of the Shuttle's onboard computers to change from 31 December to 1 January, NASA did not want to risk additional complications by flying it into the new year, and it therefore had to be back on the ground by 31 December.

Service mission STS-3B (STS-109)
This was the second half of the third service mission, and was eventually flown in March 2002. Five EVAs were accomplished on this mission. The first EVA replaced one of the two second-generation solar arrays with a third-generation array, installed a new diode box assembly, and prepared the work area for subsequent activities on the mission. The following day, the second EVA completed the work begun during the first day by replacing the second third-generation array and its diode box assembly and replacing reaction wheel assembly 1. The third EVA was utilised for replacement of the power control unit in Bay 4 and inspection of the exterior hand-rails to be used in the next EVA. The fourth EVA was utilised to replace the Faint Object Camera with the new Advanced Camera for Surveys, to install the electronics support module in the vehicle's aft shroud, and to complete the remaining power control unit clean-up tasks. During the final EVA the astronauts installed a NICMOS cryogenic cooler and its cooling system radiator. With the redeployment into orbit at the end of the mission, plans were underway for a fourth and final service mission some time between 2004 and 2006, before a final close-out mission around 2010, thus ending a twenty-year operational career. However, the loss of *Columbia* in February 2003 raised questions concerning the future of the HST servicing missions, and the fate of the telescope itself continues to be debated.

HST EVA experiences and observations
The complete story of the HST service EVA programme is beyond the scope of this book. However, the benefits can be summarised as follows.

The human element The use of experienced astronauts and the recycling of former HST EVA crew-members to later missions helped in training for the missions, as the experiences in orbit often differed from those encountered during training. Familiarity with the HST and the procedures for a servicing mission helped in a smooth transition from one crew to the next, and the assignment of experienced EVA astronauts in support roles also helped in the liaison of flight crews and ground crews. On STS-61, for example, Greg Harbaugh worked as back-up EVA astronaut for the mission and also as EVA CapCom. The effectiveness of combined crew efforts was clearly demonstrated on HST flights. When the EVA astronaut was riding the arm of the RMS, the operators could anticipate what was required, and could execute manoeuvres with very little communication. The HST EVA crews also pioneered a system of multiple EVAs and rotation of work by teams of astronauts. Each team would in turn provide EVA or IVA support on one day, work as support or EVA crew on the second day, and then provide EVA or IVA support on the third day. This not only extended the EVA capability, but also allowed one crew to rest whilst the other worked, and confined EVAs to a batch of days during a limited-duration Shuttle mission. This philosophy was adopted for early Shuttle construction missions at the ISS.

Timelines On earlier missions (for example, STS-41G and STS-51A), the crews suggested that air-lock and hatch operations would be easier if the tools were to be stored outside in the payload bay. Jeff Hoffman, however, disagrees with this idea.

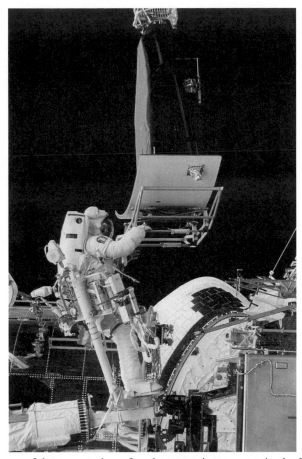

The careful manoeuvring of replacement instruments in the HST.

On the HST missions he preferred the tools to be installed inside the mid-deck so that they could be attached to the suits *prior* to entry into the air-lock. This could save as much as an hour of EVA time, which could then be spent working rather than tooling up. Hoffman has often stated that time is one of the most critical consumables often overlooked on EVA, and that if a crew can prepare more effectively before commitment to the air-lock or EVA, then the time outside can be occupied much more productively.[14] Adherence to the timeline was important for each EVA, due to the defined limit of consumables that could be used on each EVA and on the mission, and in protecting contingency EVA margins. Often, when the astronauts were ahead of the timeline, an unexpected problem or event would slow the work, and they would fall behind schedule; but the ability to perform tasks in advance would produce extra time for the next EVA, even if problems arose. Sometimes, however, it was necessary to postpone a task until the next EVA or even the next service mission – as happened when the HST encountered thermal extremes which warped its door and twisted its solar arrays.

Orbital wear and tear The availability of return visits to the HST over several years offered the opportunity to perform surveys of the deterioration of the facility during EVA, and photodocumentation by the orbiter's crew. Craters caused by orbital debris impacting across the HST were observed and reported, and sizeable hits were recorded on the aft shroud, the antenna, the solar arrays and the thermal blankets. Cracks and tears in the thermal blankets resulted in the replacement or repair of some of the damage on subsequent visits. Most of the damage was located on the sunward side of the structure. Combined with the data from Skylab, Solar Max, the return of the Long Duration Exposure Facility satellite, the repair and recovery of comsats, the Shuttle flights to Mir, and other Shuttle flights, this provided further information on the wear and tear that could be expected on the ISS.

Orbital replacement One of the most important decisions and investments in the repair and servicing of the HST was the development of manual replacement units. These units were to provide the capability to extend, upgrade and improve the telescope's mission, and clearly demonstrated an improved programme of orbital servicing and repair that was one of the key objectives of the Shuttle in the 1970s. It was first demonstrated on the Solar Max repair mission in 1984.

Summary
In little more than eight years, four HST service missions ecnompassed eighteen EVAs totalling more than 129 hours by fourteen different astronauts working in eight teams of two. The HST was thus returned to operational status, and its research capabilities were expanded and improved. In addition, the ability of the astronauts to work in confined places, with an array of tools and aids and with very small items of hardware, demonstrated the development, confidence and ability to plan, train for and execute complex tasks in back-to-back EVAs over several days. It also reveals the extent of advancement in the reliability and dexterity of pressure garments since the early days of EVA in 1965.

SHUTTLE EVAS: AN APPRAISAL

The ISS-related Shuttle EVAs are summarised in the next chapter, but it will not be possible to present a complete review of Shuttle EVA operations until after the programme has ended. It was early projected that the Shuttle would support space station construction and operations; but before firm commitment to such a programme it was used for a series of satellite service, retrieval, repair and recovery tasks. The programme of two demonstration EVAs on STS-6 and STS-41B demonstrated skills, techniques, procedures and hardware that were subsequently utilised over the next two decades to support the development of space station tasks, the repair of Solar Max, the retrieval of Palapa and Westar, the repair and relaunch of Leasat/Syncom and Intelsat, and to extend the orbital lifetime and scientific research objectives of the HST. Between April 1983 and December 1997, forty-one Shuttle-based EVAs were completed, of which twenty-nine were associated with satellite servicing, repair or retrieval operations. The Shuttle has had its fair share of critics and opponents, and there have been difficulties, erroneous decisions and setbacks; but there have also been remarkable successes, many achievements, and numerous records surpassed. When the Shuttle programme is finally completed, and when the records and history books are updated, one of the most outstanding success stories from the programme must be the capability to perform a variety of EVA tasks in the payload bay of the orbiter. Shuttle-based EVAs have led to the successful deployment of several payloads which have achieved their mission objectives, and since 1998 have provided the platform to create the ISS and to begin a new era of EVA techniques: the creation of orbital complexes.

REFERENCES

1 Band, Daniel, *Space Shuttle EVA Opportunities*, NASA JSC-11391, mid-1970s.
2 Shayler, David J., *Apollo: The Lost and Forgotten Missions*, Springer–Praxis, 2002, pp. 294–295 and 299–300.
3 Private correspondence from Jack Lousma, 15 September 2003.
4 Private correspondence from Bob Crippen, 20 August 2003.
5 STS-2 Press Kit, September 1981, p. 3
6 STS EVA Report No. 1, STS-5, Astro Info Service Publications, January 1984; STS-5 EVA Mobility Unit Anomaly, JSC-18698, NASA JSC, January 1983.
7 Private correspondence from Paul Weitz, 28 August 2003.
8 STS EVA Report No. 2, STS-6, Astro Info Service Publications, April 1984.
9 STS-41C Flight Crew Report, CB Memo, 24 May 1984.
10 STS-41G Flight Crew Report, CB Memo, 20 December 1984; Astro Info Service interviews with Dave Leestma and Kathy Sullivan.
11 STS-51A Flight Crew Report, CB Memo, 28 March 1985; Astro Info Service interviews with Joe Allen.
12 STS-51D Technical Crew Debrief, NASA JSC-20536, May 1985; Astro Info Service interviews with Jeff Hoffman.
13 *Hubble Space Telescope, Media Reference Guide*, Lockheed Missile and Space Company, *c*.1990.
14 Astro Info Service interviews and correspondence with Jeff Hoffman.

Space complexes

The orbital operations of Mir (1986–2001) and the International Space Station (from 1998) created a new phase of EVA operations which became an integral element in the expansion and operational support of both stations as long-term orbital bases.

Huge space stations were in the minds of space planners long before man actually ventured into space, but the Space Age was twenty-five years old before construction of the first large-scale space platform began. After years of study and flight experience gained from the Salyut series, the new Soviet space station core module called Mir (based on Salyut) was launched in February 1986. Over the next decade, additional modules were launched to expand the complex. Although plagued by difficulties and setbacks, the station remained operational for fifteen years, and was constantly manned for ten years. The Mir EVA programme benefited from experience gained during the Salyut programme, and itself became a valuable source of experience in operating a long-term space complex in preparation for the next: the International Space Station.

The Americans also had plans to orbit a huge space station, but lacked the finance to do so. Studies of Saturn V-based space stations, with crews of twelve, fifty, a hundred, or even two hundred, featured in most post-Apollo plans in the late 1960s and early 1970s, but Presidential approval for a space station – called Freedom – was not forthcoming until 1984. Over the next decade, countless changes in the design and configuration sent the project massively over budget, and almost led to its cancellation. But in 1993 – after yet another redesign by an international partnership including a 'new Russia' – the International Space Station emerged from the mess that was Freedom. It would be another five years before the first part of the ISS reached orbit, and over the next four years the vehicle gradually increased in complexity and size to allow permanent occupation, if not full scientific activity. EVA operations at this evolving space station is an integral and demanding part of the programme, and the success of the EVA programme smoothed the creation of the station to a point in 2002 when the configuration was close to core completion. Delivery of the remaining solar arrays and scientific modules was planned for between 2003 and 2006, but the loss of *Columbia* in February 2003 seriously delayed the completion of construction, grounded the Shuttle fleet, and threatened abandonment and cancellation of the whole programme. At the time of writing (2003) it is too early to fully review the construction phase of the ISS, as there still remains much to do. This section therefore briefly reveals the evolution of ISS EVA operations from the days of Freedom to the end of 2002, and reviews what remains to be achieved.

THE MIR PROGRAMME, 1986–2001

The core module (base block) of Mir, launched in 1986, featured a node with five docking ports. The forward port could receive visiting spacecraft, or new modules that could be relocated to one of the peripheral docking ports as required. The rear port could accept a visiting crew ferry spacecraft or small module, and also included the connections required to resupply the station via the Progress freighters. The first add-on module – the Kvant astrophysical module – was docked to the aft port on 11 April 1987, after an unplanned EVA to free an item of debris that had prevented the planned docking on 5 April. The next module – Kvant 2 – was permanently docked to the station on 8 December 1989. This was an expansion module that also included facilities for EVAs out of an enlarged EVA hatch. The scientific and technological module Kristall, which was permanently docked on 17 July 1995, also included an androgynous docking port to accept the Soviet Buran shuttle, although it was never used for this purpose. Instead, an American docking module, designed to accept American Shuttle dockings, was attached on 15 November 1995. The Spektr module was permanently attached on 2 June 1995 and finally the Priroda module on 27 April 1996, completing the main configuration that would not radically change for the rest of the orbital life of the station. Mir was planned to be a five-year operational mission, and was finally deorbited on 23 March 2001, fifteen years after launch of its core module. It was a remarkable achievement.

EVA from Mir, 1987–2000
In just over eight years, between December 1977 and May 1986, the Soviet Union completed sixteen separate EVA operations at the Salyut 6 and Salyut 7 space stations. But the EVA operations at Mir would dwarf all previous operations. Between April 1987 and May 2000, Russian and international cosmonauts would conduct seventy-two EVAs and seven IVAs from Mir, and a further two from a docked American Shuttle: a total of eighty-one EVA/IVA operations in thirteen years.

Exiting Mir
The first EVAs from Mir were conducted in 1987 – not from an EVA hatch as on Salyut, but through an unoccupied peripheral docking port on the forward node. It was not until Kvant 2 arrived that a dedicated EVA facility became available for frequent use for EVA operations at Mir, although it was not without its problems.

During an interview at Star City in June 2003, Mir cosmonaut Yuri Usachev explained the process of preparing for an EVA from Mir.[1] Updates to the EVA timeline were received from the ground about a week prior to the planned EVA, and most of that week was spent in preparing the pressure garments, EVA equipment and air-lock area in the Kvant 2 module. The cosmonauts reviewed the EVA plan and practiced elements of the activities planned, often while wearing unpressurised suits.

EVA procedures differed if there was only a two-person crew (in which case both would complete the EVA) or if there was a three-person crew on board. For two-person EVAs, the cosmonauts would prepare the Mir in the 'vacated station mode', with experiments and systems turned off and items stowed. Monitoring of station systems was therefore only possible from the ground. If there were more than two cosmonauts onboard Mir, then the station continued full operation, with the third cosmonaut acting as a 'caretaker' and EVA observer from the various modules of Mir while the other two cosmonauts were outside.

The sequence of exit began with the cosmonauts dressed in their biomedical underwater and cooling suits, floating into the air-lock area and sealing the inner hatches. After a final check of equipment, they helped each other don their Orlan suits, closed and sealed the rear hatch door, and began to pressurise the suits and lower the internal atmosphere of the air-lock. At this time the crew prepared their equipment close to the hatch that they would use for their EVA, arranged as close to sequence of use as possible. One cosmonaut would control the suit and air-lock pressurisation levels, while the second attended to the unlocking and opening of the hatch once it was safe to do so.

The node of Mir on the 1-g mock-up at TsPK. (Astro Info Service collection.)

During his residence on Mir in 1996, Usachev and his Commander Yuri Malenchenko were joined by American astronaut Shannon Lucid, who remained inside the station to monitor the EVAs. Lucid stated that when considering the amount of EVA preparation and post-EVA operations, she had not thought it possible for the same crew to conduct EVAs every four days, but she realised that in zero g this work was alleviated by the reduction in gravity, as well as careful preparation, planning and experience. Preparation and exit from Mir occupied, on average, an hour to 90 minutes, but in an emergency situation it was possible to repressurise the air-lock in no more than seven minutes. Usachev found that an increased calorie intake prior to EVAs helped maintain his condition during the tiring EVAs, during which his body liquid loss was, on average, 2–3 kg per EVA.

Once outside, the two cosmonauts progressed through their assigned timeline. For the 'two Yuris', the EVAs conducted during 1996 were their first, and no matter how much they had trained there was still the natural human concern about trusting the technology. They had their short restraint tethers and manual hooks, which they unclipped alternately to secure them to the next location and gradually move across the station, and they also had a 100-m tether attached by bolts to the suit or spacecraft structure. But both of them wondered what would happen should this become separated so that they floated off into space. They pondered on whether they

EVA handrails on the Mir 1-g mock-up at TsPK. (Astro Info Service collection.)

could reattach the separated bolted end by throwing it towards Mir 'lasso style' to determine whether it would snag some part of Mir; but fortunately this was not necessary, as their EVA progressed normally and their confidence in the system grew. They did, however, notice the strange phenomenon of loose particles in space, which, when touched or pushed, would float away but slowly return.

At the end of the EVA, the first cosmonaut out was normally the first to return to the air-lock. After re-entering the main part of the spacecraft, the first tasks were refreshment and a good wash, followed by the chore of cleaning and storing the suits and equipment and preparing for the next EVA. Personal hygiene and cleanliness, and careful disinfection of soiled areas on the suits, prevented contamination and fungal growth which might be cause by the humid conditions onboard the station. To prevent musty smells, the insides of the suits were thoroughly dried and cleaned, as were the elements of the feed water tank, the sublimator and the suit ventilation systems.

The Mir experience
EVA operations at Mir can be grouped in several categories, all of which have provided important lessons in the long-term operation of crews outside space stations. These factors will also directly influence the development of the ISS and the long-duration space platforms of the future.[2]

Contingency operations
During 1986 the Mir core module was manned twice by the same crew – Leonid Kizim and Vladimir Solovyov – who also transferred across to Salyut 7 between their visits to Mir to complete the final Salyut EVAs and retrieve the remaining experiments. No EVAs were performed on their two short visits, and they spent fifty days onboard Mir followed by fifty days on Salyut and another twenty days on Mir before returning to Earth and leaving the new station vacant again.

The launch of the core module fell behind schedule, and as a result it was decided to complete some of the pre-launch electrical tests at the launch site instead of at the factory, thus saving several months in pre-launch preparations. At the same time, the add-on scientific modules were also running behind schedule, but since it was far cheaper to 'store' the Mir core in orbit than to keep it on the ground it was decided to combine a planned launch of a final crew to Salyut 7 with a checkout of the new station, to completing a series of advance tasks to set up the new station. As a result, the first Mir-designated crew – Romanenko and Laveikin – did not arrive on the station until February 1987, a year after it had been launched. By early April they were ready to receive the Kvant astrophysics module, which would dock automatically with the rear port.

For safety, the cosmonauts moved to the Soyuz TM Descent Module in case the unmanned Kvant should strike the station and they were forced to make an emergency undocking and landing. It was a wise decision, because when the Kvant attempted to dock with Mir on 5 April, the rendezvous and docking system failed. It flew past Mir at a distance of just 120 m – much to the surprise of the two cosmonauts. Four days later, during a second attempt, soft docking was achieved without incident, but hard docking failed. For some reason, the docking probe retraction system failed to engage the capsule latches to pull Kvant into the docking ring of Mir's aft port. Looking out of the viewing ports of Mir, the cosmonauts were unable to find anything obviously wrong.

On the ground, a contingency EVA plan was developed to allow the cosmonauts to go outside and investigate the problem. This was only the nineteenth EVA of the Soviet programme and, as with Salyut 6, the first Mir EVA was linked to securing the future operational ability of the station. Ironically, Romanenko was involved in both events. On 11 April the two cosmonauts prepared to make the first exit from Mir, but with the Kvant 2 module and its specialised EVA hatch still eighteen months away from launch, and as the rear port was blocked by the soft-docked Kvant module, the EVAs had to be conducted out of one of the four berthing ports in the forward node.

The cosmonauts opened one of the ports and proceeded along the 13-m length of the core module towards the Kvant. To assist in their operation, another pair of cosmonauts was stationed in the Hydro Laboratory at Star City, to duplicate their colleagues' movements in space. As they reached the working area, the Orlan suit worn by Laveikin registered a momentary pressure drop caused by an incorrect switch setting. For a moment this caused some concern; but the problem was overcome, and the EVA proceeded. Ground controllers extended the docking probe on the Kvant to provide the cosmonauts with more room to examine the problem. When they peered between the two spacecraft they found an 'extraneous white object', jammed inside the docking mechanism, that was preventing the clean docking. Laveikin had some difficulty in removing the debris, but he finally managed to extract the object, which was later identified as a twisted piece of white cloth (earlier reports had indicated that this was an unused toilet bag) from the trash loaded onto the Progress M-28 freighter that had undocked on 26 March. The debris was cast off into space by the cosmonauts, who then moved away from the docking area while flight controllers commanded the Kvant to complete the docking with the station. There were no further problems, and so the two cosmonauts returned to the Mir node – their task accomplished and the docking completed. Kvant remained at this port for the rest of the operational lifetime of Mir.

During an interview in July 1989,[3] Romanenko stated that conducting EVAs from the node was a late decision in the mission planning for their flight (probably due to the delay in launching the dedicated Kvant 2 air-lock module. He stated that the cosmonauts found working on that area – especially the exit and entry – very difficult and restricting. The work at the rear of the station was much easier than expected, though cramped and confined, with Romanenko having to look over Laveikin's shoulder during the operation. Of his experience of conducting an EVA for which he had not previously trained on the ground, Romanenko stated: 'In all missions there are difficulties and some element of risk, but with adequate generic and on-orbit training and the constant support of ground control, these can be overcome.'

The next contingency EVA repairs at Mir, in 1990, took place not on the station itself but on the Soyuz ferry craft that had delivered Anatoli Solovyov and

Aleksandr Balandin in February of that year. It had been noticed early in the flight to the station that some thermal blankets had become detached from the DM of the Soyuz TM-9, and it was later discovered that these were three of six attached to the Soyuz DM. Since this could seriously affect the thermal levels of the spacecraft and internal systems during its dormant period docked to Mir, the station was orientated to keep the Soyuz out of sunlight as much as possible, while plans were made for an EVA to inspect and possibly repair the panels before the ordering of a replacement unmanned Soyuz to replace the damaged Soyuz. Without the thermal blanket, internal temperatures fell, and condensation formed on the internal electronics. The extremes of temperature also threatened the heat shield and pyrotechnics, and the loose blankets obscured the orientation sensors required for re-entry.

This team had not conducted extensive EVA training, as there were no planned exits during this residency; and since neither had performed an EVA before, they reviewed video film of colleagues practicing the planned EVA in the Hydro Laboratory at Star City. To aid in their preparations, the TM was moved from the rear to the front docking port to ease their access during the EVA. The crew's inexperience of EVA operations became evident when they opened the outer hatch. Instead of opening a 1–2-mm gap to allow residual air to escape while the hatch was held by retaining hooks, they turned the hand-wheel too far before full depressurisation, causing it to release the hooks early and allowing the hatch to swing back, with considerable force, against its hinges.

Due to their inexperience, translation across the Kvant and Mir took longer than planned; and when they arrived at the Soyuz they found that some of the blankets had shrunk, and so it was not possible to simply refit them. Using specially developed ladders to reach over the Soyuz OM to the DM (where again there was no EVA hand-rail provision), the cosmonauts found no obvious damage to the vehicle. However, they were unable to reach all the blankets. They folded and secured two of the blankets, but the third proved most 'disobedient', and had to be pinned back. They then retreated to the air-lock to complete their residence. In August they completed a successful re-entry and landing, without further problems concerning the integrity of the Soyuz TM.[4]

Space construction work

Mir was planned for expansion on orbit with additional modules that could be automatically docked and manoeuvred to their final positions. But it could also be expanded by EVA in a series of space construction activities. The first of these was to install additional solar arrays to increase available electrical power. Due to the need to save launch weight, the base block was launched with only two solar arrays attached, and the first resident crew – Romanenko and Laveikin – was therefore assigned the task of installing the third array during two EVAs in June 1987. The arrays were stored in the recently docked Kvant astrophysical module, and the crew spent several days preparing for the spacewalk.

First, the hinged lattice girder structure was deployed to support the folded solar arrays, which were attached during the second EVA. The cosmonauts found that they needed to use the additional volume of the OM of the Soyuz TM-2 to provide them with the leverage to manoeuvre the solar array packages out of the station through the tight confines of the Mir node. In order to save time and additional EVAs, the crew used more than one hatch in the node to manoeuvre the packages and themselves out of the station. Had they not done so, Romanenko explained, 'the number of EVAs required to deploy the solar panel could have clearly doubled [to four], which would not have been productive.' The operation had taken the two cosmonauts about five hours to complete in two EVAs. The next resident crew – Vladimir Titov and Musa Manarov – replaced a section of this solar array with a new section during their first EVA in February 1988. Before doing this, however, they had completed an on-orbit refresher course which included videos of their own underwater simulations.

During their EVA in December 1988, Jean-Loup Chrétein and Alexander Volkov deployed a folded hexagonal structure called the European Robotic Arm (ERA), which was designed to evaluate the technology of using folded deployment systems for large antennae on communication satellites. The 240-kg ERA was initially deployed by command from inside the station, but it refused to deploy correctly. Volkov suggested to ground control that he could kick it, but this ploy was rejected. Instructed to discard the structure if it did not open as planned, the cosmonauts waited until Mir passed out of the range of communications, whereupon Volkov used his EVA boot several times to 'encourage' the ERA to deploy to its full extent. At the end of the experiment, the unit was cast free of Mir.

In January 1991 the resident crew on Mir installed a telescoping boom on a Mir base block launch shroud attachment. This was designed to assist with what transpired to be a long process of relocating the two 500-kg collapsible solar arrays from the Kristall module to their new position on Kvant. This was necessary to maximise their efficiency as new modules were added that would have blocked them from the Sun. The 45-kg boom, called Strela (Arrow), was packaged in 6-m long boxes. When deployed to its maximum length it measured 14 m, and could 'carry' up to 700 kg. Apart from moving the solar arrays, it would also be used by cosmonauts as a mobile hand-rail to and to relocate bulky equipment. One cosmonaut operated a hand-crank at the base of the boom, while the other cosmonaut 'rode' the other end. During a second EVA the supports were readied for the relocation of the solar arrays. The total amount of time spent on the two EVAs was about 12 hours.

Two EVAs (totalling 11 hours) in April and June 1993 were used to install new solar array drives on the Kvant module to assist in the retraction of the arrays. Delays in the second EVA were due to the loss of one of the two Strela boom handles and the resultant wait for its replacement on the next Progress launch. Two years later, the first array was moved during a programme of three EVAs totalling 18 hours. The first Strela could only reach the −Z side of the Mir station, and so to complete the solar array manoeuvres a second boom had to be installed during an EVA in February 1996. The second solar array was eventually relocated, during two EVAs lasting more than 12 hours, in November 1997 – almost seven years after the start of the operation to relocate them.

In July 1991, four EVAs were required to construct a segmented 14.5-m girder called Sofora, which extended from the Kvant module and was designed to support a thruster package to augment the space station's onboard attitude control systems. The series began with the preparation of the Kvant area by attaching ladders for additional hand-holds, a support platform for the structures, and by using the Strela boom to relocate equipment. The construction consisted of twenty segments assembled in batches of three, eleven and six over the following three EVAs. There was also unplanned activity at the end of the final EVA of this series with the installation of a Soviet flag on top of the metal framework. In September 1993 a new team of cosmonauts was assigned to install a propulsion package on top of the Sofora girder. This 700-kg unit was delivered by Progress M-14, and was deployed automatically, by ground control, from a modified tanker section of the freighter, to allow the EVA crew easy access to the unit. It was installed over three EVAs, and the cosmonauts used an inbuilt hinge a third of the way up the length of the Sofora to 'bend' the structure so that it would be easier to relocate the propulsion package and install the cabling along the length of the structure. The USSR had by then been dismantled, and the cosmonauts therefore removed the Soviet flag, which had been torn to shreds as a result of impacts by micrometeoroids and orbital debris. In all, four cosmonauts had completed seven EVAs in two sets totalling 38 hours. In 1998 a resident crew completed a series of five EVAs to detach the old propulsion package from the top of Sofora, casting it off into space and replacing it with a new unit.

A new truss, called Rapana, was installed on a platform behind Sofora. This 26-kg cylindrical framework – thought to be related to the planned Mir 2 space station – had 'memory alloy joints' that expanded when heated, causing the structure to unfold automatically from its storage container. In three minutes it extended to 5 m. The Rapana was dismantled and stowed in April 1998, during a further programme of EVAs at Sofora.

Experiments and evaluations

During the EVA in June 1987, Romanenko and Laveikin evaluated their ability to move around outside the station while using only tethers, and not foot-restraints. This provided them with additional freedom to move as required, but they still had to cling to the spacecraft with one hand. The cosmonauts also continued the practice, begun on Salyut, of retrieving material exposure cassettes installed outside the station. These were installed and retrieved during many of the Mir EVAs, and offered a continuing use of Mir's exterior to evaluate and develop a manually operated exterior science platform, by constantly changing the samples and experiments deployed in accordance with previous results and new developments.

A selection of five technological experiments was installed on the Echantillons space exposure rack that Chrétien attached to Mir's hand-rails. They were retrieved during an EVA by Soviet cosmonauts in January 1990. The TREK cosmic ray collector was installed in June 1991 and retrieved during an EVA in July 1995. Two days later, the Belgian MIRAS spectrometer was installed. This was assembled from two parts, and attached to a boom at the end of the Spektr module. Plans for on-orbit repair to extend its one-year operational life were cancelled when the Russian space agency presented the EVA service charge figure to the Belgian government. In September 1994 an ESA radiation sensor was mounted on the exterior of the Mir base block, and the Travers Synthetic Aperture Radar antenna was manually deployed at Priroda in July 1996. During the final expedition to Mir in 2000, the crew evaluated the Germatisator Sealing experiment to determine its effectiveness in sealing punctured hulls (such as Spektr). Tests of an improved electron beam welding gun, of the type used on Salyut 7 in 1984, were planned for Mir. These tests were assigned to STS-87, but were removed from the mission in favour of NASA ISS EVA tool development tests.

Numerous material exposure experiments were deployed, exposed and returned over varying durations outside of Mir over the years, on experiments from Russia, Europe and America. These experiments included solar array samples, paints, optical glass, lightweight carbon fibres, various electronic components, thermal blanket materials, and a variety of connectors. The Kozma experiment consisted of cassettes exposing metal-foil arrays to the flow of neutral interstellar atoms for several hundred hours at a time. The first background test was undertaken during August–October 1995, and new cassettes, exposed during January–May 1996, were finally retrieved in June 1996.

MMU tests

During February 1990, cosmonaut Aleksandr Serebrov became the first Russian to fly an MMU. In test-flying the Soviet YMK unit, he moved to a distance of 33 m from Mir, but remained tethered to the station. Four days later his Commander, Alexandr Viktorenko, also flew the unit, this time to 200 m, but still remained tethered. During these highly successful test-flights the cosmonauts evaluated the unit, and even demonstrated a 'victory roll'. On the second flight, Viktorenko carried the Spin-6000 cosmic radiation measurement unit, which measured radiation levels in the vicinity of the station. The YMK unit was originally designed for operations at the Buran space shuttle and Mir 2, and was subsequently stored inside the Kvant 2 air-lock until February 1996, when it was

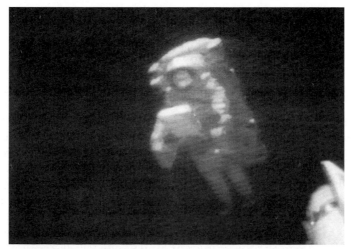

The flight of the Soviet MMU from Mir during February 1990.

taken outside and stored on the exterior support facility to provide extra space inside the air-lock. It was never used again. Like the American MMU, the YMK was a little ahead of its time, and had few useful objectives.

Repair and inspection

Being such a large and long-duration facility, Mir provided the first opportunity for extended inspections and examination, and inspection of the exterior of the station – both in the evaluation of hardware, sub-systems and materials, and in evaluations for impending or previously attempted EVAs – continued throughout the operational lifetime of the complex.

Exterior photography was first conducted by the Okean crew (1988), who also dusted off several portholes with a small hand-held brush. Several inspections and repairs of appendages were included in a number of Mir EVAs, and because of wear and tear, system failures and occasional accidental breakages, new tasks were added to later EVAs. TV surveys of the exterior surfaces of the station were completed under the Panorama experiment, which revealed the condition of various materials and components after exposure to the space environment (solar heating/shading, repeated Earth day/night orbital cycles, micrometeoroid bombardment, man-made damage and wear and tear along repeated EVA traverse paths).

The first 'extensive' repair EVAs took place in June 1988, when the cosmonauts attempted to repair the TTM (*Russian*, Telescope, Shadow and Mask – Coded Mask Telescope) on the Roentgen Observatory located inside Kvant. This was a joint Dutch–British experiment, and on-orbit servicing was not incorporated in the design of the instrument. The crew carried with them a new detector and specialised tools and equipment on portable carriers – amounting to more than 40 kg – across 25 m from one end of the station core module to the other. Despite having no foot-restraints in the area in which they were working, they managed to cut through twenty layers of thermal insulation, cut off bolts and remove restraint clips from the faulty detector. However, there were more clips than expected, and a check of the original drawings and the simulator model again revealed that the flight model differed from the model prepared for EVA simulations.

The cosmonauts were instructed to ignore the extra clips, and they continued their work. Some 25-mm connecters were held down by small screws, which were impossible to remove while wearing an EVA glove; and as this was in an area not planned for EVA servicing, they were told to cut through the screws. The cosmonauts also found resin on some of the screws, which came as a surprise to those who built the instrument and were monitoring the EVA from Mission Control. The resin had apparently been applied on a small repair that had not been recorded during construction. Using a saw, the cosmonauts managed to remove sufficient resin to allow inspection of the inside, and they then began to remove the detector; but the specially designed tool snapped. This was frustrating for both those in space and on the ground, and it ended the repair attempt for that EVA. The crew then temporarily replaced the thermal coverings, and completed other EVAs tasks. A replacement tool had to be added to the inventory for the next Progress launch, and this further delayed the repair. Four months later, however, a second attempt was successful.[5]

When Anatoli Solovyov and Aleksandr Balandin conducted their first EVA to inspect and repair the Soyuz spacecraft in July 1990, they inadvertently damaged the Kvant 2 air-lock hatch by opening it before the pressure had been vacated. Their work-load had also been delayed, so, having been outside for more than five hours, and nearing the six-hour limit of their Orlan DMA suits, they were disturbed to find that the outer air-lock hatch was buckled and would not close. Fortunately, Kvant 2 was designed with the option to use the module's instrument science compartment as a contingency air-lock, leaving the Special Air-lock Compartment exposed to the vacuum of space while options for repair were studied on the ground. At the time, Russian officials stated that this additional air-lock volume would allow items of larger mass to be taken out into space on EVA, although this was never demonstrated during the lifetime of Mir.

The work was not finished at the Soyuz TM site, and a week later a second EVA was conducted to retrieve ladders and tools used to fix the thermal blankets to the spacecraft, which had been left in place to allow return to the air-lock within the suit safety limits. Here the cosmonauts also televised the damage to the hatch by relaying images of damaged hinges. Engineers on the ground devised a means of closing the faulty hatch to test its integrity, and despite difficulty with leverage, the cosmonauts managed to close the hatch. After being pressure-tested for 24 hours, the air-lock compartment showed no signs of leakage through the outer hatch. Confident of its integrity pending more permanent repairs, the cosmonauts opened the interconnecting hatches again and continued their mission. The hatch would be repaired by the next crew.

In October 1990 the new crew – Gennadi Manakov and Gennadi Strekalov – investigated the hinge in more detail, and discovered that it was beyond repair on this mission. However, they installed a special latch to ensure its correct closure and continual use, and in January 1991 the replacement crew – Viktor Afanasyev and Musa Manarov – repaired the hatch by fitting it with a new hinge mechanism. The hinge was not originally designed for in-flight serving, and the repairs took four hours to complete in a very complex and difficult operation. As a final check the hatch was closed and reopened, and the cosmonauts then continued with their other EVA tasks.

The final major EVA repairs at Mir, in 1997, were related to the damaged Spektr module. Internally, cables were rerouted to the module, and a series of EVAs was utilised to locate the source of the leak from the pressure compartment and to reroute external cables connecting the solar arrays on the module, to readjust them and regain as much power as possible.

Internal EVAs

The facility to dock large modules at Mir's front node enabled expansion of the research capability. The new modules arriving at the front port would later be moved, by the Lyappa arm, to a peripheral docking location. The Konus docking drogues therefore had to be relocated from one port to another to allow movement of the modules from the front port to a peripheral port. To accomplish this task, cosmonauts wearing Orlan suits sealed hatches between the modules and the main base block, and depressurised the forward node to relocate the docking equipment. Three such IVAs were completed during 1995, but the first move of Konus docking equipment took place, as part of Viktorenko and Serebrov's second EVA, during the final use of the forward node for exit into space in January 1990, and was not classed as an IVA.

Three other IVAs were connected to the isolation of the Spektr module and the recovery of power systems following the collision of Progress in June 1997. The suits' sublimator heat exchangers could not be used due to the vacuum in the Spektr, and Zvezda therefore instead devised a 10-m hose extension to support the suits' cooling facilities. On the first of these IVAs in August 1997, a new 'hermaplate' sealing hatch was fitted, and some of the severed cables were reconnected. Before sealing off the module, the cosmonauts carried out a brief internal inspection to retrieve personal items and other equipment, and also reported to ground controllers about the condition of the module. During a second internal EVA two months later, rerouted power cables were reconnected; and during a third, in September 1998, connections to the hermaplate were completed. Spektr would never be manned again; but the series of EVAs and IVAs had bypassed the damage caused by the collision, and the rest of Mir could continue to be used.

Another EVA, planned for March 1998, was eventually carried out as an internal EVA at the Kvant 2 hatch, which was still causing problems. A leak was traced to one of ten securing bolts, but during the attempt to repair the bolt the cosmonauts succeeded only in breaking three wrenches. Tougher wrenches arrived on the next Progress, after which the repair was completed.

International EVAs at Mir

A cooperative programme between the French and the Russians resulted in several long-duration residencies on Mir by French spationauts. In 1988, Jean-Loup

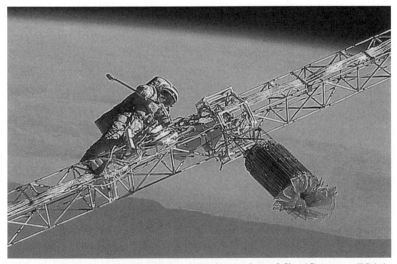

The deployment of the European experiment from Mir. (Courtesy ESA.)

Chrétien became the first person other than a Russian or an American to perform an EVA, and in 1999 Jean-Pierre Haignere completed a second French EVA from the station. German astronaut Thomas Reiter also completed two exits from Mir, in October 1985 and February 1996.

The joint US–Russian Shuttle–Mir project (1995–98) offered the opportunity to send American astronauts on long-duration missions to a space station for the first time since the Skylab missions in the early 1970s. It also facilitated the docking of a Shuttle to a space station, which was originally planned for the (later cancelled) space station Freedom, and was important for the impending International Space Station, in which the Russians would play a large part in construction and operation. This programme also offered the chance for several American space station EVAs and for externally mounted experiments to be flight-tested in preparation for ISS operations. It also provided NASA with comparative data for the series of Shuttle-based EVAs being conducted under a test programme of EVA experiments prior to the commencement of activities at the ISS.

The first American EVA at Mir took place during the STS-76 Shuttle–Mir docking mission in March 1996, when Linda Godwin and Rich Clifford performed an EVA from the docked *Atlantis* to install the Mir Environmental Effects Payload (MEEP) panels on the exterior of the DM (which had been delivered by STS-74 the previous November and was located permanently on the end of the Kristall module). However, the astronauts were unfamiliar with the architecture outside the Kristall module and were not trained for EVAs at the module, and as the Russians wanted to avoid possible damage they were not allowed to transfer across the DM/Kristall threshold. Godwin and Clifford used new tether hooks that could fit over Mir hand-rails on the DM, and foot-restraints that would accept both Orlan and Shuttle EMU EVA boots; but they had to rely on helmet-mounted lights and *Atlantis's* payload-bay lights for illumination on the night-side pass of each orbit, as Mir had little exterior EVA lighting. Because Mir had to be pointed towards the Sun, the astronauts' hands became extremely warm, and they were glad to see the sunset, which would alleviate the heat on the back of their hands. The astronauts were impressed with the size of the facility (the combined mass of *Atlantis* and Mir was 237,494 kg); but the core module had been in space for ten years, and there were signs of orbital deterioration of the thermal blankets and on the surfaces of the station, although there was less wear on the newer elements.

In October 1997, American astronaut Scott Parazynski and Russian cosmonaut Vladimir Titov conducted a second EVA from the docked *Atlantis*, during which they retrieved the MEEP experiment from the exterior of the DM. This was the first that time a Russian (and a Mir EVA veteran) had conducted an EVA from the Shuttle while wearing the EMU, and it was part of the original programme of cooperative Shuttle–Mir flights.

Three EVAs by long-duration Mir resident Americans were also completed as part of the Shuttle–Mir programme. On the first of these, on 29 April 1997, Jerry Linenger became the first American to conduct an EVA in a Russian Orlan suit, from a Russian station with a Russian cosmonaut, Vasily Tsibliyev. During the EVA, the pair evaluated the new Orlan DM spacesuits planned for use on the ISS, and

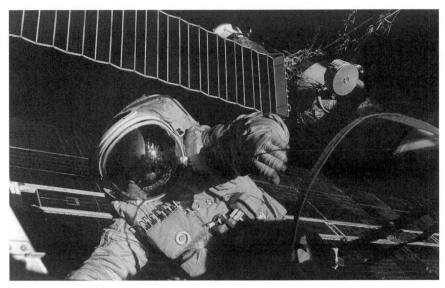

Jerry Linenger performs the first American EVA using the Orlan suit at Mir.

deployed and retrieved experiments. In September 1997, Michael Foale became the second American to conduct an EVA with a Russian cosmonaut, Anatoli Solovyov – this time to rotate the solar array outside Spektr and to inspect for signs of leakage from the damaged pressurised compartment. In January 1998, Solovyov again accompanied an American astronaut – Dave Wolf – on EVA, during which they conducted experiments with a spectroreflectometer supplied by NASA. These cooperative EVAs were, on the whole, successful; but they also highlighted differences in training and operations that would have to be addressed for cooperative EVAs at the ISS.

Summary
The final hatch opening from Mir took place on 12 May 2000, when the Enisei crew – Sergei Zaletin and Aleksandr Kaleri – performed an excursion of 4 hrs 52 min to test the Germatisator Sealing experiment hardware, examine a failed solar battery, and carry out the final Panorama external inspection of the station during their 73-day mission. This was the twenty-eighth resident crew to man the station since 1986, and for ten years, between September 1989 and August 1999, cosmonauts had continuously lived and worked onboard the station. This had a significant effect on the EVAs planned for, and operations conducted on, the station, because tasks that could not be completed by one crew could be assigned to a follow-up crew. It also allowed an increase in on-orbit training for tasks not simulated to a great degree on the ground before the mission.

The Mir EVA programme provided valuable baseline data for operating both prolonged and series EVA operations at an orbital base, and although the Russians had hoped that this experience would be utilised at their Mir 2 complex, it would instead provide a useful reference for EVAs conducted at the ISS – especially from Russian segments – and for repair and servicing activities by resident crews upon the completion of the main construction tasks by visiting Shuttle crews.

The annual EVA programme at Mir was as follows:

Year	EVA	IVA (hrs:min)	Total	
1986	–	–	00:00	Launch of Mir
1987	3	–	08:48	
1988	4	–	19:43	
1989	–	–	–	Temporary vacation of Mir
1990	8	–	32:14	Kvant 2 air-lock available
1991	10	–	52:57	
1992	6	–	24:36	
1993	7	–	24:11	
1994	2	–	11:07	
1995	7	3	39:00	
1996	9	–	43:04	
			06:02	American EVA from docked STS-76 *Atlantis*
1997	4	2	33:03	Collision of Progress
			05:01	American/Russian EVA from docked STS-86 *Atlantis*
1998	8	2	51:43	
1999	3	–	17:48	
2000	1	–	04:52	
2001	–	–	–	Deorbit of Mir
Totals	72	7	374:09	

A great deal was learned from Mir EVA activities that had direct application to operations at the ISS, for planning other long-term space bases and, in time, for EVA operations in deep space. The interaction and participation of American astronauts also revealed more about the physiology and execution of Russian EVA operations at a long-term space platform, compared to the short and intense Shuttle EVAs with which they were accustomed.

Planning EVA planning evolved throughout the Mir programme and included evaluating the tasks to be accomplished, the distances from the air-lock to the work-station, the provision of tools, support aids and other equipment, the prior experience of the cosmonauts undertaking the planned EVAs, a reserve of about 20% safety margin for life support systems, contingency plans, and several emergency scenarios. It was also found that up to 30% of EVA time was wasted in carrying out tasks in support of the EVA: traversing to and from work-sites, establishing work platforms, relocating tools, and laying ladders and hand-rails. Disorientation was also experienced, and the Russians therefore marked EVA paths over the surface of the station to reveal the best and most productive translation route to a particular location, although at least one American astronaut (Linenger) had trouble identifying them.

Training As well as additional training in the Hydro Laboratory at Star City, cosmonauts also increased their level of on-orbit training by using videos of ground simulations, working in the modules to practice specific elements of repair or service, and utilising the support of ground teams during EVAs of a more complex nature. Russian EVA training for long-duration missions is general rather than task-specific, as on the American Shuttle EVAs. This became very clear during the EVAs conducted by the Americans outside Mir, where they had learned how to operate the suit and the equipment to support it, but not the task of the EVA in detail or in pacing their activities. They were almost encouraged to go outside and carry on with the job by themselves.

On-orbit servicing and preparation Preparations for each EVA took a considerable amount of time, and this caused some concern over the ability to prepare equipment for EVA and still continue a viable research programme, especially with only a two-person main crew manning the station. Improvements to the Orlan suits have helped reduce the preparations, but it still takes several days to prepare all the equipment, suits and crews to conduct the EVA, and a post-EVA recovery period is still required. The condition of each flight-suit and the interval between its use on EVA cycles affected the amount of maintenance it required. A new suit delivered to the station, or the reuse of a suit after a long period of inactivity, required more work (around 6.5 hours) than maintenance between a series of EVAs by the same crew (averaging around 1.5–2 hours). As new crews took over the running of Mir, the Orlan suits required adjustment to fit the new occupants. There was also a requirement for a three-hour period of post-EVA maintenance such as the drying and cleaning of the suits, which was usually completed the day after the EVA.

Malfunctions The sheer number of EVAs conducted at Mir over several years inevitably resulted in some malfunctions of hardware such as leaking coolant loops and malfunctioning water-temperature monitoring devices. The results of the long-term use of suits and their prolonged exposure to orbital flight provided designers with valuable information when preparing upgrades for subsequent programmes, and it also presented the opportunity to flight-test components and systems to evaluate the suits' design and performance under real flight conditions. During an EVA by Anatoly Artsebarski in July 1991, his Orlan suit overheated and temporarily blinded him with perspiration, so that he had to be guided back to the hatch area by his colleague, Sergei Krikalev. In February 1992, Alexander Volkov's suit developed problems that forced early termination of the planned EVA; and in September 1993, Vasily Tsibliyev's suit overheated and again shortened the EVA. In July 1995, problems with Anatoli Solovyov's suit prevented him from exiting the air-lock, and he therefore remained in the hatch area to support Nikolai Budarin, who went about completing as many as possible of the tasks originally assigned to both of them.

Equipment On several occasions the cosmonauts' efforts were frustrated by faulty or broken equipment, and the resultant rescheduling of the tasks required a change of plans or a delay to a later EVA, and cost both time and consumables. Following his EVA in April 1997, Linenger commented that during his translation on the Strela boom, it acted like a flexible fishing pole, swinging him back and forth 'like a yo-yo'. He concluded that the Strela was not an effective method for precision positioning.[6] There were also fewer dedicated hand-holds, foot-restraints and tether location points across Mir than on American spacecraft (such as the HST).

EVA routes Linenger also recalled that his impression of the exterior of Mir was that it was a tangle of solar arrays, disused experiments and experiment mounts, unused equipment, booms, antennae and other appendages, and that care had to be taken not to float into them, especially those with razor-sharp edges. This scenario was not duplicated on the underwater mock-ups, and the 'road signs' placed on earlier EVAs were apparently unclear to him.

Ground and flight hardware differences Differences between what was used in training and what was actually included in the flight hardware caused some difficulties and necessitated a number of changes to the EVA timeline. However, these anomalies have arisen several times on American EVA activities over the years.

Timeline There was little flight documentation – such as cuff checklists and EVA planning timelines – to support EVAs at Mir. The donning of suits was summarised on a cardboard flip-up book attached to the inner hull of Kvant 2, but unlike with the American schedules, EVA timelines were certainly not detailed. It was often necessary to use helmet lights during night-time passes to compensate for the time lost as a result of the repeated traverses required to set up work-stations.

Communications Due to the lack of constant orbital communication with Mir there were long periods when it was out of range, during which time the cosmonauts carried on their activities following their own decisions and intuition. When there was a third cosmonaut on board, there was little communication between the IVA and EVA crews (although the Americans on Mir communicated with the ground as they observed and photographed their colleagues on EVA).

Wages and advertisements In 1992, reports were circulated that each EVA from Mir cost, on average, 1 million roubles an hour. (In 2003, 1 million roubles = £13,350.) This encouraged both the cosmonauts and the ground controllers to quickly complete all assigned EVA tasks, despite the increased risks; but it is neither sensible nor safe to cut corners in an attempt to save cash – particularly in spaceflight, where such an unforgiving environment exists, and especially during an EVA, when the crew-members are beyond the protection of the spacecraft. With the change from the Soviet Union to the 'new Russian Federation', a new pay system was introduced. The cosmonauts were employed on a contractual basis, and were awarded bonuses for additional work, including successful completion of EVAs. In 1993, unconfirmed reports claimed that for three EVAs, two cosmonauts could 'earn' up to 1 million roubles. In 1996, in an effort to generate further revenue for the cash-starved Russian space programme, a deal was struck with PepsiCo to video-tape the deployment of a mock-up Pepsi can during an EVA. The 1.2-m replica was delivered by Progress M-31 in May 1996, and the cosmonauts constructed it by using aluminium struts and nylon sheeting. The commercial filming, which took place over two EVAs, showed the mock-up can between the two cosmonauts outside the space station.

Suit disposal In July 1991, the worn-out Orlan DMA no.10 unit, used nine times by different cosmonauts, was taken outside empty and released into space. (*Pravda* suggested that it would have been far better to return the suit and sell it for profit to a museum, at a time when both the space programme and the country was seeking commercial opportunities to fund future survival). Because suits were stored on orbit – sometimes for several years – it was very difficult to examine a worn suit, and assessments relied on in-flight photography and comments by the cosmonauts. Calls for improvements in comfort, the replacement of systems, and the repair of faulty equipment, all had to be evaluated on merit, and suit no.18, after it was finally used in December 1995, was returned to Earth onboard the Shuttle.[7]

In summary, the activities of cosmonauts outside the Mir complex over a thirteen-year period significantly increased the databse of experience. Due to extended evaluation and development, the prolonged use of Orlan suit technology on an orbital space complex was well proven, and upgrades to the system and suits throughout the Mir programme proved sufficient to assign the suits and support facilities to early ISS operations. The reliability and flexibility of the Orlan system complemented the complexity and endurance of the Shuttle EMUs to offer state-of-the-art pressure-suit technology and flight-proven equipment for early operations at the ISS. At the same time, investigations continued into improved methods of supporting prolonged EVA operations at the station and the evaluation of future EVA programmes on other worlds.

EVA AT THE ISS: CONSTRUCTION ON A GRAND SCALE

Studies into the creation of a large space station originated at NASA soon after its foundation in 1958, and during the 1960s and 1970s, dozens of space station designs were proposed, evaluated and rejected. In most of these proposals the space stations were composed of several elements, launched separately and brought together in orbit and attached by teams of EVA construction workers. One of the earliest concepts was that a large space complex should be fabricated and expanded item by item, due mainly to the limitations of the launch vehicles available or on the drawing board.

During the 1970s, design studies of space stations focused on the lift and payload capacity of the Shuttle, and depicted a combination of Shuttle RMS operations and Shuttle-based EVA construction. Some of these structures were huge, and others were smaller, and were intended to be crew-tended rather than permanently occupied. However, difficulties with the development of Shuttle launches, in selling its commercial capabilities, and in promoting its national and international opportunities, also led to increased costs.

During the 1960s, NASA accepted studies of space station hardware based on Apollo lunar technology as a way of expanding the programme beyond that of a few

An early (1970s?) design for a pressurised free-flying space construction facility. (Courtesy Grumman.)

lunar landings. The Apollo/Saturn-based space station programme was promoted outside NASA as an extension of the Apollo lunar mission and an application of previously developed hardware, so as to secure funding to develop the missions. If the space agency had referred to the Apollo Extension Program or the Apollo Applications Program as a 'new' space station programme, the funds would have dried up, and this could well have seriously affected the budget of the primary goal of Apollo: the Moon. Therefore, what was originally promoted as an extension of the lunar programme, and not a full follow-on programme, eventually became Skylab.

In January 1984 President Reagan challenged NASA to create a large space station by 1994. This, however, proved more difficult and more expensive than was at first thought. The story of changing designs, cost over-runs and a floundering programme are well-documented, but in all of these designs, EVA features as a key element in creating and expanding the station, as well as in maintaining its operational readiness throughout its lifetime.

EARLY STUDIES AND SIMULATIONS

With President Reagan's authorisation to develop what was to become the ISS, EVA skills featured high in programme planning. Previously, America's Earth-orbital spacewalking experience had been accrued on five Gemini flights, during one Apollo mission, and by the three Skylab space station crews a decade earlier. Even the Soviets had completed only eight Earth-orbital EVAs since 1965, and it was therefore obvious that experience of extensive, complicated and repetitive EVA construction was required. Even the Space Shuttle – promoted as the base from which this new space station EVA programme would take place – had supported only one three-hour demonstration EVA. It was all very well to generate studies on paper and stunning artwork, but after NASA sold the idea of a space station it had to build it; and to do so, construction-based EVAs had to be a primary objective on several Shuttle flights in order to develop the techniques and to form the cadre of astronauts who were to build the space station and determine the best method of construction. As the ink dried on the Presidential authorisation for the space station, it was clear that there remained much to learn before a Shuttle crew could begin construction; but at least the project had begun.

In 1978 the systems laboratory at the Massachusetts Institute of Technology began researching human factors in space structural assembly techniques. It was clear that in previous studies there was very little emphasis on the productivity of humans carrying out EVA structural assembly, despite the high priority assigned to multiple spacewalking operations in achieving the construction of the proposed structure.[8] Over the next two years, MIT carried out relevant studies, and concluded with a proposal for a programme of neutral buoyancy tests in the large water tank at the MSFC in 1979 and 1980.

Flight Demonstration Programme

In October 1983 the Office of Space Flight at NASA HQ established the Flight Demonstration Program, which would support a series of developments that provided 'young' engineers at NASA with direct experience in the development of flight hardware, the development of business for the STS programme, and the stimulation of customer use of the unique capabilities of the Shuttle. Many (although not all) of these demonstrations were related to EVA, in satellite servicing and resupply, and the construction of large space structures. The first of these was the orbital refuelling programme flown on STS-41B and STS-41G, the second was Experimental Assembly of Structure of EVA (EASE), and the third was Assembly Concept for Construction of Erectable Space Structures (ACCESS). EASE was a flight experiment to validate MIT's research and to provide a quantitative comparison between neutral buoyancy simulations and orbital structure assembly. In November 1985, EASE was combined with ACCESS as a flight demonstration objective on STS-61B.

Underwater evaluations

Between March and September 1979 a programme of seventeen underwater tests was completed in the water tank at MSFC. These tests focused on the potential utilisation of manned EVAs for assembling erectable trusses, which had been evaluated as the best materials to form the basis of a variety of space structures.[9] Additional tests involved the underwater construction of a thirty-six-element tetrahedral truss and antenna, a trapezoidal truss platform, and the evaluation of new training and simulation equipment, including a Personal Underwater

Early underwater tests of space construction techniques.

Manoeuvring Apparatus (PUMA), designed at MIT, to simulate use of the MMU during neutral buoyancy tests. The PUMA was controlled by two hand-controllers (one on each arm of the unit), and was similar to the flight MMU which was then under development. It was designed with all operational controls on one controller, to allow use of a free hand in transporting elements of the hardware from the storage area to the work-site, and also featured four manoeuvring motors at each corner of the unit, for mobility in the water. However, it could not be determined whether productivity or efficiency was enhanced by its use.

The seventeen assembly tests were conducted using a regular six-column tetrahedral cell – the 'fundamental element' of a planned large tetrahedral truss structure, with each strut 5.4 m long. The programme was completed by two spacesuited test subjects working in a mock-up of the Shuttle payload bay. Tests 1–11 consisted of a series of manual assembly modes varying in assembly time from 26 minutes to 87 minutes, depending on the task and the number of procedures required to complete the work. These simulations also included modifications in the procedure, affecting the tasks assigned to the next test run. Manual simulations without modifications were carried out on Tests 12–14, which each lasted around 29–30 minutes. Tests 15–17 were conducted with no modification to procedures, but were RMS-assisted, and lasted for about 15 minutes. Throughout these and other underwater tests, procedures were constantly modified to enhance the performance of the test subjects. The portable foot-restraints and the RMS mock-up proved their value by freeing the subjects' hands for assembly tasks and by reacting to their generated forces as they moved around the pool.

Development of the components of the EASE/ACCESS hardware benefited from these simulations, and the information gleaned proved beneficial for STS-61B. Much experience was gained in securing feet and hands, the dexterity of handling large and small objects, repetitive and complex tasks, and in verification that the operation had been accomplished step by step. The original plan – which involved the tethering of each strut prior to removal from the stowage canister until it was secured in position – proved very cumbersome and time-consuming, and required extensive handling, which tired the test subjects. With little changes to the hardware, it was agreed that handling by the astronauts would be less hindered by using a staggered strut storage facility to stow the struts during their final attachment, rather than have several tethered. With so many metal surfaces sliding, rotating, and interfacing directly in a number of close-tolerances joints, material selection and the surface coating of the struts had to be carefully evaluated in order to reduce excessive wear and friction. The material selected for the node guides, which slide on the 12.5-mm stainless steel rods attached to the ends of the guide rails on the assembly fixture, was 7075-T73 aluminium. A rigid test programme was conducted both in the water tank and in 1-g mock-ups, to minimise structural stress on the truss segments that controlled and restricted tolerance build-up in construction, and for expected thermal extremes on orbit. Vibration tests were also carried out to simulate launch loads, and other tests were used to select the optimum length of each strut and the number of strengthening cross-members. This data was multiplied in theoretical simulations and models of structures which were too large for the water tank, but which, at the time, were to be similarly constructed in space during the 1980s and 1990s.

These studies confirmed that 'astronaut EVA assembly of large space structures is well within man's physical capabilities in spite of the limitations of the pressure suit. The time to complete the assembly of the tetrahedral cell – 29 minutes at five minutes per column – may be significantly improved by using machines, such as the remote manipulator, to reduce the number of EVA tasks to yield a three-minute per column assembly rate.'

Pioneering orbital construction tests

On 14 December 1983 a Request for Flight Assignment form was received at JSC to manifest EASE/ACCESS to a specific Shuttle flight (STS-25) in the summer of 1985. However, over the ensuing two years a range of manifesting problems led to changes in this plan that resulted in a reassignment on STS-31 (later STS-61B), which was finally launched in November 1985. The WETF at JSC was too small to accommodate the large three-dimensional structure, and training had to be completed in the larger NBS pool at MSFC. In addition, the WETF did not have a functioning RMS simulator, and that at MSFC was a working model without 'flight-like' controls. An MPRESS mock-up and a 'cardboard crew-man', for RMS operator training, were therefore installed in the Manipulator Development Facility (MDF) in Building 9 at JSC.

Crew training for EASE/ACCESS commenced eight months before launch, with a cadre of nine astronauts working on development runs for EVA procedures on the mission. These included the EVA flight crew – Jerry Ross and Sherwood Spring –

who handled most of the development work, supported by John Blaha, Gordon Fullerton, Ron McNair, Bryan O'Connor, Ellison Onizuka. Robert Springer and George Nelson. This training clearly defined the timeline for the first and second planned EVAs, defined the method of using the Manipulator Foot-Restraint as a remote work platform, resolved safety concerns about the jettisoning hardware, and defined and simulated the roles of the IV crew-members in areas of photodocumentation, IV choreography and RMS operations, and cross-training in back-up roles. This division of labour and procedures was the first step in defining orbital construction techniques from the Shuttle that have continued to current ISS construction almost two decades later. Total crew EVA training for STS-61B (with EV1 and EV2 only, and including contingency runs) amounted to 55.25 hours, IV TV photographic training totalled 10.5 hours, and RMS training totalled 33.25 hours. Extensive experience was also gathered on the planning of construction EVAs into a Shuttle mission profiles, the launch processing of construction hardware at the Cape, and a two-way communication loop to and from the flight crew and the management and technical teams via the STS-61B EVA officer and CapCom.

The STS-61B crew conducted the first EVA tests on Flight Day 4. It had taken 58 minutes to assemble the 3.4-m ACCESS truss, cell by cell, in the water tank, and a period of two hours was allocated for assembly on EVA. However, It took only 55 minutes, and so the structure was dismantled and the test was repeated. The 3.6-m

EASE/ACCESS demonstration EVAs during STS-61B.

three-sided pyramidical EASE was to be assembled six times, but the astronaut completed eight construction assemblies. They noted that towards the end of the EVA their hands suffered from tiredness due to their working against the pressure in the gloves. Flight Day 5 was a rest day, and on Flight Day 6 the second EVA was used to continue assembly tests with the EASE/ACCESS equipment (flown under the DTO 0817 programme). After assembling the ACCESS girder, the astronauts simulated the attachment of a cable run by attaching a tether to the structure whilst supported by the RMS. They simulated structural repairs by replacing beams, mass handling of both structures, and working alone or in a team of two, with or without the RMS, in an evaluation of what was planned to be a standard operation on Space Station Freedom. In their post-flight debriefong they commented that ACCESS worked well, but that EASE required too much free floating, which was probably not the best way to build a space station. They had tried to assign the MMU to their flight to assist in their EVA tasks for work beyond the reach of the RMS, but were unsuccessful.

Lessons learned

From ACCESS, all basic EVA space construction tests were successfully accomplished on orbit. They provided a good comparison with simulations in the water tank, although the flight timelines were slightly longer than the best time logged underwater. This was attributed to the fact that the flight hardware was 'new', and had tighter tolerances than the well-worn equipment used during training under water over many months. The average assembly time for the truss – which was almost 14 m long – was 25.5 minutes at an assembly rate of 3.6 struts per minute. The experiment was judged by the EVA astronauts to be very hand-intensive – especially for the lower-positioned astronaut, who had 70% of the work-load with no rest. The pressure points in the glove contributed to the numbness – a factor that was considered when the Shuttle EMU glove assembly was modified. What was remarkable was that on the first attempt more than five hundred *untethered* pieces, each with a mass of around 0.5 kg, were assembled. Manipulation of the truss was found to be much easier in space than in the tank, where the viscosity of the water affected mobility, and both astronauts reported that provided the rate was kept low, the handling of larger masses should not present problems. Although the orbiter was flying on automatic attitude control during the EVA, neither astronaut felt any vernier firings. Riding the RMS on increased rates – double those used on previous missions – was 'comfortable', and working on the top of the 14-m truss was not disorientating. From EASE it was emphasised in the post-flight results that this was not intended to be a structural experiment. It was purely an EVA experiment to provide a measurement of flight results against ground simulations and studies, and to provide further data points for continued research, at MIT, into various constructional shapes and procedures in support of space station research.

In summarising the EASE/ACCESS experiment, the overriding conclusion was that 'the success of ACCESS confirmed the feasibility of EVA space assembly of erectable trusses, and played a role in the decision to baseline Space Station Freedom as a 5-metre erectable structure.'

EVA and Space Station Freedom

The changing face of Space Station Freedom, and its budgetary conflict and uncertain future, have been reported in a variety of publications (see Bibliography). After a decade of planning, the change from Space Station Freedom to the International Space Station was, in part, a reflection of the immense EVA tasks that the original design required. The selection of a baseline 5-m erectable truss as the overall framework upon which to attach the various laboratory modules and other facilities, systems and utilities, was recognised as a 'formidable challenge' by those involved in planning and scheduling. In order to accomplish this, several contractors (some using current or former astronauts, including many of the Skylab crew-members) and NASA centres set about evaluating the question of EVA at Space Station Freedom, which had to be constantly revised as the design (and budget) changed over the period of 1984–94.

EVA research and development in areas of human factors engineering resulted in a four-day conference, held at NASA Ames Research Center during 3-6 December 1985, to summarise the *first* year's research related to the creation of Space Station Freedom.[10] This in-depth study reviewed a variety of topics and subjects related to crew performance and, in particular, EVA scheduling, operations and tasks. Some of the areas covered included work-stations; equipment and tools; pressure suits (care, maintenance and selection, including a review of the Shuttle EMU and Russian Orlan configurations and advanced suits); total EVA mission time; EVA scenarios; Department of Defense EVA requirements; and constructional and maintenance work-loads expected at the ISS. In addition, the conference also discussed the impact of the space station EVA programme against other Shuttle-based EVA requirements for servicing and repairing large and small satellites, man-tended servicing in low Earth geostationary and polar orbit facilities, and testing and evaluating new and developmental EVA hardware and procedures. There was also a growing concern over the extensive amount of crew training time and the lack of training facilities to support multiple EVA crew training year on year over a sustained period. In addition, the expense of flight hardware required to support such a vast EVA programme, and the manpower required not only by the crew but by contractors and support personnel, produced some staggering figures in the summer of 1985.

From concept and projection studies completed at Langley, it was estimated, based on data supplied by JSC, that if all the planned and candidate EVA hours were to be added to the operational and flight opportunity hours at the station and on other programmes, then in 1992 the annual total time spent on EVA could reach 2,864 hours, and by 2001 this could increase to 4,013 hours! This was calculated with '100% efficiency with no contingencies, no equipment downtime, no crew sickness, and no mistakes'. Minimum estimates were 1,872 EVA hours per year at six hours a day for six days a week for two two-man crews (3,744 man-hours), rising to a top-of-the-range estimated growth figure of 5,616 EVA hours for six two-man crews (11,232 man-hours). This was clearly a major and worrying amount of EVA time, and was based on the success of all EVAs. This was, of course, a paper study produced several years before expected flight operations, but it reflected the complexity of the design of the space station and the expected flight growth of the Shuttle at the time (1985).

In-flight EVA activities would clearly have to be gradually augmented to provide not only operational experience but also a trained cadre of astronauts who would complete multiple EVA tasks with flight-proven equipment and procedures. A detailed EVA training programme would help prepare for space station construction tasks, and provide experience in preparing crews for other EVA tasks expected during the 1990s. As a result of these and other studies, in February 1986 – two weeks after the loss of *Challenger* – the Space Station Structures and Dynamics Technical Integration Panel met to establish the necessary data for an integrated structure flight experiment to be assigned to a Shuttle flight (as was EASE/ACCESS), and to define hardware that could be demonstrated in-flight to validate design procedures, tools, fabrication processes and ground-test techniques prior to multiple production of a reliable and primary structure for the space station. This Structural Assembly Verification Experiment (SAVE) resulted from a three-month concept study by NASA Langley and Boeing.[11]

This experiment – which was planned as a two-day EVA programme, with each EVA lasting approximately 5.5 hours – was to have been performed by two astronauts. The first EVA would be utilised for the construction of eight bays, with two additional bays each side of the top assembly, forming a T-shaped structure stretching out of the payload bay; while the second EVA would complete the assembly with eight bays, elevating the structure to sixteen bays, after which structural dynamics, photodocumentation, strain-gauge readings and thermal quality would have been conducted prior to ejection from the payload bay, using the RMS. This was an extension of the ACCESS experiment, and according to mission documentations it was 'feasible' to construct twenty bays in two six-hour EVAs.

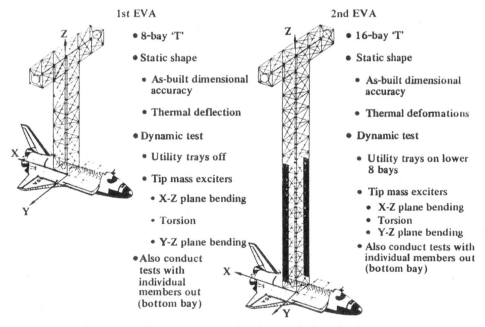

SAVE assembly techniques planned as a post-EASE/ACCESS demonstration.

In 1986 the experiment was planned for launch on 9 June 1991 (18 months before the first Shuttle flight to the space station), but when the plans for the mission were examined, the true scale of the experiment became evident. Each of the twenty truss structures were 5 m in length when deployed, so that when all sixteen were assembled the tower would stretch 80 m above the payload bay – well beyond the reach of the Shuttle's single RMS. Indeed, the report suggested the use of the RMS to speed up assembly time, and more study was required to determine whether this was possible. Finally, however, as a result of the changes to Space Station Freedom based on further analysis and design reviews, this experiment was terminated in the early 1990s. Had it flown, and been successful, it would have been a dramatic sight. One of the working concerns was the plan to conduct structural tests on the eight-bay and sixteen-bay configurations 'with individual members [struts] taken out of the bottom bay'. It can only be imagined what safety concerns the Astronaut Office would have expressed about the swaying of an 80-m structure above the heads of the astronauts, with base elements deliberately removed to determine what would happen!

Dozens of reports, studies and evaluations were conducted into the various aspects of the impact of EVA at Freedom, including the use of robotic and free-flying aids, and the concept of a mobility work-station,[12] which investigated mobile translation devices to assist the EVA crews in construction and servicing across the truss length of the station to ease the work-load and provide access to outreaching points on the station, far beyond the pressurised modules. These ideas were subsequently amended and adapted for the ISS to form the Crew and Equipment Translation Assembly (CETA) and to develop the Canadian Mobile Servicing System and 'walking' Space Station Robot Arm.

Another major study was the Man–Systems Integration Standards document produced by Boeing and a Government/Industry Advisory Group, published in four volumes in 1987. This study considered various aspects of man–system integration and design requirements, and various example solutions for the development of 'future' manned space systems – primarily space station. It included sections on anthropometrics and biomechanics, human performance capabilities, natural and induced environments, crew safety, heath management, design architecture, work stations, activity centres and sub-systems, hardware and equipment, maintenance design, management of the facility, and EVA.[13]

However, by far the most extensive (and publicly discussed) document was the two-volume report produced by the External Maintenance Task Team (EMTT), established in 1989 to review the increasing number of EVAs which the design of Freedom was accumulating. This report, published in July 1990, discussed an apparent discrepancy between official NASA estimates and those from a NASA study group into the EVA issue.[14] The use of robotics for external maintenance at Freedom had long been planned, but did not address the amount of external activity that could be undertaken. In October 1989, NASA directed a team, headed by Dr C. Bryant Cramer, to provide an estimate of expected EVA maintenance time at the station. The findings indicated that about 432 hours of EVA repair would be required each year; and when contingencies and 'uncertainties' were added to an overhead margin, it was estimated that 1,732 hours of EVA – 2.8 two-man EVAs

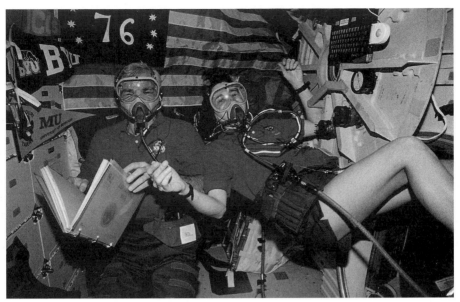

Pre-breathe operations prior to Shuttle-based EVAs, *c.*1996.

every week – could be required every year simply to maintain the condition of the station. This study was, of course, primarily concerned with *external maintenance* of the station over the duration of the programme, *without* the additional assembly and construction tasks required to create the station in orbit during the first few years of flight operations. At the time, NASA had allocated just 132 hours *a year* – one two-man EVA every *month* – to external maintenance. Clearly, there was a discrepancy. In December 1989, therefore, the EMTT was established, under the joint chairmanship of Charles R. Price, Chief of the Robotics Systems Division branch at JSC, and astronaut William F. Fisher, with a team of nine officials representing all relevant Space Station Freedom programme organisational elements and technical disciplines, supported by a group of thirty-three contributors including seven astronauts (Michael Foale, Mike Gernhardt, Linda Godwin, Tamara Jernigan, David Leestma, David Low and Jerry Ross).

The EMTT examined the ten primary components of Freedom, in areas of design, assembly, expected system failures, number of orbital replacement units, and maintenance levels. Simulations were also conducted to determine as accurately as possible both human and robotic tasks, and from these a list of recommendations was compiled. Overall, the team determined that the size of the station, and the number of parts required to construct it, multiplied the number of hours required to maintain it. Although certainly sizeable, the effort necessary to maintain the station was not insurmountable provided that joint human and robotic systems were employed. The stumbling block was the huge amount of EVA associated with maintenance and assembly (then planned for 1995–99), and further study was required in this area. In addition, a large amount of spares would be required on orbit for in-flight replacement of failed components.

EVA maintenance actions per year, over an estimated 35-year lifetime of Freedom, was found to average around 507 – equating to an annual average of 625 EVA work-site hours. This amounted to 3,276 hours of EVA time – 273 two-man EVAs annually, or 5.3 two-man EVAs every week. This was far greater than estimated in the Cramer report, which questioned NASA's official estimates. The report also evaluated EVA requirements between launch of the first element and permanent occupation (the first thirty months), and first occupation to completion of the assembly (twenty-four months): 1,752 maintenance actions requiring 11,517 hours of EVA – 960 two-man EVAs, or 4.1 two-man EVAs, *every week* for about 4.5 years. Then, depending on the external maintenance planned, a low year (2003) could see 189 two-man EVAs (3.7 a week) rising to a peak year (2005) of 10.4 two-man EVAs a week as the station becomes older and components begin to deteriorate. This was not what NASA wanted to hear from its own workforce, but it provided another element to the growing range of concerns that Freedom was too big, too complicated, and too expensive. The result was another review of designs, extended cooperation with international partners that now included Russia, and a final authorisation by a new President, Bill Clinton. Although EVA would still feature highly in planning for the ISS, it was by no means as complex as for the cancelled Freedom design that it replaced.

Developing the technique

During 1986–91 there was a lull in Shuttle EVAs, due mainly to the loss of *Challenger* and the absence of EVAs in the recovery and return to flight of the Shuttle fleet, including a number of scientific and deployment missions. The first EVAs from the Shuttle took place during the STS-37 mission in April 1991. The first was a contingency EVA to assist in the deployment of the Compton Gamma Ray Observatory, and the second evaluated the CETA, planned for use on Freedom. New glove assemblies were also evaluated, although they proved disappointing despite improved performance levels during ground simulations. During their post-flight debriefing, the crew (which included Jerry Ross, who had performed the EASE/ACCESS experiments and had spent the intervening years working on Freedom EVA studies) emphasised that during the five years between STS-61B and STS-37, a significant amount of talent, experience and skills had been wasted in the areas of equipment design, the building and testing of hardware, the planning of EVAs for mission, and in the training of crew to accomplish them. There was also a serious lack of experience at Mission Control and on flight crews. They also recognised the growing concerns for the amount of EVA expected at Freedom, and the apparent lack of mission planning to develop the skills and techniques to achieve this efficiently and safely: 'While this mission helped to regain some of this expertise, there were also many indications that additional EVAs are required to establish the robust level of EVA capabilities that will be necessary at all levels to support assembly and operation of SSF.' In their recommendations, the crew suggested that mission plans should include one or two EVA missions each year to gather experience, and to rigorously flight-test EVA systems and hardware well ahead of assembly during SSF missions. They also suggested that consecutive-day EVAs, while possible, would not be avoidable for more complex operations at the space station.

As a result of this and other studies, NASA instigated a series of Detailed Test Objectives for developing EVA procedures prior to space station operations under the DTO 1210 EVA Operations Procedures/Training programme. These would expand the knowledge of EVA by means of planning and practise. They would then apply this new knowledge to future EVAs, which would produce a far better understanding of human performance, as well as refine the timeline of the EVAs compared to training simulations. This could then be applied to both space station hardware and service missions such as at the HST.

The STS-49 mission was flow between the STS-37 EVA and the first of the DTO 1210 EVAs. After retrieving and redeploying the Intelsat satellite it supported a fourth EVA dedicated to the Assembly of Space Station by EVA Methods (ASEM) – a McDonnell Douglas experiment aimed at further evaluation of Freedom trusses (on a much smaller scale than SAFE). The ASEM was related to truss assembly experiments under the Freedom programme, but by the time the experiment was manifested to the flight, NASA had already decided to launch station trusses in preassembled and preintegrated segments to reduce the requirement for EVA construction of each truss. However, the ASEM was still completed to provide additional experience and comparative data for STS-61B EASE/ACCESS, although it had different design characteristics. For the astronauts it became a frustrating exercise in excessive use of their arms, underlying the decision to use prefabricated trusses.

On 25 November 1992, NASA announced the decision to assign EVAs to impending Shuttle flights, to gain experience, to train astronauts and flight controllers, and to develop space construction techniques.[15] Over the next five years, a series of EVAs was accomplished to support ISS and HST EVA operations.

STS-54 (1993) Evaluation of carrying a large mass (another astronaut); demonstration of the use of large tools on EVA; and testing of the ability to alight bulky objects.

STS-57 (1993) The first exit from the Shuttle air-lock as part of the payload bay tunnel extension, which was moved from the mid-deck to provide more room; evaluation of a large mass (another astronaut) whilst riding the RMS. The crew experienced low temperatures when outside the payload bay and facing deep space (out of sunlight), which caused shivers and numb and painful hands.

STS-51 (1993) Evaluation of the provisional stowage assembly; evaluation of the glove-warming experiment using payload bay lights; testing of tethers for low and high torque; comparisons of WETF and flight EVA activities.

STS-64 (1994) These were the first untethered EVAs since STS-51A and the use of the MMU in 1984. This time the astronauts used the first SAFER (Simplified Aid For EVA Rescue) devices in tests broken into four elements: familisation, programmed jet tests for gathering engineering data, tumbling tests, and a series of precision manoeuvring tests. Each astronaut took turns to complete the tests while

Cold-soak thermal tests on the end of the RMS as part of the EVA development programme conducted in the 1990s in preparation for ISS EVA operations.

the second rode the Manipulator Foot-Restraint on the RMS. The Shuttle crew stood by to 'rescue' the SAFER astronaut by manoeuvring the Shuttle to 'scoop him up'. It was determined that the SAFER used less nitrogen than previously thought, which was more welcome discovery than the realisation that the electronic cuff checklist – planned to replace the cuff checklist used since Apollo 12 – did not perform as planned.

STS-63 (1995) This was the first of a series of EVAs under the new EVA Development Flight Test (EDFT-01) programme, which was more focused in preparing for the ISS programme. The EVA crew evaluated thermal improvements to the Shuttle EMU, incorporating a liquid-cooling and ventilation garment bypass switch that allowed a reduction of cooling water flow but without any reduction in suit ventilation, wearing thicker underwear, and better insulated EVA gloves. EVA objectives included mass-handling tasks with the 1,363-kg Spartan-204 free-flyer; determination of the thermal properties of the EMU, either on the RMS or foot-restraints with the payload bay pointed at the Sun in daylight passes or deep space during night-time passes. For fifteen minutes, Michael Foale stood on the RMS, 9.1 m above the payload bay, to compare data collected by sensors on the Manipulator Foot-Restraint and in the glove, and to provide objective observations. During the tests, the temperature recorded on Foale's gloves dropped to –6° C, and Bernard Harris's feet became cold when he contacted the structure of the orbiter at a temperature of –148° C. The astronauts were requested to register their comfort on a sale of 1–8, and the EVA was terminated early after they reported that it was 3 – 'unacceptable cold'.

STS-69 (1995) EDFT-02 featured the installation of thermal sensors on the RMS, and work on removing ISS developmental micrometeoroid/orbital debris shields and insulation blankets. Each astronaut completed evaluation of thermal conditions during repetitive handling of tools, further cold soak evaluation 9.1 m above the payload bay for 45 minutes, and a variety of maintenance and assembly tasks using power tools, fasteners, electrical conduits and tethers. Fingertip heaters warmed their hands, and they reported that they were quite comfortable.

STS-72 (1996) EDFT-03 tested ISS equipment for demonstration of a simulated assembly task by attaching a rigid umbilical diagonally across the payload bay. Also evaluated was the Lockheed Martin Portable Work Platform – a combination of restraint aid and temporary storage facility for equipment transfer – a portable foot-restraint work-station, and an articulating portable foot-restraint. During a second EVA the crew conducted further thermal soak tests on an astronaut in a Shuttle EMU in sunlight and in darkness, with further improvements to keep the astronaut warm in the suit systems. Tests were also carried out on clamps and cable trays being developed for the ISS.

STS-60 (1996) EDFT-04 included a planned evaluation of a crane for moving bulky orbital replacement units for the ISS. However, this test was cancelled when the EVA hatch could not be opened. It was later discovered that a small screw had worked loose and had jammed the door-opening ratchet system. Had a contingency EVA been required, the astronauts could not have exited the orbiter. NASA therefore added a set of contingency hatch-door tools for freeing the hatch mechanism from the inside of the air-lock.

STS-87 (1997) EDFT-05 consisted of the final pre-ISS EVA developmental spacewalks. This time the crew performed a contingency EVA task at the beginning of the first of two EVAs, to manually capture the Spartan-201 free-flyer and secure its berthing in the payload bay. They then successfuly completed the EDFT-05 tasks, which included evaluation of a telescoping crane boom, simulated replacement of ISS solar array batteries, and evaluation of a cable caddy, a body restraint tether, a multi-use tether, and other handling and restraint aids. This mission also demonstrated the use of the Autonomous EVA Robotic Camera Sprint (AERCam Sprint) – a prototype free-flying TV camera of the type planned for use on remote inspections of the exterior of the space station. The AERCam Sprint – 35.5 cm in diameter and weighing 15.8 kg – included two TV camera, twelve small nitrogen-powered thrusters, an avionics system (resembling a large football, and controlled automatically from the Shuttle aft flight deck), a hand-operated controller, two laptop computers, and an antenna mounted in an aft flight deck window. The power supply ensured seven hours of operational use – the duration of a nominal Shuttle-based EVA.

ISS EVA OPERATIONS, 1998–2002

The first EVAs connected to the physical construction of the ISS were accomplished during the STS-88 mission in December 1998. Over the next four years the rapid pace of EVA continued to assist in the expansion and support of ISS operations. The loss of *Columbia* in February 2003, however, not only grounded the Shuttle but also seriously affected the expansion of the station due to the suspension of the Shuttle's heavy-lift capability. This also affected the EVA programme from the space station. Station-based EVAs commenced, with the second resident crew, in 2001, and have been developed and modified as the station expands. The programme continues to be maintained as the issues pertaining to the Shuttle's return to flight are pursued.

Quest or Pirs?

All of the initial EVAs at the space station (1998–2001) could be accomplished only from Shuttle-based air-locks; but in 2001, delivery of the American Quest air-lock module and the Russian Pirs docking port, with air-lock facilities, allowed EVAs to be conducted from the ISS without a docked Shuttle. With the grounding of the Shuttle fleet this capability has proved a valuable asset, and has enabled resident caretaker ISS crews to perform EVA should the need arise, whilst expansion of the station by Shuttle EVA crews has been suspended until at least the autumn of 2004.

Quest is the joint air-lock module that can support EVAs by crew-members wearing American EMU or Russian Orlan pressure garments. It is 6 m long and 4 m in diameter, and is divided into two sections. The equipment lock is the EVA

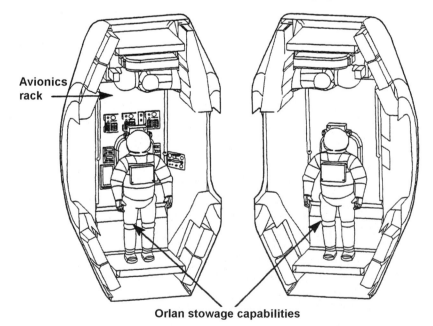

EVA equipment provisions inside the ISS joint air-lock (Quest).

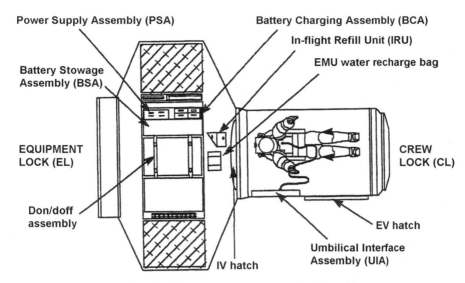

Power Supply Assembly (PSA)

Battery Charging Assembly (BCA)

In-flight Refill Unit (IRU)

EMU water recharge bag

Battery Stowage
Assembly (BSA)

EQUIPMENT
LOCK (EL)

CREW
LOCK (CL)

Don/doff
assembly

EV hatch

Umbilical Interface
Assembly (UIA)

IV hatch

Details of the internal arrangements inside Quest.

preparation and pre-breathe area, and is used for equipment storage; whilst the crew lock is the air-lock/hatch passage used to begin and end the EVAs required at the American elements. It was delivered by STS-104 in July 2001, and is located at the starboard common berthing mechanism on the Unity node.

Pirs arrived at the ISS via a Progress M cargo vehicle in September 2001, and was located on the Zvezda module nadir port. It was a third docking port for Soyuz TM and Progress resupply craft, and had facilities to support Orlan-based EVAs at the Russian segment. The docking compartment was 4.85 m long and 2.1 m in diameter.

The wall of EVA
By 1998 the first elements of the station were ready for launch. In 1999, problems

Mission Control Center, Houston, during ISS EVA operations dealing with the 'wall of EVA'.

with Shuttle propulsion systems and electrical wiring sub-systems, and with the launch of the Russian Zvezda SM, delayed the preparations for permanent residency on the station, although the first four years of EVA operations progressed smoothly. The early station EVAs – to establish the ISS in orbit – presented an operational and technical hurdle. The number and complexity of the EVAs could not be avoided, and in NASA this became known as the ISS 'wall of EVA'. EVA Flight Director Glenda Laws explained the term during an interview in 2002:[16] 'You are going at [ISS EVA operations] hard ... really hard ... Everybody knew [that when] we started building the space station. [We have] a little chart that shows the number of EVAs per year and the projected number of EVAs per year, and it's really low and flat until you get to last year [2001] and this year [2002], and then it just hits a steep curve, and we climb right up this curve, and that's what we call the wall of EVA. And then it stays peaked for many, many years, and what we are learning now is it's not going back down again for a long time because of all the maintenance stuff that comes up.' It was also becoming a clear conflict between operational necessity and budget reality. If NASA was at the bottom of the 'wall of EVA' at the end of 2002, then they slipped off in early 2003 as a result of the loss of *Columbia*. There is no doubt that there is an urgent need to again climb the wall as soon as is it is safe and prudent to do so.

Shuttle-based construction EVAs
In June 1997, NASA announced a cadre of fourteen astronauts who would be trained for an extensive programme of initial Shuttle–ISS assembly flights. Many of them had already been training for several years, and had completed EVAs on earlier flights.[17] Between December 1988 and November 2002, thirty-nine EVAs were completed by Shuttle crews, either from the orbiter or from the Quest air-lock on the station.

The following summarises the first phase of Shuttle EVA operations during this period. The initial five – STS-88 to STS-106 – were associated with the installation of connections between the Unity module and the Zarya module, the installation of tools, EVA support hardware and restraints on the exterior of the station, for later EVAs, and the completion of connections between Zvezda and Zarya. The next seven – STS-92 and STS-97 – mainly focused on the preparation and installation of the Z1 truss, the P6 solar array, and PMA-3. The three EVAs of STS-98 were related to the installation of the Destiny laboratory module, the attachment of a spar communication antenna, and the releasing of a cooling radiator. The two EVAs on STS-102 focused on the transfer of previously installed hardware and equipment and the installation of an external stowage platform. Two EVAs during STS-100 were connected with the installation of the Canadian-built RMS on the station, whilst three STS-104 EVAs installed the Quest docking module and a pair of oxygen and nitrogen tanks. Further installation of hardware – including science experiment packages – was completed during two EVAs on STS-105 and during one EVA on STS-108. During 2002, thirteen EVAs on STS-110 to STS-113 were completed in support of the expansion of the truss and solar arrays and associated electrical and fluid line connections, and in completing the installation of the Canadarm 2 and the Mobile Base System and its cart transporter.

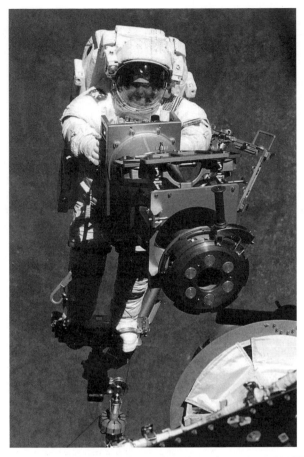

An STS-96 EVA to deliver supplies and EVA equipment to the ISS in 1999.

Although every EVA encounters small difficulties and failures, the only major mishap occurred during the first EVA of STS-98 in February 2001, when a coolant line, being connected by Robert Curbeam, leaked a small amount of ammonia crystals. This leak was quickly stopped, after which the escaped crystals dispersed and vapourised. Mission safety rules required Curbeam to 'sunbathe' for thirty minutes to allow solar heating to vapourise any residual crystals that might be in the folds of his suit or behind associated equipment; and as a precaution his colleague, Thomas Jones, brushed off his own suit. The re-entry to the Shuttle air-lock also required partial pressurisation and venting to flush out any residual particles that might have entered the air-lock. The orbiter crew then wore oxygen masks for twenty minutes as the air-lock was opened and the orbiter environmental control system flushed the air to provide further cleansing. Contaminant measurements were also carried out. This operation extended the EVA by a further 100 minutes; but there were no adverse affects, and it provided an operational demonstration of emergency EVA procedures developed over many years.

The delivery of the Quest airlock on STS-104 in 2001.

ISS-based EVAs

The first ISS-based EVA by a resident crew was delayed from ISS-1 due to configuration work-load, and was completed by the ISS-2 crew. It was a short internal EVA to install a docking cone in the docking node of Zvezda to accept the Pirs docking compartment. The ISS-3 crew completed four EVAs from Pirs to install a variety of instruments on the exterior of Zvezda and connecters between Pirs and Zvezda. This crew also conducted the first unplanned EVA from the ISS, during which they cut away a rubberised O-ring seal, left from the previous undocking of a freighter craft, that was preventing the hard docking of the next unmanned Progress at the aft port of Zvezda. The ISS-4 crew completed three EVAs to relocate the Strela crane and to began the installation of six thruster deflectors at the rear of Zvezda. They also completed a range of housekeeping and advance tasks in preparation for later EVAs. The ISS-5 crew continued this work with the installation of the first six of twenty-three debris protection shields on Zvezda's exterior, and also equipment and Russian experiments in the exterior of the station.

In January 2003 the ISS-6 crew continued the outfitting and activation of the P-1 truss and the relocation of a tool-box from the Z1 truss to the CETA. The following month, the loss of *Columbia* necessitated a major review of ISS orbital operations, resulting in the reduction of resident ISS crews from three to two, and the use of the Soyuz TM spacecraft to deliver and return resident crews until the resumption of Shuttle flights. This affected the on-orbit scientific and EVA programme, and seriously delayed further expansion of the ISS by means of the Shuttle system. In April 2003 the ISS-6 crew completed a reconfiguration of the ISS power systems by supplying a secondary power source for one of the control moment gyros that was to be replaced on the (delayed) STS-114. During the ISS-7 residency, no EVAs were conducted, although the crew completed IVA training and fit and function tests of their Orlan suits. In November 2003 the ISS-8 crew completed another IVA training run involved tests of a retreat, while wearing Orlan suits, from the Pirs DC-1 air-lock to the Soyuz crew return vehicle docking to the aft port on the Pirs module. This was a first-time demonstration exercise to confirm the ability of the crew to access the OM should a problem, during an operational EVA, prevent nominal completion of an EVA and re-entry into the habitable sections of the ISS. This type of contingency EVA has never been simulated in the Hydro Laboratory, and so the crew assembled and checked out an Orlan suit, pressurised (inflated) it without a crew-member inside, and pushed it through the connecting hatches. There was marginal clearance, and the next day, after modifications to the test, Flight Engineer Kaleri suited up to attempt the

ISS EVA operations.

transfer. It required the unsuited Michael Foale to push him through the tight hatches into the Soyuz. Analysis of the results indicated that the Soyuz ferry might need to be relocated to the more extensive nadir hatch area on the node of Zarya.[18] Limited ISS EVA operations are to continue during the wait for the resumption of Shuttle flights, which is currently planned for no earlier than the fourth quarter of 2004.

Summary

In 1984 it was hoped that the station would be completed by 1994; but construction began only in 1998, and at the time of writing (2003) it is still less than 50% complete. If the Shuttle returns to flight in late 2004 (as current projections state), it will take much longer than two years to recover from the two years of lost ISS construction. Four years of ISS EVA operations have only scratched the surface of what needs to be completed. It is too early to fully review ISS operations, but to date there has been remarkable success in both Shuttle-based and ISS-based EVAs. There remain, however, a significant amount of EVA operations before the end of the construction of the facility, and after that the focus will shift to a long-term (10–15-year?) orbital maintenance programme. At the end of the ISS programme, EVA activities, over what may possibly be as many as twenty years, will constitute an important and valuable chapter in long-term operations not only in Earth orbit but also across the Solar System, where the next steps in human spaceflight and EVA operations are destined to be taken.

REFERENCES

1 Astro Info Service interview with Yuri Usachev, TsPK Moscow, 19 June 2003.
2 Shayler, David J., 'Outside Mir: Ten Years of EVA Operations', *Journal of the British Interplanetary Society*, **51**, 1998, 27–36.
3 Astro Info Service interview with Yuri Romanenko, London, July 1989.
4 Shayler, David J. and Hall, Rex D., *Soyuz: A Universal Spacecraft*, Springer–Praxis, 2003, pp. 336–339.
5 Salmon, Andy, 'The TTM Repair EVA: An Insider's View', *Zenit*, No. 22, December 1988, 8–12.
6 Portree, David S.F. and Trevino, Robert C., *Walking to Olympus*, NASA Monograph No. 7. pp. 127–129.
7 Information revealed during Astro Info Service visit to Zvezda facility, Moscow 16 June 2003; Abramov, Isaak P. and Skoog, Å. Ingemaar, *Russian Spacesuits*, Springer–Praxis, 2003, pp. 186.
8 Proceedings of a Conference [on space construction] held at NASA Langley Research Center, Hampton, Virginia, 6–7 August 1986, NASA Conference Publication CP-2490, 1987.
9 *EVA Assembly of Large Space Structure Elements*, NASA Technical Paper TP-1872, June 1981.
10 *Space Station Human Factors Research Review*, Vol. 1, EVA Research and

Development; Vol. 2, Behavioural Research; Vol. 3, Space Station Habitability and Function, Architectural Research, NASA Conference Publication CP-2426, 1988.

11 *A Space Station Structure and Assembly Verification Experiment: SAVE*, NASA Technical Memorandum TM-89004, August 1986.

12 *A Mobile Work Station Concept for Mechanically Aided Astronaut Assembly of Large Space Trusses*, NASA Technical Paper TP-2108, March 1983; *Tests of an Alternate Mobile Transporter and EVA Assembly Procedure for SS Freedom Truss*, NASA TP-3245, October 1982.

13 *Man-Systems Integration Standards*, 4 vols., NASA STS 3000, March 1987.

14 *Space Station Freedom External Maintenance Task Team Final Report*, 2 vols., July 1990.

15 NASA News, 92-066.

16 Astro Info Service interview with Glenda Laws, NASA Shuttle/EVA Flight Director, Houston, Texas, 8 May 2002.

17 Shayler, David J., 'NASA Shuttle Missions to ISS, 1998–2001', in *ISS: Imagination to Reality*, British Interplanetary Society, 2002, pp. 81–118.

18 ISS On Orbit Status Report, 18 November 2003.

Next steps

This book is not intended to be a detailed account of every EVA (which is being prepared by the author under other projects). Rather, the aim here has been to present an overview of what is required to complete a safe and successful operation in the vacuum of space. There are the years of development and testing of the hardware, the suits, and the procedures; the objectives have to be defined and the timeline set; and the EVA and IVA crews, as well as the flight controllers, have to be trained. To bring these all together, everything has to function correctly on the mission, up to and including the EVAs and throughout the rest of the flight, to truly claim that the excursions have been successful. Of course, things can go wrong, plans can be thwarted, and equipment can break down; but the provision of as much flexibility and redundancy as possible can, in most cases, result in a successful mission.

The story of EVA began in literature and films, and in the dreams of space pioneers who believed that the exploration of our Solar System was possible with the use rocket-propelled vehicles from which suitably protected crew-members could exit to explore the vacuum of space and the moons and planets. After the first fledging steps into the void of space in 1965, most of our extravehicular activity has been spent in Earth orbit. We have managed to log fifteen excursions on the lunar surface, and just three in deep space. However, for the foreseeable future, although we discuss plans to explore the plains of Mars and expect to return to the Moon, all of our human spaceflight experience will be confined to Earth orbit. We still have yet to really *explore* space. But we have started.

THE FUTURE OF EVA

In the Foreword to the author's *Apollo: the Lost and Forgotten Missions*, former astronaut Don Lind recalled that it had taken explorers more than 250 years to truly discover and explore the Pacific Ocean. In contrast, we have been placing humans in space for less than a fifth of that time period; so why should we expect more? Good timing, effort and will, and political and public support, combined with a military fear, took us from the Earth to the Moon – from Sputnik to Apollo 11 – in less than twelve years. In 2003 we celebrated a century of aviation; and in those celebrations we could have included manned flights to Mars, had we followed the same drive that took Apollo to the Moon. We all explore space every day of our lives, as the Earth moves around the Sun; and because of the concern for our environment and our planet, we must explore and understand how space 'works', and how it affects life here on Earth. Space is always close – just 80 km above our heads, and less than a ten-minute ride away, straight up. In the future, the exploration of space should encompass defined goals and objectives – an integrated plan that combines unmanned satellites with robotic explorers and manned spacecraft. An integral part of this will be EVA, by both humans and machines; and because of the expected frequency of EVAs and the conditions of where we may explore in the future, this is an invaluable development. So, what does the future promise? And where are the next steps outside of our spacecraft likely to take us?

The ISS: outpost in orbit
For at least the next decade or so, the programme of extensive EVAs will be focused at the International Space Station. Although the expansion of the facility currently relies on the Shuttle's return to flight, it is at this outpost in orbit that most of the EVAs now being planned will take place. According to information received in October 2003, Return to Assembly commences in early 2005, and the American core should be completed by the end of 2006. This will be followed by the completion of the International Partner assembly in 2010, after which the ISS facilities and components will be utilised for the duration of its operational lifetime, which should be as least another ten years.

At the time of the loss of *Columbia*, most of the assembly flights for 2003 and 2004 were focused on the installation of the remaining truss and solar arrays, and current planning indicates that after the Return to Flight missions of STS-114 (which will still deliver and install the all-important control moment gyros) and the newly inserted STS-121, the next five missions will attempt to deliver and install the remaining truss segments and solar arrays. In view of past experience, it can be estimated that three or four EVAs will be required on each mission. Approximately seventeen further missions will be required to complete the construction of the ISS; and here again, EVA will be a feature of almost every flight, to ascend the 'wall of EVA'.

The next resident crew (ISS-9), in April 2004, is expected to complete the first EVAs at the ISS for more than a year. As the rotation of ISS crews continues, and with the return to a three-person resident crew after the Shuttle resumes flying, EVAs will begin to feature as a regular part of the long-duration mission, as on Mir.

The Shuttle: return to flight
A key element in the Shuttle's return to flight is the inclusion of repair systems in the thermal protection system of the vehicle, to be tested on STS-114 and STS-121 before being flight-qualified. Tile repair kits, for replacing lost tiles, were originally evaluated for orbital flight-tests in the late 1970s, but were not flown after better systems for attaching the tiles to the Shuttle were developed. Flight data from the four orbital flight-test missions also revealed that although tiles were lost or

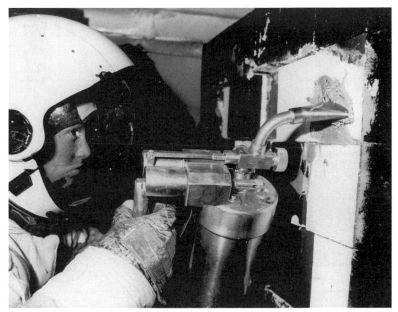

Anna Fisher evaluates the proposed (but not flown) tile repair kit during pre-STS-1 underwater EVA simulations *c*.1980.

damaged, they were (at least on those missions) lost from less critical areas. Experience gained from more than a hundred flights of the Shuttle, twenty years of EVA from the Shuttle, and new materials and techniques, have resulted in the development of a material which can be filled in and shaped to repair small indentations and missing tiles. NASA, however, is still working on how to repair the reinforced carbon high-temperature area that appears to have been the main cause of the loss of *Columbia*. To allow all-round orbiter surveys, extensions to the 500-kg RMS will probably be carried on all missions. The addition of EVA tile repair hardware weighing more than 300 kg, and other associated equipment, increases the lift-off mass of each Shuttle, and this mass will be further increased by a range of possible materials for a variety of repairs – all of which will have a deleterious effect on the payload mass deliverable to the station. Whatever the final plan adopted, EVAs, and the supply of consumables and equipment to support them, will feature prominently in every remaining Shuttle mission.

NASA plans seven Return to Flight missions by 2007, and between fifteen and twenty-five missions (five a year) until 2012. EVA will feature in the design of the Crew Exploration Vehicle, which, even in its role of crew transfer or rescue vehicle, will need a contingency EVA rescue capability. An EVA rescue from STS-107 was suggested *after* the accident, but if the Crew Exploration Vehicle is to retain a simple and less complicated design and a limited mission profile, then by including EVAs, the air-locks, suits and associated equipment will increase costs, complexity, launch weights and crew size. It will be interesting to discover what role EVA might play in any vehicle that follows the Shuttle after 2012.

The evaluation of Shuttle tile repair simulations in 2003, after the loss of *Columbia*.

KERMIt: repairing the ISS

At MSFC, research is being conducted into a patch repair kit to seal punctures and debris impact damage at the ISS. The Kit for External Repair of Module Impacts (KERMIt) consist of externally emplaced patches that can seal punctures up to 10 cm in diameter and cracks as long as 20 cm. In 1999 it was stated that larger areas of damage would be 'highly improbable' and exceedingly difficult to repair. As revealed at Mir in 1997, small impacts can be survivable given adequate time and luck; but that will not always be the case. KERMIt is a three-component system consisting of a clear lexan disk with a torodial seal on one side, a toggle bolt through the centre, and fittings for injection of adhesive sealant. It also incorporates special tools for fitting and securing the patch to the hull of the station. While on EVA, the crew examines the puncture and clears the area of debris and thermal insulation. After cleaning the local area to improve contact, a special tool is then used to determine the size and shape of the hole and to mark reference points that are used to determine which size of patch is to be used. A second EVA is then undertaken to complete the repair by inserting the toggle bolt in the hole, tightening the seal patch against the surface of the hull, and injecting a caulking-type epoxy resin to fill and seal the hole.

This type of repair, however, presents problems. Tests have shown that a 2.5-cm hole can cause an unacceptable drop in pressure within an hour, in which case the crew must evacuate the module and seal it off to preserve the rest of the atmosphere and to allow time to conduct the repair over two days. This would be a serious challenge should crew-members be injured or sub-systems be damaged. Curing takes 2–7 days, and the seal must be capable of withstanding internal pressure for at least six months. But would the crew feel comfortable or safe, knowing they were working in patched module? Certainly, any patching kit would be tested on EVA during its evaluation for operational use. It is obvious that more research needs to be completed before these types of repair kit are added to the station's equipment.

Risk assessments of impacts on the ISS have been conducted, and studies continue to be carried out into the long-term risk of impacts on astronauts on EVA at the ISS, using probability risk analysis models based on available data on the particle flux in low Earth orbit, the data from JSC tests that determine the ability of the EMU to absorb these loads, the frequency of micrometeoroids, and particles of orbital debris. This continuous accumulation of data is then used to determine future suit design, astronaut shielding options, and the number, scheduling and frequency of EVAs during construction and utilisation.

Space rescue by EVA?

After the *Columbia* accident, several news reports opined that NASA could have sent up a second Shuttle to rescue the crew. Theoretically, that might have been possible. A vehicle was being processed for STS-114, and the procedure might have been accelerated to offer a quick launch; but several factors still affect a probable rescue using EVA. The problem of keeping the crew alive and capable of supporting a rescue from a second Shuttle was discussed by space news analysts, including Jim Oberg, who reviewed a 34-page document which, as part of the work of the Columbia Accident Investigation Board, examined inspection and repair by the *Columbia* crew or rescue by *Atlantis*. It was assumed that the damage was more fully known than was being revealed, and that some type of rescue could have been mounted with adequate provision to keep the crew alive in space. A flight to the ISS was not an option, as it was flying in a different orbit. Problems with keeping the crew alive by adjusting their work/sleep pattern, the use of lithium hydroxide canisters, and power-down, reflected NASA thinking *c*.1970 and the days of Apollo 13; and given time, correct information, communications, and a great deal of luck, NASA might very well have succeeded. We shall never know.

However, flying two orbiters in proximity operations had never been attempted, and transferrence of a crew by EVA rescue, though planned in the 1970s, was never taken to operational mode. Furthermore, a shortage of EVA suits made the operation more complicated. All seven astronauts had the orange 'pumpkin' pressure garments, but although these garments were vacuum-qualified they had limited independent life support and no thermal protection. *Atlantis* could have carried extra EVA suits into orbit by to supplement the two suits carried on *Columbia* for contingency EVAs, but the transfer of the *Columbia* crew in EVAs of two, three and two would occupy several hours and rely on adequate consumables in order to depressurise the air-lock and keep the two Shuttles flying in tandem. And with *Columbia* unattended during the last transfer, this would be a tricky operation, even in the best of circumstances.

In the 1970s, during the early days of the programme, NASA often promoted

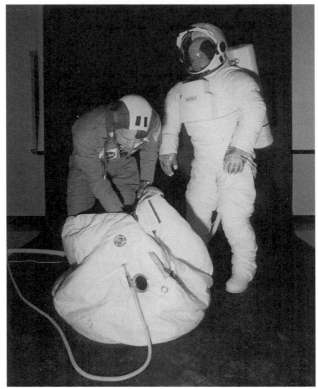

The proposed (but not flown) Shuttle crew Personal Rescue Enclosures designed to transfer stranded IV crew-members via EVA from a damaged Shuttle to a rescue vehicle (ideally, a second Shuttle orbiter).

one-person rescue capsules (spheres) for such a transfer. These could be carried by hand from one Shuttle to another, but although two were built they were never tested in space, and are now in storage. In 1985 an IAF paper suggested that there was the capability to use the MMU to transfer these rescue spheres or a stranded EVA crewman, but that it was never tested operationally.[1] Again, the MMUs have been in storage for years.

Another opportunity for the development of a true space rescue capability was lost with the Russian MMU, which had been developed for use in conjunction with the Buran shuttle and Mir 2, by using an air-lock on the Shuttle that evolved into the Pirs docking compartment on the ISS. Had there been cooperation between America and the USSR just a few years earlier, then perhaps the grounding of Buran could have been avoided. If it had been promoted as an extra redundancy to the Shuttle, with a 'space rescue' capability, and if funding had been forthcoming, then it might have been used to support ISS operations.

Rescue systems have been proposed for many years, but only the docking systems have been evaluated in flight – during the Apollo–Soyuz Test Project in the 1970s, on Shuttle–Mir in the 1990s, and now at the ISS. Even in science fiction, EVA rescue was demonstrated in the 1969 film *Marooned*, and has been the subject of countless films before and since. Also in 1969, the Soyuz 4/5 crew completed an EVA that was described as a development of the techniques of crew rescue by EVA. The only problem has been that in all the major accidents since then – Soyuz 11, *Challenger* and *Columbia* – the crew has had insufficient time and resources to exercise a 'rescue option'. In these cases, the combination of a lack of clear information, the location of the incident, the rapid sequence of events and a shortage of time prevented any hope of rescue. Rescue by EVA is a viable option if there is sufficient time to conduct it; but with the current systems of spacecraft and pressure suits, time is severely limited.

The HST: rescue or re-entry?
STS-122 is currently under review as the fourth HST service mission; but it is not expected that it will fly before 2006, if at all! If the mission proceeds, four or five EVAs will follow the flight plan of past servicing missions to upgrade, service and replace instruments to allow a continuation of science research into 2010, when the HST's operational life is expected to end. Prior to the loss of *Columbia* there were discussions concerning the fate of the HST after 2010. Another Shuttle flight could provide a package to reboost it to a higher orbit or destructive re-entry, or it could be brought back to Earth for studies on long-term exposure to the space environment over twenty years. The latter would provide a valuable test-bed facility with applications to future manned and unmanned spacecraft development. It might then be sent to a museum, to be placed on public display.

However, after the *Columbia* accident the options for the future of the telescope continue to be studied more closely. With the Shuttle fleet reduced from four to three vehicles, and commitment to the ISS, the risks involved in flying a mission to the HST increase in light of the *Challenger* and *Columbia* accidents. Some say that because of what the HST has achieved and still has to offer, it is worth the risk; but others argue that it should be fitted with an engine that can be used to boost it to a storage orbit after the end of its life in 2010, thus saving one service/retrieval mission. There is also the option of flying another service mission in 2009, to increase the operational lifetime to 2011, by which time the James Webb Space Telescope (JWST) should be operational. But the JWST is no HST, and at its planned operational point it will be far beyond the current capacity for EVA servicing and repair should anything untoward happen. Another option is to leave the HST in orbit until the JWST is at least certified for operations. The future of HST service missions remains uncertain.

Robots and EVA
We have seen the beginning of robotic systems used in EVA operations, the RMS systems on the Shuttle and the ISS, AERCam Sprint, and several proposals for telerobotic devices to aid maintenance on the ISS. In 1972 the film *Silent Running* featured three such robotic droids that could 'walk' across the exterior of the spacecraft to conduct maintenance and repairs, independent of human participation. Other science fiction films also featured robot–human cooperation for hazardous EVAs in deep space and on 'strange new worlds'. But it is not all science fiction.

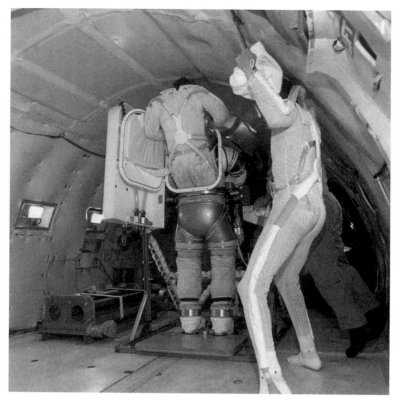

Testing new types of EVA suits. The next generation of constant-volume EVA suits have been evaluated on Earth for decades. but have yet to be flown or tested during a mission. During the next decade, the development of this technology will be a milestone in our return to the Moon and extended exploration at Mars.

On 2 July 2003, NASA issued a news release (H03-227) that reported of humans and robots working together in a 'spacewalk squad', designed to improve the productivity of astronauts working outside the ISS, on other space vehicle, or on the surface of other planets. A dexterous humanoid robot called Robonaut – a joint project of JSC and the Defense Advanced Research Products Agency – has been under development for several years, and tests of advanced pressure garments with additional robotics features, designed to replace the current types of EVA suit, have for some time been under development in America, Russia and Europe. In recent tests, astronaut Nancy Currie worked with two Robonauts in the assembly of trusses. These tests – which were carried out several times – significantly reduced the time required to complete the task, compared with the time taken by regular-suited teams of EVA astronauts. Currie reported that the tests were very successful, and that over the next five years, further evaluations could result in the use of Robonauts to reduce the current 20% EVA timeline required to set up a work-site or close it out, or to assist in manoeuvring and placing large and bulky items of hardware.

Robotic EVA systems could also be used for venturing to locations unsafe for astronauts – not only on the ISS, but also on the surface of planets, the atmosphere and pressures of which increase the difficulty in planning human EVA activities. Working in teams, Robonauts could assist EVA astronauts like a nurse in an operating theatre, passing and stowing tools by verbal command, and they could assist in large orbital assembly tasks independent of a human crew, or beyond low Earth orbit (such as at the JWST). They would not replace humans, but would complement and assist them in their EVA tasks – a true blend of robotic and human space exploration.

A Chinese EVA?

In October 2003 the successful flight of Yang Liwei, onboard the Shenzhou 5 spacecraft, generated a wealth of speculation concerning China's future plans now that it has attained manned spaceflight capability. The Chinese have for several years indicated their long-term desire to create a national space station and to send crews to explore the Moon. Both of these programmes – depending on their success and progress – will feature EVA operations to gain the most from the investment. In the short term, further evaluations of the Shenzhou vehicle will include rendezvous, docking, and probably an EVA demonstration or crew transfer (similiar to the 1969 Soyuz EVA) to qualify the EVA suit and support systems. On Shenzhou 5, no EVA suit was carried or evaluated, and Yang Liwei wore only a personal rescue pressure garment (similiar to the Russian Sokol) for the duration of his 21-hour mission. It therefore seems reasonable to assume – based on recent official Chinese statements – that rendezvous and docking are of high priority for the next Shenzhou missions, leading to an increase of orbital stay-time with a crew on board, and a demonstration of EVA capability.

Back to the Moon?

Arguments continue for and against a return to the Moon prior to going on to Mars. Undoubtedly, at some point we will return to the Moon, as it is our nearest neighbour in space and somewhat easier to reach; and we have shown that for a short time we can live and work there. The next step is the creation of a research station – a lunar base – and extended surface exploration over several weeks or months, of the type projected in the 1950s and 1960s. EVA will certainly be a key element in this process, but the suits, equipment and procedures, although based on Apollo experiences, will be developed at least 40–50 years after the last men of Apollo left the Moon. Technology has already changed so much that we will not return to the Moon in the style of Apollo, for short term exploration, but to stay. It is interesting to speculate whether the new lunar explorers will be American, Russian, European, Chinese, or of other nationalities.

And so to Mars ...

It is said that the first person who will step onto the surface of Mars has already been born and is attending school somewhere on Earth. This is to be hoped for, as it implies that at some time during the next thirty years we will see the first human exploration of the Red Planet. New developments in suit technology and methods of

Our return to the Moon will not be such a 'small step' – but will it be permanent?

reliability, so far from our home planet, will lead to dramatic changes in the way we control and operate space missions. And EVAs on Mars will be the next 'giant leap'.[2]

And after that ...?

At the height of Apollo, eyes were looking towards Mars as the 'next logical step'; and now, thirty years later, as we ponder on actually making that step, we are wondering where we shall go next.

In the Office of Exploration annual report for 1988 there is an illustration, by Pat Rawling, of an astronaut wearing 'EMU and MMU'-type equipment, but with added 'bumper fenders' to explore the martian moons, Phobos and Deimos, which have *very* low gravity (hence the requirement for the MMU and fenders). Soviet designs for a Phobos lander featured harpoons for securing the unmanned spacecraft to this tiny body, but this might not be so practical for human exploration! The use of retrorockets to stop the bounce might also disturb deep pools of dust in crevices, which might in turn hinder vision or equipment. (During the Apollo mission, lunar dust was always a problem.) When Arthur C. Clarke wrote of 'a fall of Moon dust' there was a fear of deep craters full of soft 'quicksand-like' material which would create hazards for lunar explorers. But although Apollo disproved this portrayal as far as the Moon was concerned, such hazards will be encountered on worlds yet to be

The next 'giant leap' – the human exploration of Mars – and the introduction of a new word – Marswalk.

explored. Perhaps, therefore, the Robonauts would be useful and effective in the furthest reaches of the Solar System, where radiation and intense cold (or heat, towards the Sun) will diminish the prospect of human exploration by EVA.

It is thought that the moons of Mars are probably captured asteroids, and they have often been the subject of long-range plans to set up mining systems for the extraction of minerals. Spacecraft have already been sent to asteroids, and have landed on them, and so the prospect of human investigation is not too far in the future. However, it seems fairly improbable that we shall be able to explore the inner or outer planets by EVA.

Closer to the Sun it is simply too hot, though perhaps Robonauts can be developed to perform EVA work near Venus should the returning Mars craft take that route. Surface operations on Venus are severely limited by the extremely high atmospheric pressure and the intense heat. To land an automatic probe on Venus, to take pictures and measurements, is difficult enough, but to have it retrieve a sample and then relaunch back into a path from where it can be retrieved is a completely different problem. Mercury is even closer to the Sun and is totally out of the range of human spaceflight, although robotic surface exploration and sample return might be an option in the latter decades of the twenty-first century.

Moving away from the Sun and into the depths of the Solar System, the temperature plummets and radiation increases. Because of its orbit, Jupiter's moon Io is continuously pummled by tidal forces and constantly subjected to intense radiation, and it is a sulphurous volcanic nightmare. Saturn's largest moon, Titan, is blessed with an atmosphere of methane, and other moons present a wide variety of extreme alien environments and landscapes. Pre-1970s artistic impressions of remote research stations, with EVAs conducted across the surfaces of the moons of the gas giants Saturn, Uranus and Neptune have, since the discoveries of the Voyager, Galileo and Cassini probes, proved to be far removed from reality. Feature films such as *Saturn 3, 2001: A Space Odyssey* and *2010* depict EVAs far out in the Solar System, but such scenes now belong only in the imagination.

If we are truly to explore the Solar System, new flight-propulsion methods, life support systems, shielding and exploration techniques need to be developed before we can confidently begin our journey to the stars. As recalled earlier, it took 250 years to explore the Pacific Ocean, and yet only fifty years separates Sputnik and Gagarin from the Shuttle and the ISS. So, what are our prospects for the next two hundred years in space? Although EVA will probably be limited to the regions between Venus and the asteroid belt, the operations will always evolve from the lessons first learned by the pioneers of EVA between 1965 and 2005 – from Leonov and White, through the excitement and daring of Apollo, to the space stations Salyut, Skylab, Mir and the ISS.

The lessons learned from EVA

In this book I have tried to present an account of the development of the *techniques* of EVA, and how we progressed from science fiction and theory to the Moon, and through the development of space stations to the ISS. In almost forty years of EVA operations there have been several key lessons learned that have applied to most EVAs, and it is those elements that we will probably continue to utilise throughout the next few decades, and take forward into the exploration of our solar system:

- Underwater simulation provides the best method of training for EVA on Earth.
- In-flight training accurate simulates tasks to be performed in zero g.
- Restraint and tethers are paramount in securing an EVA crew-member for effective work-loads.
- Experienced EVA crew-members, on different missions or in support roles on IVA crews or at Mission Control, aid the performance of the EVA crew.
- Ensuring that flight hardware is actually depicted in ground training models saves time and effort in overcoming differences when on EVA.
- The development of new tools, suits, equipment and procedures is important

in ensuring a constant flow of EVA skills and testing and evaluation of EVA technology prior to operational use.

- The provision of EVA in-flight hardware makes servicing and repair much easier and increases the chance of mission success.
- The use of humans and robotics, working as an EVA squad, is the way forward, although the human decision-making process should never be neglected, even when using remote robotic EVA procedures.
- Contingency procedures and redundancy capabilities provide options in reaching EVA goals despite in-flight difficulties.
- Communication of successes and failures, before, during and after EVA, can lead to greater mission success in future EVA planning and operations.
- The involvement of EVA crews in the development of techniques and procedures greatly assists in the later execution of operational EVA tasks.
- EVA remains the greatest fun on any spaceflight – according to those who have achieved it!

The door to space has been ajar for more than forty years, and we have stepped through and back many times. It is now time to leap into the void and plant our footsteps not only on the Moon again, but on the dusty surface of Mars, and elsewhere. EVA will help us to achieve our dream of planetary exploration, and with it, new techniques will enable us to voyage further and deeper into space. This may happen soon or in the distant future, but those of us who will not witness these advancements can say that we were there when EVA began the true human exploration of the Cosmos.

REFERENCES

1 Rogers, L., 'Use of the MMU for On-Orbit Rescue Operations', presented at the 36th Congress of IAF, Stockholm, 7–12 October 1985, IAA-85-332.
2 Shayler, David J., Salmon, Andy, and Shayler, Michael, *Marswalk One*, Springer–Praxis, (due in 2004). This book includes discussion on martian EVAs – the first 'small steps' on the Red Planet.

Appendix 1

THE SPACEWALKERS

Selected milestones in the development of EVA techniques, 1965–2003

1965	Mar 18	Soviet cosmonaut Alexei Leonov becomes the first person to complete a walk in space (12 min).
	Jun 4	Astronaut Ed White becomes first American to walk in space (21 min); McDivitt exposed to space vacuum in first dual EVA operation.
1966	Jun 5	Gene Cernan becomes first person to exceed 2 hours on one EVA.
	Jul 20	Mike Collins becomes first person to perform two separate spacewalks on the same mission (total time, 1 hr 29 min).
	Nov 14	Buzz Aldrin sets a new endurance of 5 hrs 30 min on three EVAs.
1969	Jan 16	Cosmonauts Yevgeny Khrunov and Alexei Yeliseyev complete first EVA transfer from Soyuz 5 to docked Soyuz 4.
	Mar 6	Dave Scott and Rusty Schweickart perform dual EVA from two separate spacecraft; McDivitt exposed to vacuum inside LM in first three-person EVA operation.
	Jul 20	Neil Armstrong and Buzz Aldrin (Apollo 11) complete first EVA on lunar surface (2 hrs 32 min).
	Nov 20	Pete Conrad and Al Bean (Apollo 12) perform their second lunar-surface EVA.
1971	Jul 30	Dave Scott (Apollo 15) performs first (and so far only) stand-up EVA on the Moon.
	Jul 31	Dave Scott (Apollo 15) becomes first person to drive over the surface of the Moon.
	Aug 2	Dave Scott and Jim Irwin (Apollo 15) become the first to complete a third surface exploration on one mission.
	Aug 5	Al Worden performs first deep-space EVA.
1972	Dec 13	Gene Cernan and Jack Schmitt complete their third (fourteenth and final) Apollo EVA on the surface of the Moon (7 hrs 15 min).
1973	May 25	First space-station-related EVA (stand-up) performed by Paul Weitz from hatch of Apollo CM.

	Jun 7	First EVA conducted from a space station, performed by Pete Conrad and Joe Kerwin.
1974	Feb 3	Final Skylab EVA by Jerry Carr and Ed Gibson.
1977	Dec 20	First Soviet EVA since 1969; first EVA from a Salyut space station (Salyut 6).
1979	Aug 15	Ryumin and Lyakhov perform first unplanned EVA (from Salyut 6).
1982	Jul 30	Valentin Lebedev and Anatoli Berezovoi perform first EVA from Salyut 7.
1983	Apr 7	Story Musgrave and Don Peterson perform first Shuttle EVA (from *Challenger*).
1984	Feb 7	Bruce McCandless performs first untethered EVA by flying MMU 300 feet from Shuttle orbiter.
	Apr–Aug	Leonid Kizim and V. Solovyov carry out extensive repairs to Salyut 7.
	Jul 25	Svetlana Savistskaya becomes first woman to perform EVA (3 hrs 35 min).
	Oct 11	Kathryn Sullivan becomes first American women to conduct an EVA (3 hrs 27 min).
	Nov	Joe Allen and Dale Gardner, using MMUs, rescue two stranded satellites.
1985	Apr 16	Jeff Hoffman and Dave Griggs perform first unplanned EVA from Shuttle.
1987	Apr 11	First EVA from Mir (unscheduled) by Yuri Romanenko sand Alexandr Laveikin.
1988	Dec 9	Jean-Loup Chrétien becomes first Frenchman, and first non-Russian/nonAmerican, to perform EVA.
1990	Feb 1	Alexandr Serebrov flies first (tethered) Soviet MMU from Mir.
1991	Apr 7	Jerry Ross and Jay Apt complete first Shuttle-based EVA for 64 months.
1992	May 13	Hundredth EVA; first three-person EVA by Pierre Thuot, Rich Hieb and Tom Akers (8 hrs 29 min), setting a new record.
1993	Dec 4	Jeff Hoffman and Story Musgrave perform First Hubble Space Telescope servicing EVA.
1995	Oct 20	Thomas Reiter become first German to perform EVA.
1996	Mar 27	Linda Godwin and Rich Clifford perform first Shuttle-based EVA at a space station (Mir).
1997	Apr 29	Jerry Linenger becomes first American to perform EVA in a Russian suit (Orlan) from a Russian space station (Mir).
	Oct 1	Vladimir Titov becomes first Russian to perform EVA from the Shuttle (docked to Mir).
	Nov 24	Takio Doi becomes first Japanese citizen to perform EVA.
1998	Dec 7	Jerry Ross and Jim Newman perform first ISS-related EVA from the Shuttle.

1999	Dec {?}	Claudie Nicollolier{?} becomes first Swiss to perform EVA (from the Shuttle).
2001	Apr 22	Chris Hadfield becomes first Canandian to perform EVA.
	Jun 8	Jim Voss and Yuri Usachev perform first ISS-based IVA.
	Oct 8	Vladimir Dezhurov and Mikhail Tyurin perform first ISS-based EVA without docked Shuttle present.

Single EVA duration records, 1965–2003

1965	Mar 18	12 min	Alexei Leonov
	Jun 4	21 min	Ed White
1966	Jun 5	2 hrs 10 min	Eugene Cernan
	Nov 12	2 hrs 29 min	Buzz Aldrin
1969	Jul 20	2 hrs 32 min	Neil Armstrong and Buzz Aldrin
	Nov 19	3 hrs 56 min	Pete Conrad and Al Bean
1971	Feb 5	4 hrs 48 min	Al Shepard and Ed Mitchell
	Jul 31	6 hrs 33 min	Dave Scott and Jim Irwin
	Aug 1	7 hrs 12 min	Dave Scott and Jim Irwin
1972	Dec 12	7 hrs 37 min	Gene Cernan and Jack Schmitt
1992	May 13	8 hrs 29 min	Pierre Thuot, Rich Hieb and Tom Akers
2001	Mar 11	8 hrs 56 min	Jim Voss and Susan Helms

Appendix 2

WORLD EVA LOG, 1965–2003

In recording EVAs and IVAs, one of the most difficult points of agreement is qualification and categorisation in official records; and even official details vary, depending on the source. Records can detail the times from depressurisation to repressurisation, from transferrance from the suit systems and then back to the spacecraft systems, or from the opening or closing of a hatch.

In the table which follows, crew-members are sequenced in order of exit or by command structure, as it is not always clear who exited first or if the flight plan was followed. All the official EVAs are listed, together with accepted IVAs during which suits were worn for internal activities in a vacuum, even though the external EVA hatches might have been closed.

On the Gemini flights the Pilots completed EVA, but the Command Pilots, although remaining in the spacecraft, were also exposed to the vacuum. On the Apollo flights, McDivitt wore a pressure suit but did not exit the LM during the Apollo 9 EVA, and Irwin remained inside the pressurised LM during Scott's stand-up EVA on Apollo 15; and although the CM Pilots on Apollo 15–17 each completed a trans-Earth EVA and each LM Pilot performed a stand-up EVA in the CM hatch, each Commander was also exposed to the vacuum without exiting the hatch. On Skylab 2, Conrad and Kerwin were both exposed to the vacuum during the Skylab stand-up EVA. These details are not included in official lists.

During Mir operations and on the ISS, recorded internal EVAs were carried out in the node of the station to relocate docking equipment, and several IVA operations were conducted in support of safing the station following the collision in 1997. During the contingency demonstration by the ISS-8 crew in November 2003, one of the crew, in a pressurised suit, moved from the Pirs module to the Soyuz OM; but the internal compartments were not depressurised, and to assist his colleague the other crew-member remained unsuited. This demonstration, however, is not officially listed as an IVA.

EVA	Start date of EVA	Duration hrs:min	EVA crew-members	Mission/ spacecraft	Notes
001	1965 Mar 18	0:12	Leonov	Voskhod 2	First person (Soviet) to perform EVA; experienced difficulty in returning to the air-lock.
002	1965 Jun 4	0:21	White	Gemini 4	First US EVA; used zip-gun to manoeuvre; McDivitt exposed to space vacuum; did not exit (IVA).
003	1966 Jun 5	2:10	Cernan	Gemini 9A	Tried to put on AMU, but failed due to overheating of suit; Stafford on IVA; first EVA to exceed 1 hour.
004	1966 Jul 19	0:39	Collins	Gemini 10	Stand-up EVA; photographs of stellar background; cut short due to impurities in air supply; Young on IVA.
005	1966 Jul 20	0:50	Collins	Gemini 10	Removed Agena 8 experiment package; used HHMU; lost stills camera from open hatch; Young on IVA.
–	–	–	–	Gemini 10	Equipment jettison; Collins opened hatch and pushed out unwanted equipment; Young's hatch closed.
006	1966 Sep 13	0:33	Gordon	Gemini 11	Straddled neck of Gemini to attach 100-foot tether to Agena; dubbed 'Space Cowboy'; Conrad on IVA.
–	–	–	–	Gemini 11	Equipment jettison; Gordon opened hatch and pushed out unwanted equipment; Conrad's hatch closed.
007	1966 Sep 14	2:08	Gordon	Gemini 11	Stand-up EVA for photography of stellar background in UV; Conrad on IVA.
008	1966 Nov 12	2:29	Aldrin	Gemini 12	Stand-up EVA for astronomical photography; equipment prep for later EVAs; Lovell on IVA; EVA record.
009	1966 Nov 13	2:09	Aldrin	Gemini 12	Evaluation of restraints and tethers during completion of EVA work-task tests; Lovell on IVA.
010	1966 Nov 14	0:59	Aldrin	Gemini 12	Final Gemini EVA; completed astronomical photography objectives; Lovell on IVA.
011	1969 Jan 16	0:37	Khrunov/Yeliseyev	Soyuz 5/4	First Soviet EVA for four years; first EVA crew transfer between two spacecraft; related to lunar programme.
012	1969 Mar 6	1:07	Schweickart/Scott, D.	Apollo 9/LM/CM	First test of Apollo lunar EVA suit in Earth orbit; Scott conducted stand-up EVA from CM hatch; Schweickart demonstrated exit on to LM porch and partial crew transfer to LM; McDivitt remained inside LM on IVA.

No.	Date	Duration	Crew	Mission	Description
013	1969 Jul 20	2:32	Armstrong/Aldrin	Apollo 11/LM	First moonwalk; EVA record; set up experiments and took samples; Tranquillity Base EVA world's most televised event to date.
–	1969 Jul 20	–	–	Apollo 11/LM	Equipment jettison; Armstrong and Aldrin remained inside LM on IVA.
014	1969 Nov 19	3:56	Conrad/Bean	Apollo 12/LM	Second moonwalk; set up first ALSEP; geological sampling near landing site; EVA record.
015	1969 Nov 20	3:49	Conrad/Bean	Apollo 12/LM	Third moonwalk; geological field trip on foot; retrieved parts from unmanned Surveyor III.
–	1969 Nov 20	–	–	Apollo 12/LM	Equipment jettison; Conrad and Bean remained inside LM on IVA.
016	1971 Feb 5	4:48	Shepard/Mitchell	Apollo 14/LM	Fourth moonwalk; deployed second ALSEP and completed short geological field trip on foot; EVA record.
017	1971 Feb 6	4:35	Shepard/Mitchell	Apollo 14/LM	Fifth moonwalk; geological field trip on foot to Cone Crater using MET to carry equipment.
–	1971 Feb 6	–	–	Apollo 14/LM	Equipment jettison; Shepard and Mitchell remained inside LM on IVA.
018	1971 Jul 30	0:27	Scott, D.	Apollo 15/LM	Stand-up EVA from LM top hatch to photographically survey Hadley Rille landing site; Irwin remained in LM on IVA.
019	1971 Jul 31	6:33	Scott, D./Irwin	Apollo 15/LM	Sixth moonwalk; set up third ALSEP geological traverse on first LRV set a new EVA record.
020	1971 Aug 1	7:12	Scott, D./Irwin	Apollo 15/LM	Seventh moonwalk; geological traverse on LRV; round trip to Hadley–Apennine Front; new EVA record.
021	1971 Aug 2	4:50	Scott, D./Irwin	Apollo 15/LM	Eighth moonwalk; geological traverse on LRV to Hadley Rille.
–	1971 Aug 2	–	–	Apollo 15/LM	Equipment jettison; Scott and Irwin remained inside LM on IVA.
022	1971 Aug 5	0:39	Worden/Irwin	Apollo 15/CM	First deep-space EVA; Worden retrieved SIM-bay film cassettes; Irwin filmed and assisted during stand-up EVA in CM hatch; Scott exposed to vacuum, but did not exit CM.
023	1972 Apr 21	7:11	Young/Duke	Apollo 16/LM	Ninth moonwalk; deployed fourth ALSEP; drove second LRV on geological traverse to Flag Crater.

No.	Date	Duration	Crew	Mission/Vehicle	Description
024	1972 Apr 22	7:23	Young/Duke	Apollo 16/LM	Tenth moonwalk; LRV geological traverse to Stone mountain; new EVA record.
025	1972 Apr 23	5:40	Young/Duke	Apollo 16/LM	Eleventh moonwalk; LRV geological traverse to House Rock; largest boulder visited by Apollo crews.
—	1972 Apr 23	—	—	Apollo 16/LM	Equipment jettison; Young and Duke remained inside LM on IVA.
026	1972 Apr 25	1:24	Mattingly/Duke	Apollo 16/CM	Second deep-space EVA; Mattingly retrieved SIM-bay film cassettes from SM; Duke filmed and assisted during stand-up EVA in CM hatch; Young exposed to vacuum but did not exit CM.
027	1972 Dec 11	7:12	Cernan/Schmitt	Apollo 17/LM	Twelfth moonwalk; deployed fifth and final ALSEP; drove third LRV on short geological traverse.
028	1972 Dec 12	7:37	Cernan/Schmitt	Apollo 17/LM	Thirteenth moonwalk; LRV geological traverse; discovered 'orange soil'; set new EVA record.
029	1972 Dec 13	7:15	Cernan/Schmitt	Apollo 17/LM	Fourteenth and last moonwalk; LRV geological traverse; performed ceremonies marking end of Apollo.
—	1972 Dec 13	—	—	Apollo 17/LM	Equipment jettison; Cernan and Schmitt remained inside LM on IVA.
—	1972 Dec 13	—	—	Apollo 17/LM	Second equipment jettison; Cernan and Schmitt remained inside LM on IVA.
030	1972 Dec 17	1:06	Evans/Schmitt	Apollo 17 CM	Third deep space EVA; Evans retrieved SIM-bay film cassettes; Schmitt filmed and assisted during stand-up EVA in CM hatch; Cernan exposed to vacuum but did not exit CM; final Apollo EVA.
031	1973 May 25	0:37	Weitz	Skylab 2/CM	Stand-up EVA from open CM hatch; attempts to deploy stuck solar array; assisted from inside CM by Kerwin and Conrad on IVA.
032	1973 Jun 7	3:25	Conrad/Kerwin	Skylab 2/OWS	Successfully deployed stuck solar wing, saving station; tethered Conrad catapulted into space by motion of array.
033	1973 Jun 19	1:44	Conrad/Weitz	Skylab 2/OWS	Retrieval and replacement of ATM film cassettes; inspected solar array and previously deployed parasol sunshade.
034	1973 Aug 6	6:31	Garriott/Lousma	Skylab 3/OWS	Erected twin pole assembly over parasol to improve thermal conditions inside Skylab OWS.

	Date	Duration	Crew	Mission	Notes
035	1973 Aug 24	4:30	Garriott/Lousma	Skylab 3/OWS	ATM film cassette retrieval and replacement.
036	1973 Sep 22	2:45	Bean/Garriott	Skylab 3/OWS	ATM film cassette retrieval and replacement.
037	1973 Nov 22	6:33	Gibson/Pogue	Skylab 4/OWS	ATM film cassette retrieval and replacement; routine repair tasks.
038	1973 Dec 25	7:01	Carr/Pogue	Skylab 4/OWS	ATM film cassette retrieval and replacement; first EVA on Christmas Day; observed Comet Kohoutek.
039	1973 Dec 29	3:38	Carr/Gibson	Skylab 4/OWS	ATM film cassette retrieval and replacement; further observations of Kohoutek.
040	1974 Feb 3	5:19	Carr/Gibson	Skylab 4/OWS	Final ATM film cassette retrieval; final Skylab EVA.
041	1977 Dec 20	1:28	Grechko	EO1/Salyut 6	Stand-up EVA examined forward docking port after failure of Soyuz 25 to dock; Romanenko conducted IVA; first Soviet EVA in almost nine years; first EVA from a Salyut space station.
042	1978 Jul 29	2:05	Ivanchenko/Kovalenok	EO2/Salyut 6	Removed and replaced samples from exterior of Salyut 6; Kovalenok performed stand-up EVA.
043	1979 Aug 15	1:23	Ryumin/Lyakhov	EO3/Salyut 6	Unscheduled EVA to free KRT-10 telescope antenna from aft docking port.
044	1982 Jul 30	2:33	Lebedev/Berezovoi	EO1/Salyut 7	Collected and replaced samples on exterior of Salyut.
–	1982 Nov	–	Lenoir/Allen	STS-5/OV102	Planned EVA cancelled with crew in air-lock.
045	1983 Apr 7	4:17	Musgrave/Peterson	STS-6/OV099	First Shuttle demonstration EVA; evaluated new EVA suits and restraint systems.
046	1983 Nov 1	2:50	Alexandrov/Lyakhov	EO2/Salyut 7	First in a series of EVAs adding extra panels to solar arrays; installed additional panels to central array.
047	1983 Nov 3	2:55	Alexandrov/Lyakhov	EO2/Salyut 7	Added second panel to central array.
048	1984 Feb 7	5:55	McCandless/Stewart	STS-41B/OV099	First untethered EVAs; McCandless flew first MMU 300 feet from orbiter; Stewart also test flew MMU.
049	1984 Feb 9	6:17	McCandless/Stewart	STS-41B/OV099	Further MMU flights; also evaluated procedures for satellite repairs planned for later missions.
050	1984 Apr 8	2:57	Nelson/Van Hoften	STS-41C/OV099	Nelson attempts to capture Solar Max by flying MMU; unsuccessful; later captured by Shuttle RMS.
051	1984 Apr 11	6:16	Nelson/Van Hoften	STS-41C/OV099	Repaired Solar Max in payload bay; redeployed; Van Hoften flew untethered MMU in payload bay.

No.	Date	Duration	Crew	Mission	Description
052	1984 Apr 23	4:15	Kizim/Solovyov, V.	EO3/Salyut 7	First of a series of six EVAs; transported ladder and EVA tools to work area, and prepared work site.
053	1984 Apr 26	4:56	Kizim/Solovyov, V.	EO3/Salyut 7	Installed new propellant valve by cutting into station's skin.
054	1984 Apr 29	2:45	Kizim/Solovyov, V.	EO3/Salyut 7	Installation of new conduit; replacement of thermal covering of station.
055	1984 May 3	2:45	Kizim/Solovyov, V.	EO3/Salyut 7	Installation of second conduit; verification of both conduits; fuel leak pinpointed.
056	1984 May 18	3:05	Kizim/Solovyov, V.	EO3/Salyut 7	Installation of a second set of solar array extensions to main array.
057	1984 Jul 25	3:35	Savitskaya/Dzhanibekov	T12/Salyut 7	First female to perform EVA; tested multipurpose welding gun.
058	1984 Aug 8	5:00	Kizim/Solovyov, V.	EO3/Salyut 7	Leaking fuel pipe sealed; retrieved samples from solar arrays for evaluation on Earth.
059	1984 Oct 11	3:27	Leestma/Sullivan	STS-41G/OV099	First US female to perform EVA; completed satellite refuelling demonstration in payload bay of Shuttle.
060	1984 Nov 12	6:00	Allen/Gardner	STS-51A/OV103	Used MMU to retrieve rogue Palapa Comsat.
061	1984 Nov 14	5:42	Allen/Gardner	STS-51A/OV103	Used MMU to retrieve rogue Westar Comsat.
062	1985 Apr 16	3:00	Hoffman/Griggs	STS-51D/OV103	First unscheduled (contingency) US EVA; crew attached 'fly-swatter' to RMS attempting to activate Leasat.
063	1985 Aug 2	5:00	Savinykh/Dzhanibekov	EO4/Salyut 7	Attached third and final set of additional solar panels to main arrays; evaluated new Orlan EVA suits.
064	1985 Aug 31	7:08	Van Hoften/Fisher	STS-51I/OV103	Manual capture of Leasat deployed on 51D; commenced repairs in payload bay.
065	1985 Sep 1	4:26	Van Hoften/Fisher	STS-51I/OV103	Completed Leasat repair and redeployed by hand from end of RMS.
066	1985 Nov 29	5:30	Spring/Ross	STS-61B/OV104	Space construction tests using EASE and ACCESS in payload bay for future space station activities.
067	1985 Dec 1	6:30	Spring/Ross	STS-61B/OV104	Space construction tests with EASE and ACCESS related to Space Station Freedom programme.
068	1986 May 28	3:50	Kizim/Solovyov, V.	EO5/Salyut 7	Collected experiments from exterior of station; evaluated a beam builder for future space construction tasks.
069	1986 May 31	5:00	Kizim/Solovyov, V.	EO5/Salyut 7	Completed additional space construction tests; used improved URI welding gun; final EVA from a Salyut station.

No.	Date	Duration	Crew	EO/Location	Description
070	1987 Apr 11	3:40	Romanenko/Laveikin	EO2/Mir node	Unscheduled EVA to remove a foreign object from rear Mir docking port which prevented hard docking of Kvant astrophysical module; first EVA from Mir.
071	1987 Jun 12	1:53	Romanenko/Laveikin	EO2/Mir node	Added an extra set of solar panels to the Mir exterior.
072	1987 Jun 16	3:15	Romanenko/Laveikin	EO2/Mir node	Installed a second set of additional solar panels to exterior of Mir.
073	1988 Feb 26	4:25	Titov, V./Manarov	EO3/Mir node	Replaced elements of solar array panels erected by EO2 crew.
074	1988 Jun 30	5:10	Titov, V./Manarov	EO3/Mir node	Attempted repair of the X-ray telescope on Kvant 1; terminated due to broken wrench.
075	1988 Oct 20	4:12	Titov, V./Manarov	EO3/Mir node	Completed repair of the TTM telescope on Kvant.
076	1988 Dec 9	5:57	Volkov, A./Chrétien	EO4/Mir node	First non-US, non-USSR EVA; first French EVA (Chrétien); erected French ERA structure (by kicking the container) and French experiments.
077	1990 Jan 8	2:56	Viktorenko/Serebrov	EO5/Mir node	Deployment of two star sensors on exterior of Mir and retrieved samples from hull of station.
078	1990 Jan 11	2:54	Viktorenko/Serebrov	EO5/Mir node	Retrieved French experiments deployed in December 1988 and installed new experiment packages.
079	1990 Jan 26	3:02	Viktorenko/Serebrov	EO5/Kvant 2	First use of specialised air-lock on Kvant 2 module; prepared docking device for use with Soviet MMU; removed Kurs antenna; installed new TV system and tested improved EVA suit.
080	1990 Feb 1	4:59	Viktorenko/Serebrov	EO5/Kvant 2	First flight of Soviet MMU; Serebrov flew (tethered) up to 30m from Mir.
081	1990 Feb 5	3:45	Viktorenko/Serebrov	EO5/Kvant 2	Viktorenko flies (tethered) MMU 45 m from Mir and performs a 'Victory Roll' in celebration.
082	1990 Jul 17	7:00	Solovyov, A./Balandin	EO6/Kvant 2	Attempted repair of damaged Soyuz thermal blankets; a damaged outer Kvant 2 hatch meant using a back-up method to re-enter Mir.
083	1990 Jul 26	3:31	Solovyov, A./Balandin	EO6/Kvant 2	Stowed external ladders on Mir for future use; completed temporary repairs to damaged Kvant 2 outer hatch.
084	1990 Oct 30	3:45	Manakov/Strekalov	EO7/Kvant 2	Completed a partially successful repair of Kvant 2 outer hatch.

No.	Date	Duration	Crew	Mission	Description
085	1991 Jan 7	5:18	Afanasyev/Manarov	EO8/Kvant 2	Completed repairs to Kvant 2 outer hatch; installed support structure for a crane to relocate solar arrays from Kristall to Kvant 1 module.
086	1991 Jan 23	5:33	Afanasyev/Manarov	EO8/Kvant 2	Installation of first crane jib near the forward multiple docking adapter node on Mir core module.
087	1991 Jan 26	6:20	Afanasyev/Manarov	EO8/Kvant 2	Installation of second crane jib on Kvant 1 for future relocation of solar arrays.
088	1991 Apr 7	4:38	Ross/Apt	STS-37/OV104	First US EVA for 64 months; unscheduled EVA to deploy stuck Gamma Ray Observatory high-gain antenna.
089	1991 Apr 8	6:11	Ross/Apt	STS-37/OV104	Completed EVA experiments related to future space station construction, including CETA mobility tests.
090	1991 Apr 25	3:34	Afanasyev/Manarov	EO8/Kvant 2	Inspection of faulty Kurs antenna discovering missing receiver dish; replacement of exterior TV camera.
091	1991 Jun 25	4:58	Artsebarsky/Krikalev	EO9/Kvant 2	Replacement of damaged Kurs antenna.
092	1991 Jun 28	3:24	Artsebarsky/Krikalev	EO9/Kvant 2	Installation of US cosmic ray detector experiment on exterior of Mir.
093	1991 Jul 15	5:55	Artsebarsky/Krikalev	EO9/Kvant 2	Commenced construction of 15-m Sofora girder on the exterior of Kvant 1.
094	1991 Jul 19	6:20	Artsebarsky/Krikalev	EO9/Kvant 2	Continued construction of Sofora.
095	1991 Jul 23	5:34	Artsebarsky/Krikalev	EO9/Kvant 2	Continued construction of Sofora.
096	1991 Jul 27	6:49	Artsebarsky/Krikalev	EO9/Kvant 2	Completed construction of Sofora; Artsebarsky's suit overheats and he is blinded by perspiration; he is guided back to the hatch by Krikalev.
097	1992 Feb 20	4:12	Volkov, A./Viktorenko	EO10/Kvant 2	Installation of new equipment on exterior of Kvant 2; Volkov's suit experienced problems that forced an early termination of the EVA.
098	1992 May 10	3:43	Thuot/Hieb	STS-49/OV105	An attempt to capture the Intelsat VI satellite fails.
099	1992 May 12	5:30	Thout/Heib	STS-49/OV105	Second attempt to capture Intelsat VI satellite fails.
100	1992 May 13	8:29	Thout/Heib/Akers	STS-49/OV105	First three-person EVA; trio captured Intelsat VI by hand, then attached a new kick motor and redeployed it, the hundredth EVA was also the longest in EVA history.
101	1992 May 14	7:45	Thornton, K./Akers	STS-49/OV105	Tested Space Station Freedom construction techniques and crew rescue procedures; Thornton set a new EVA duration record for a female.

	Date	Duration	Crew	Mission/Station	Description
102	1992 Jul 8	2:05	Viktorenko/Kaleri	EO11/Kvant 2	Completed external repairs to the Kvant 2 module.
103	1992 Sep 3	3:56	Solovyov, A./Avdeyev	EO12/Kvant 2	Commenced work to install new VDU propulsion system onto the Sofora girder.
104	1992 Sep 7	5:08	Solovyov, A./Avdeyev	EO12/Kvant 2	Continued installation of VDU on Sofora girder; lowered USSR flag from outside Mir.
105	1992 Sep 11	5:44	Solovyov, A./Avdeyev	EO12/Kvant 2	Continued installation of VDU on Sofora girder.
106	1992 Sep 15	3:33	Solovyov, A./Avdeyev	EO12/Kvant 2	Antenna attached to Kristall module in order to assist planned Buran and US Shuttle docking radar.
107	1993 Jan 17	4:28	Harbaugh/Runco	STS-54/OV105	Demonstration and tests of techniques under development for construction of Space Station Freedom and HST repair activities.
108	1993 Apr 19	5:25	Manakov/Poleshchuk	EO13/Kvant 2	Commenced process of transferring solar arrays on exterior of Mir.
109	1993 Jun 18	4:18	Manakov/Poleshchuk	EO13/Kvant 2	Commenced configuration of Mir exterior for a series of EVAs by next main crew.
110	1993 Jun 25	5:50	Low/Wisoff	STS-57/OV105	Crew attached EURECA antennae; evaluated SS Freedom EVA tasks and HST repair methods.
111	1993 Sep 16	4:18	Tsibliyev/Serebrov	EO14/Kvant 2	First of three EVAs to install Rapana mast on exterior of Mir; also checked hull for Perseid meteoroid damage.
112	1993 Sep 16	7:05	Walz/Newman	STS-51/OV103	Further tests of HST repair tools and techniques.
113	1993 Sep 20	3:13	Tsibliyev/Serebrov	EO14/Kvant 2	Continued deployment of Rapana mast; also installed sample packages for later retrieval.
114	1993 Sep 28	1:52	Tsibliyev/Serebrov	EO14/Kvant 2	Planned EVA to complete installation of Rapana and film exterior of station; the EVA was shortened when Tsibliyev's suit overheated.
115	1993 Oct 22	0:38	Tsibliyev/Serebrov	EO14/Kvant 2	Installation of new instrument block on Kvant 2; completed exterior filming of station.
116	1993 Oct 29	4:12	Tsibliyev/Serebrov	EO14/Kvant 2	Inspected solar arrays and exterior antenna; checked Sofora mount and retrieved materials samples in order to determine future operational lifetime of Mir.
117	1993 Dec 4	7:54	Hoffman/Musgrave	STS-61/OV105	First HST service mission; first service EVA included the replacement of malfunctioning gyroscopes.

	Date	Time	Crew	Mission	
118	1993 Dec 5	6:36	Thornton, K./Akers	STS-61/OV105	Second HST service EVA to remove old solar arrays and install new ones.
119	1993 Dec 6	6:47	Hoffman/Musgrave	STS-61/OV105	Third HST service EVA; installed new camera.
120	1993 Dec 7	6:50	Thornton, K./Akers	STS-61/OV105	Fourth HST service EVA; installed COSTAR and a new computer; cut loose solar panel.
121	1993 Dec 8	7:21	Hoffman/Musgrave	STS-61/OV105	Fifth HST service EVA; new control systems installed.
122	1994 Sep 9	5:06	Malenchenko/Musabayev	EO16/Kvant 2	Inspected docking port after collision by Progress M-24; repaired thermal blanket torn by Soyuz TM-17; attached new solar panels; configured Mir for planned US Shuttle docking.
123	1994 Sep 13	6:01	Malenchenko/Musabayev	EO16/Kvant 2	Samples retrieved from Rapana; maintenance work completed on Sofora truss; maintenance on exterior of Kvant 2 and solar panels.
124	1994 Sep 16	6:51	Lee/Meade	STS-64/OV103	First tests of SAFER, free-flying astronaut rescue jetpack (smaller MMU).
125	1995 Feb 9	4:39	Harris/Foale	STS-63/OV103	Evaluation of the astronauts ability to translate large objects; related to future Alpha Space Station tasks; (EDFT-3); unsuccessful test of cold temperature EVA gloves.
126	1995 May 12	6:15	Dezhurov/Strekalov	EO18/Kvant 2	Exterior of Mir was prepared for the transfer of solar panels allowing for docking of the US Shuttle.
127	1995 May 17	6:30	Dezhurov/Strekalov	EO18/Kvant 2	Commenced moving solar panels but failed to complete initial move.
128	1995 May 22	5:15	Dezhurov/Strekalov	EO18/Kvant 2	Completed the move of the first solar array started on previous EVA.
129	1995 May 28	0:21	Dezhurov/Strekalov	EO18/Mir node	Internal EVA (first of several at Mir) wearing full EVA suits in forward transfer compartment to relocate Konus equipment to allow docking of Spektr module on June 1.
130	1995 Jun 1	0:24	Dezhurov/Strekalov	EO18/Mir node	Second internal EVA returning the Konus docking cone to its pre-May 28 location.
131	1995 Jul 14	5:34	Solovyov, A./Budarin	EO19/Kvant 2	Inspection of a leaky docking collar; moved two solar arrays allowing later transfer of Kristall module.
132	1995 Jul 19	3:08	Solovyov, A./Budarin	EO19/Kvant 2	Budarin commences installation of MIRAS infrared spectrometer; Solovyov unable to exit hatch due to a suit problem and remains in transfer compartment.

#	Date	Duration	Crew	Mission	Description
133	1995 Jul 21	5:50	Solovyov, A./Budarin	EO19/Kvant 2	Completed installation of MIRAS spectrometer.
134	1995 Sep 15	6:46	Voss/Gernhardt	STS-69/OV105	Tests of EVA thermal gear; continued tests on Space Station Alpha techniques (EDFT-2).
135	1995 Oct 20	5:16	Avdeyev/Reiter	EO20/Kvant 2	ESA astronaut Reiter (first German to perform EVA) erects European experiment on outside of Mir.
136	1995 Dec 8	0:29	Gidzenko/Avdeyev	EO20/Mir node	Third Mir internal EVA; relocation of −Z docking cone to +Z port for Priroda module docking.
137	1996 Jan 15	6:09	Chiao/Barry	STS-72/OV105	Space station hardware evaluation, the third EVA Development Flight Test (EDFT-3) exercise; this time umbilical lines, utility boxes and work platforms were used.
138	1996 Jan 17	6:54	Chiao/Scott, W.	STS-72/OV105	Continuation of the EDFT programme begun on previous EVA; in addition a 30 minute 'cold soak' of Scott's EMU was completed to test its thermal properties.
139	1996 Feb 8	3:06	Gidzenko/Reiter	EO20/Kvant 2	Retrieved the exposure facility ESEF that was deployed during EO20 EVA1 on 20 October 1995; the crew also moved the Ikarus MMU to a permanent location outside of Kvant 2 to allow more room inside the air-lock; the MMU remained there until the destruction of Mir in March 2000; failed to remove an antenna from one of the solar arrays.
140	1996 Mar 15	5:51	Onufrienko/Usachev	EO21/Kvant 2	Installation of a second Strela crane on the Mir base block; they also set cables ready for new solar panels on Kvant.
141	1996 Mar 27	6:02	Godwin/Clifford	STS-76/OV104	First Shuttle-based EVA by American astronauts docked to Mir; the crew attached MEEP dust collectors to the exterior of Mir docking module but did not traverse over to Mir; they also evaluated common foot restraints and tether hooks intended for ISS EVA operations.
142	1996 May 20	5:20	Onufrienko/Usachev	EO21/Kvant 2	Relocated solar battery from exterior of Mir docking module to exterior of Kvant 1.
143	1996 May 24	5:43	Onufrienko/Usachev	EO21/Kvant 2	Installed Russian/American solar panel on to the exterior of Kvant 1.
144	1996 May 30	4:20	Onufrienko/Usachev	EO21/Kvant 2	Installation of MOMS-2 camera and an EVA handrail to the exterior of Mir.

No.	Date	Duration	Crew	Mission	Description
145	1996 Jun 6	3:34	Onufrienko/Usachev	EO21/Kvant 2	Installed the SKK-11 experiment on the outside of Mir and replaced the KOMZA experiment package. They also filmed part one of a sponsored Pepsi commercial.
146	1996 Jun 13	5:46	Onufrienko/Usachev	EO21/Kvant 2	Crew installed and deployed Ferma-3 girder and repaired the Travers antenna; they also completed filming the second part of the Pepsi commercial.
–	1996 Nov 28	0:48	Jones/Jernigan	STS-60/OV102	Crew failed to complete planned EVA due to stuck hatch on air-lock module; in a vacuum condition within the air-lock for 48 minutes.
147	1996 Dec 2	5:57	Korzun/Kaleri	EO22/Kvant 2	Linked power cables from solar battery to the main electrical bus on Mir.
148	1996 Dec 9	6:36	Korzun/Kaleri	EO22/Kvant 2	Completed the linkage of power cables begun on previous EVA.
149	1997 Feb 13	6:42	Lee/Smith	STS-82/OV103	Sixth HST servicing EVA; replacement of older High Resolution Spectrograph and Faint Object Spectrograph with the new Space Telescope Imaging Spectrograph (STIS) and Near Infrared Camera and Multi-Object Spectrometer (NICMOS).
150	1997 Feb 14	7:27	Harbaugh/Tanner	STS-82/OV103	Seventh HST serving EVA; replacement of the Far Guidance System (FGS) and out of date recorders; installation of the Optical Control Electronics Enhancement Kit (OCE-EK); the crew also noted insulation damage on the telescope.
151	1997 Feb 15	7:11	Lee/Smith	STS-82/OV103	Eighth HST servicing EVA; replacement of the Data Interface unit (DIU) and the installation of a new solid state data recorder.
152	1997 Feb 16	6:34	Harbaugh/Tanner	STS-82/OV103	Ninth HST servicing EVA; replacement of Solar Array Drive electronics (SADE); installation of covers for magnetometers; commenced the repair of insulation noted on Feb 14 EVA.
153	1997 Feb 17	5:17	Lee/Smith	STS-82/OV103	Tenth HST servicing EVA; this was an additional EVA added to the flight plan in order to attach thermal insulation blankets to the exterior of the telescope.

No.	Date	Duration	Crew	Mission	Description
154	1997 Apr 29	4:48	Tsibliyev/Linenger	EO23/Kvant 2	First US/Russian EVA; first American to use a Russian EVA suit (Orlan M); retrieval of experiment packages from exterior of Mir.
155	1997 Aug 22	3:16	Solovyov, A./Vinogradov	EO24/Mir node	Fourth Mir IVA; connected power cables from the damaged Spektr module to the Mir base block.
156	1997 Sep 6	6:00	Solovyov, A./Foale	EO24/Kvant 2	Second US/Russian EVA; crew searched for evidence of puncture in hull of Spektr but found none; completed a manual realignment of solar arrays.
157	1997 Oct 1	5:01	Parazynski/Titov, V.	STS-86/OV104	Third US/Russian EVA; first from Shuttle; retrieval of the MEEP from exterior of Mir.
158	1997 Oct 20	6:38	Solovyov, A./Vinogradov	EO24/Mir node	Fifth Mir IVA; completed connections of Spektr power cables.
159	1997 Nov 3	6:04	Solovyov, A./Vinogradov	EO24/Kvant 2	Disconnected the old Kvant solar array and replaced it; hand launched mini replica of Sputnik, marking the fortirth anniversary of the Space Age (October 4).
160	1997 Nov 6	6:17	Solovyov, A./Vinogradov	EO24/Kvant 2	Transferred solar panel; temporarily installed a cap for possible Spektr leak repair; a hatch leak delayed the EVA closeout procedures.
161	1997 Nov 24	7:43	Scott, W./Doi	STS-87/OV102	First Japanese (Doi) to perform EVA; unplanned Spartan retrieval; further tests of ISS EVA hardware.
162	1997 Dec 3	4:59	Scott, W./Doi	STS-87/OV102	Second Japanese EVA; Continued the originally planned ISS EVA hardware/procedures tests delayed from EVA1.
163	1998 Jan 8	4:04	Solovyov, A./Vinogradov	EO24/Kvant 2	Attempted repair of leaking air-lock hatch.
164	1998 Jan 14	6:38	Solovyov, A./Wolf	EO24/Kvant 2	Fourth US/Russian EVA; inspection of Mir exterior and repair of the faulty EVA air-lock hatch.
–	1998 Mar 3	1:15	Musabayev/Budarin	EO25/Kvant 2	Failed EVA attempt; sixth Mir IVA; Budarin broke tool trying to open the EVA hatch.
165	1998 Apr 1	6:40	Musabayev/Budarin	EO25/Kvant 2	Installation of handrails on the outside of Mir in preparation for a planned repair of Spektr solar array mounting.
166	1998 Apr 6	4:23	Musabayev/Budarin	EO25/Kvant 2	Strengthening of Spektr solar array mounting; the EVA was shortened by mission control error in regard to the orientation system on Mir.

#	Date	Duration	Crew	Mission	Description
167	1998 Apr 11	6:25	Musabayev/Budarin	EO25/Kvant 2	Crew dismantled and discarded the exterior control engine (VDU) unit.
168	1998 Apr 17	6:32	Musabayev/Budarin	EO25/Kvant 2	Dismantled the Strela boom and truss #3 and commenced the installation of a new VDU unit.
169	1998 Apr 22	6:21	Musabayev/Budarin	EO25/Kvant 2	Completion of the installation of the new VDU unit on top of a tower on the exterior of Mir; hundredth space station-related EVA.
–	1998 Sep 15	0:30	Avdeyev/Padalka	EO26/Mir node	Seventh Mir IVA; connection of cables on Spektr hatch thermaplate within forward docking node.
170	1998 Nov 10	5:54	Avdeyev/Padalka	EO26/Kvant 2	Set up six new external experiments; conducted a maintenance and repair programme; hand launched second replica mini Sputnik satellite.
171	1998 Dec 7	7:21	Ross/Newman	STS-88/OV105	First ISS-related EVA; Shuttle orbiter based; connected electrical cables from Zarya to Unity; installation of handrails and support equipment for future EVAs.
172	1998 Dec 9	7:02	Ross/Newman	STS-88/OV105	Second ISS EVA; installation of two antennae on Unity and removal of launch restraint pins on Node hatches; also deployment of stuck TORU antenna on Zarya.
173	1998 Dec 12	6:59	Ross/Newman	STS-88/OV105	Third ISS EVA; stowed bag of EVA tools on Unity for future EVA crews; freed second stuck TORU antenna on Zarya and conducted a survey of exterior of station hull.
174	1999 Apr 16	6:19	Afanasyev/Haignere	EO26/Kvant 2	Second French EVA; performed CNES experiments; deployment by hand of third mini-replica of Sputnik.
175	1999 May 28	7:55	Jernigan/Barry	STS-96/OV103	Fourth ISS EVA; Relocation of stowed US Orbital Transfer Device and Russian Strela cranes from Shuttle payload bay to Unity; attached a further set of EVA tools to the station; longest EVA for a female.
176	1999 Jul 23	6:07	Afanasyev/Avdeyev	EO26/Kvant 2	An attempt to deploy Russian/Georgian reflector antenna.
177	1999 Jul 28	5:22	Afanasyev/Avdeyev	EO26/Kvant 2	Deployment of Russian/Georgian reflector; deployed and retrieved exposed sample cassettes.
178	1999 Dec	–	Smith/Grunsfeld	STS-103/OV103	Eleventh HST servicing EVA.
179	1999 Dec	–	Foale/Nicollier	STS-103/OV103	Twelfth HST servicing EVA; first Swiss EVA.
180	1999 Dec	–	Smith/Grunsfeld	STS-103/OV103	Thirteenth HST servicing EVA.

#	Date	Duration	Crew	Mission	Description
181	2000 May 12	4:52	Zaletin/Kaleri	EO28/Kvant 2	The final EVA from the Mir complex; used Germatisator Sealing Experiment; completed a panoramic inspection; examined a failed solar battery.
182	2000 May 21	6:44	Williams/Voss	STS-101/OV104	Fifth ISS EVA; secured OTD; installation of final elements of Strela crane and replaced antenna on Unity; completed a number of 'get ahead' tasks.
183	2000 Sep 10	6:14	Lu/Malenchenko	STS-106/OV104	Sixth ISS EVA; fifth US/Russian EVA; connection of electrical and communication cables between Zvezda and Zarya; also installed magnetometer.
184	2000 Oct 15	6:28	Chiao/McArthur	STS-92/OV103	Seventh ISS EVA; deployed two antenna assemblies and installed EVA tool box; electrical cable connection for Z1 truss.
185	2000 Oct 16	7:07	Wisoff/Lopez-Alegria	STS-92/OV103	Eighth ISS EVA; preparation of Z1 truss for attachment of solar arrays; released launch latches securing PMA-3.
186	2000 Oct 17	6:48	Chiao/McArthur	STS-92/OV103	Ninth ISS EVA; installed second EVA tool-box; continued programme of reconfigurations and connections of electrical cables; installation of two DC converter units on Z1 truss.
187	2000 Oct 18	6:56	Wisoff/Lopez-Alegria	STS-92/OV103	Tenth ISS EVA; Tested SAFER back-pack; completed EVA wrap up tasks.
188	2000 Dec 3	7:33	Tanner/Noriega	STS-97/OV105	Eleventh ISS EVA; Installed P6 truss segment and deployed two solar arrays (after one stuck); connected power and data cables.
189	2000 Dec 5	6:37	Tanner/Noriega	STS-97/OV105	Twelfth ISS EVA; inspection of partially deployed solar array; connected power and data cables and coolant lines from P6; replacement of S Band assembly at P6.
190	2000 Dec 7	5:10	Tanner/Noriega	STS-97/OV105	Thirteenth ISS EVA; Repaired P6; installation of a small antenna and a sensor on a radiator; also installed a Floating Potential Probe and a centre-line camera.
191	2001 Feb 10	7:34	Jones/Curbeam	STS-98/OV104	Fourteenth ISS EVA; Monitored relocation of PMA-2; connected power and data cables; Curbeam's suit was contaminated with leaking ammonia requiring a 'bake' in sunlight to clean.
192	2001 Feb 12	6:50	Jones/Curbeam	STS-98/OV104	Fifteenth ISS EVA; EVA crew assisted in repositioning of PMA-2 on Destiny Lab; installed insulation covers in payload bay of *Discovery*; installed SSRMS base.

No.	Date	Duration	Crew	Mission	Description
193	2001 Feb 14	5:25	Jones/Curbeam	STS-98/OV104	Sixteenth ISS EVA; attachment of S-band antenna; released cooling radiator; inspected connections; test flew SAFER; this marked the hundredth US EVA since Gemini 4 in June 1965.
194	2001 Mar 11	8:56	Voss/Helms (EO2)	STS-102/OV103	Seventeenth ISS EVA; relocated Early Communications antenna from Unity to PMA attachment; installation of Lab Cradle Assembly on Destiny; installed SSRMS cable tray; set new EVA endurance record.
195	2001 Mar 13	6:21	Thomas/Richards	STS-102/OV103	Eighteenth ISS EVA; Configuration of stowage platform; completed the connection of cables commenced in previous EVA; made minor adjustments to solar array brace.
196	2001 Apr 22	7:10	Parazynski/Hadfield	STS-100/OV105	Nineteenth ISS EVA; first Canadian EVA (Hadfield); installation of UHF antenna and video command and power cables between the SSRMS and Destiny Lab robotic workstation; installation of Canadarm 2 (SSRMS).
197	2001 Apr 24	7:40	Parazynski/Hadfield	STS-100/OV105	Twentieth ISS EVA; second Canadian EVA; rewired and rerouted power and data cables for Canadarm 2.
198	2001 Jun 8	0:19	Usachev/Voss	EO2/Zvezda	Twenty-first ISS EVA, first IVA; sixth US/Russian EVA; replacement of flat plate in Zvezda nadir docking port with a docking cone; first IVA at ISS without a Shuttle docked.
199	2001 Jul 14	5:59	Gernhardt/Reilly	STS-104/OV104	Twenty-second ISS EVA; Configuration of Quest air-lock module for transfer to Unity module; connection of heater cables.
200	2001 Jul 17	6:29	Gernhardt/Reilly	STS-104/OV104	Twenty-third ISS EVA; transfer of two oxygen tanks and one nitrogen storage tank from *Atlantis* payload bay to exterior of Quest.
201	2001 Jul 20	4:02	Gernhardt/Reilly	STS-104/Quest	Twenty-fourth ISS EVA; first EVA from Quest air-lock; transferred one nitrogen tank from *Atlantis* payload bay to exterior of Quest; examined gimbal assembly on top of solar array truss; inauguration of Quest fell on thirty-second anniversary of first moonwalk during Apollo 11 (1969).
202	2001 Aug 16	6:16	Barry/Forrester	STS-105/OV103	Twenty-fifth ISS EVA; crew installed Early Ammonia Servicer (EAS) on P6 and the Material ISS Experiment (MISSE) on Quest.

No.	Date	Duration	Crew	Mission/Module	Notes
203	2001 Aug 18	5:29	Barry/Forrester	STS-105/OV103	Twenty-sixth ISS EVA; installation of six EVA hand-rails; relocation of a further two on Destiny; also strung heater cables for future SO truss.
204	2001 Oct 8	4:58	Dezhurov/Tyurin	EO3/Pirs	Twenty-seventh ISS EVA; first from Pirs and first all-Russian ISS EVA; erected antenna and docking targets on exterior of Pirs module; first EVA from ISS without Shuttle docked to it; hundredth Soviet/Russian EVA since Voskhod 2 in March 1965.
205	2001 Oct 15	5:52	Dezhurov/Tyurin	EO3/Pirs	Twenty-eighth ISS EVA; placed Kravka sample detector on Zvezda; also placed MPAC and SEED experiments on Zvezda; revoked Russian flag and installed commercial logos.
206	2001 Nov 12	5:04	Dezhurov/Culbertson	EO3/Pirs	Twenty-ninth ISS EVA; seventh US/Russian EVA; Routed and fixed cables from Pirs to ISS interior; checked and photographed small section of solar battery on Zvezda; installed handrails on Pirs; tested extension of Strela cargo crane.
207	2001 Dec 3	2:46	Dezhurov/Tyurin	EO3/Pirs	Thirtieth ISS EVA; first unplanned EVA at ISS; removed rubber O-ring obstruction from Pirs docking equipment left behind during 22 Nov undocking of Progress M-45; this prevented 28 Nov docking of Progress M1-7 which was achieved later on Dec 3.
208	2001 Dec 11	4:12	Godwin/Tani	STS-108/OV105	Thirty-first ISS EVA; installed insulation blankets on beta gimbal assembly located on the top of the P6 truss; unsuccessful attempt to free stuck solar array cable.
209	2002 Jan 14	6:03	Onufrienko/Walz	EO4/Pirs	Thirty-second ISS EVA; eighth US/Russian EVA; installed ham radio antenna; completed the assembly of the Strela unit.
210	2002 Jan 25	5:59	Onufrienko/Bursch	EO4/Pirs	Thirty-third ISS EVA; ninth US/Russian EVA; installation of six thruster plume deflectors; installed four new experiment packages and retrieved one older one; attached tether guides and a ham radio antenna.

211	2002 Feb 20	5:47	Walz/Bursch	EO4/Quest	Thirty-fourth ISS EVA; connected cables from Destiny to Z1 truss; removed tools and handrails that had be used on earlier EVAs; this EVA was conducted on the fortieth anniversary of John Glenn's Mercury 6 flight.
212	2002 Mar 4	7:01	Grunsfeld/Linnehan	STS-109/OV102	Fourteenth HST servicing EVA; replacement of one of the telescope's two second generation solar arrays; replaced a Diode Box assembly; completed other prep work for later EVAs on this mission.
213	2002 Mar 5	7:16	Newman/Massimino	STS-109/OV102	Fifteenth HST servicing EVA; replacement of second array with a new unit and its Diode Box Assembly; replacement of Reaction Wheel Assembly-1.
214	2002 Mar 6	6:48	Grunsfeld/Linnehan	STS-109/OV102	Sixteenth HST servicing EVA; replacement of the telescope's Power Control Unit in Bay 4; Linnehan conducted an inspection of HST's exterior handrails to be used during the fourth and fifth EVAs.
215	2002 Mar 7	7:30	Newman/Massimino	STS-109/OV102	Seventeenth HST servicing EVA; replacement of the Faint Object Camera with the new Advanced Camera for Surveys; completed Power Control Unit clean up tasks.
216	2002 Mar 8	7:20	Grunsfeld/Linnehan	STS-109/OV102	Eighteenth HST servicing EVA; installed the Cryogenic Cooler and its cooling system radiator around the NICMOS experiment.
217	2002 Apr 11	7:48	Smith, S./Walheim	STS-110/OV104	Thirty-fifth ISS EVA; transfer and attachment of S-Zero truss.
218	2002 Apr 13	7:30	Ross/Morin	STS-110/OV104	Thirty-sixth ISS EVA; completed attachment of S-Zero truss; attached redundant power cable.
219	2002 Apr 14	6:27	Smith, S./Walheim	STS-110/OV104	Thirty-seventh ISS EVA; connection of mobile transporter; routed power connections to SSRMS through S0 truss.
220	2002 Apr 16	6:37	Ross/Morin	STS-110/OV104	Thirty-eighth ISS EVA; installed work lights and air-lock spur for future EVA work on S0 truss.
221	2002 Jun 9	7:14	Chang-Diaz/Perrin	STS-111/OV105	Thirty-ninth ISS EVA; third French EVA; transfer of power/data/grapple fixture to solar array; attached space debris shields; preparation of Mobile Base System (MBS) for installation.
222	2002 Jun 11	5:00	Chang-Diaz/Perrin	STS-111/OV105	Fortieth ISS EVA; fourth French EVA; hard mate of the MBS and connected power, data and electronics cables.

No.	Date	Duration	Crew	Mission	Description
223	2002 Jun 13	7:17	Chang-Diaz/Perrin	STS-111/OV105	Forty-first ISS EVA; fifth French EVA; replacement of failed SSRMS wrist roll joint; preparation of P6 truss for future relocation.
224	2002 Aug 16	4:25	Korzun/Whitson	EO5/Pirs	Forty-second ISS EVA; tenth US/Russian EVA; Installed first six (of planned 23) debris panels onto Zvezda. Installation of Russian Kromka experiment postponed due to late start of EVA.
225	2002 26 Aug	5:21	Korzun/Treschev	EO5/Pirs	Forty-third ISS EVA; installation of exterior frame on Zarya to house components for future EVA assembly tasks. Installed new materials samples on a pair of NASDA (Japanese) experiments housed on outside of Zvezda. The cosmonauts also installed devices to simplify routing of tethers in future assembly EVAs; installed two Ham radio antenna on Zvezda and installed Kromka hardware that measures residual emissions from Zvezda jet thrusters.
226	2002 Oct 10	7:01	Wolf/Sellers	STS-112/Quest	Forty-fourth ISS EVA; attached power, data and fluid lines between S0 and newly installed S1 truss; deployed second S-band comm system; installed first of two external camera systems; released launch restraints on the truss's mobile EVA workstation (CETA) and released launch lock holding S1's radiators in place for launch.
227	2002 Oct12	6:04	Wolf/Sellers	STS-112/Quest	Forty-fifth ISS EVA: set up the second external camera system and released further radiator beam launch locks; removed insulation on quick disconnect fittings near Z1 and P6 junctions to install Spool Positioning Devices; released starboard side launch restraints on CETA and attached Ammonia Tank Assembly cables.
228	2002 Oct 14	6:36	Wolf/Sellers	STS-112/Quest	Forty-sixth ISS EVA; replaced Interface Umbilical Assembly on Mobile Transporter; installed two jumpers to allow flow of coolant between S1 and S0 Trusses; released the large metal rod used as a launch restraint for S1 and stowed it (a drag link); installed Spool Positioning Devices on ammonia lines.

229	2002 Nov 26	6:45	STS-113/Quest	Lopez-Alegria/ Herrington	Forty-seventh ISS EVA; made connections between P1 and S0 Truss; released launch restraint on CETA; installed Spool Positioning Devices onto ISS; removed drag link on P1; installed wireless video system on External Transceiver Assembly on to Unity node to support EVA helmet camera operations.
230	2002 Nov 28	6:10	STS-113/Quest	Lopez-Alegria/ Herrington	Forty-eighth ISS EVA; installed fluid jumpers at S0/P1 attachment point; removed P1's starboard keel pin; installed a second wireless video system External Transceiver Assembly on to the P1; removed port keel pin; relocated CETA cart from P1 to S1 Truss allowing Mobile Transporter to move along P1 to assist in future missions.
231	2002 Nov 30	7:00	STS-113/Quest	Lopez-Alegria/ Herrington	Forty-ninth ISS EVA; installed further SPDs (total for the three EVAs over 40); reconfigured external electrical harnesses that route power through Main Bus Switching Units; attached Ammonia Tank Assembly lines.
232	2003 Jan 15	6:51	EO6/Quest	Bowersox/Pettit	Fiftieth ISS EVA; continued the outfitting and activation of P1 truss; released remaining radiator launch locks allowing full deployment; removal of debris on sealing ring of Unity's Earth facing docking port; tested P6 Truss ammonia reserve; unable to complete installation of a light fixture on CETA; reassigned to a future EVA; cut away thermal cover strap that was interfering with the rotation of Quest air-lock hatch and delayed the start of this EVA.
233	2003 Apr 8	6:26	EO6/Quest	Bowersox/Pettit	Fifty-first ISS EVA; reconfigured ISS power systems to provide a secondary power source for one of the CMGs; secured thermal control system quick-disconnect fittings; released stuck latch on the CETA cart light support system.

Bibliography

I have pursed an interest in the history of spacewalking, EVA operations, spacesuit technology and EVA hardware development since 1969 and the first lunar landing. However, it would not be feasible to list all the source material and references collected over the past thirty-five years, of which only a small portion was used in the compilation of this book, which itself is part of a larger study of the history of EVA. During 1988–2002, fifteen visits to the USA encompassed tours and research at NASA JSC, Houston, NASA KSC, Florida, NARA, Fort Worth, and Rice University, Houston. I have also carried out extensive research using the private collection of Skylab astronaut Jerry Carr, in Arkansas. During these visits, several dedicated trips included research at NASA JSC History Archive, currently located at the University of Clear Lake, Houston (Apollo, Skylab, the Shuttle and the ISS), and at NARA, Fort Worth (Gemini). Visits to the Yuri Gagarin Cosmonaut Training Centre, and the Zvezda facility, Moscow, in June 2003, also proved very fruitful.

PERSONAL ARCHIVES

The Astro Info Service collection consists of research archives, mission documentation and projects amassed and compiled since 1968, and contniues to be expanded. This collection includes personal interviews and correspondence with the following astronauts and cosmonauts during 1988–2003, many of which have been utilised for this book.

Joe Allen	Gregori Grechko	Jerry Ross
Jerry Carr	Bernard Harris	Viktor Savinykh
Gene Cernan	Jim Irwin	Kathy Sullivan
Leroy Chiao	Jack Lousma	Yuri Usachev
Walter Cunningham	Bruce McCandless	Alexandr Volkov
Bill Fisher	Bruce Melnick	Paul Weitz
Michael Foale	Story Musgrave	Dave Wolf
Ed Gibson	Bill Pogue	John Young
Dick Gordon	Yuri Romanenko	

BOOKS AND ARTICLES

1956–	*Spaceflight* (British Interplanetary Society)
1965–1995	*Astronautics and Aeronautics* (NASA, various volumes)
1965–	*Aviation Week and Space Technology*
1965	*Gemini: America's Historic Walk in Space*, United Press International/Prentice–Hall
1967	*Summary of Gemini Extravehicular Activity*, Reginald M. Machell (*ed.*), NASA SP-149
1969	*Project Gemini, Technology and Operations: A Chronology*, James M. Grimwood, Barton C. Hacker and Peter J. Vorzimmer, NASA SP-4002
	Apollo 11 Technical Crew Debrief, NASA Mission Operations Branch, Flight Crew Support Division, NASA Manned Spacecraft Center, Houston, Texas, 2 vols.
	Apollo 12 Technical Crew Debrief, NASA Mission Operations Branch, Flight Crew Support Division, NASA Manned Spacecraft Center, Houston, Texas
	Apollo Program (Phase 1 Lunar Exploration) Mission Definitions, 1 December 1969, NASA Manned Spacecraft Center, MSC-01266, (from the Jerry Carr collection)
	The Apollo Spacecraft: A Chronology, NASA-SP-4009, 4 vols.
1970	*First on the Moon: A Voyage with Neil Armstrong*, Michael Collins and Edwin E. Aldrin Jr., with Gene Farmer and Dora Jane Hamblin, Michael Joseph
1971	*Suiting Up for Space: The Evolution of the Sspace Suit*, Lloyd Mallan, John Day Company
	Apollo 14 Technical Crew Debrief, NASA Mission Operations Branch, Flight Crew Support Division, NASA Manned Spacecraft Center, Houston, Texas
	Apollo 15 Technical Crew Debrief, NASA Mission Operations Branch, Flight Crew Support Division, NASA Manned Spacecraft Center, Houston, Texas
1972	*Apollo 16 Technical Crew Debrief*, NASA Mission Operations Branch, Flight Crew Support Division, NASA Manned Spacecraft Center, Houston, Texas
1973	*Apollo 17 Technical Crew Debrief*, NASA Mission Operations Branch, Flight Crew Support Division, NASA Manned Spacecraft Center, Houston, Texas, 2 vols.
	Return to Earth, Buzz Aldrin and Wayne Warga, Random House
	To Rule the Night, James B. Irwin and William A. Emerson Jr., Holman
	Skylab: A Guidebook, NASA EP-107
1974	*Carrying the Fire*, Michael Collins, Farrar Straus Giroux
	MSFC Skylab Mission Report: Saturn Workshop, Skylab Program Office, NASA MSFC, Alabama, NASA Technical Memorandum, TM X-64814, (from the Jerry Carr collection)

1975	*Apollo Program Summary Report*, NASA JSC-09423
	Apollo Expeditions to the Moon, NASA SP-350
1977	*On the Shoulders of Titans: A History of Project Gemini*, Barton C. Hacker and James M. Grimwood, NASA SP-4203
	Skylab: A Chronology, Roland Newkirk, Ivan Ertel and Courtney Brooks, NASA SP-4011
1979	*Chariots for Apollo: A History of Manned Lunar Spacecraft*, Courtney Brooks, James Grimwood and Loyd Swenson Jr., NASA SP-4205
1980	*I Walk in Space*, Alexei Leonov, Malysh Publishers, Moscow
	Handbook of Soviet Manned Spaceflight, Nicholas L. Johnson, Science and Techology Series, Vol.48, AAS Publications
1981	*Red Star In Orbit*, James E. Oberg, Harrap
1983	*Space Transportation System: EVA Description and Design Criteria*, NASA JSC-10615 Rev A
1984	*STS EVA Reports*, No. 1, STS-5, and No.2, STS-6, Astro Info Service
1985	*Don Lind, Mormon Astronaut*, Kathleen Maughan Lind, Deseret Books
1986	*An Astronaut's Diary*, Jeff Hoffman, Caliban Press
1987	*Before Lift-Off: The Making of a Space Shuttle Crew*, Henry S.F. Cooper, Johns Hopkins University Press.
1988	*Lift-off: The Story of America's Adventure in Space*, Michael Collins, Grove Press
	Diary of a Cosmonaut, Valentin Lebedev, Gloss Company (US edition)
	The Soviet Manned Space Programme, Phillip Clark, Salamander Books
	Race into Space, Brian Harvey, Ellis–Horwood
1989	*Where No Man Has Gone Before: A History of Apollo Lunar Exploration Missions*, William David Compton, NASA SP-4214
	Footprints, Douglas MacKinnon and Joseph Baldanza, Acropolis books
1990	*Moonwalker*, Charles and Dotty Duke, Oliver Nelson
	Almanac of Soviet Manned Spaceflight, Dennis Newkirk, Gulf Publishers
1993	*To a Rocky Moon: A Geologist's History of Lunar Exploration*, Don E Williams, University of Arizona Press
1994	*US Space Gear: Outfitting the Astronaut*, Lillian D. Kozloski, Smithsonian Institution Press
	Catalog of Apollo Experiment Operations, Thomas A. Sullivan, NASA Reference Publication
	A Man on the Moon: The Voyages of the Apollo Astronauts, Andrew Chaikin, Michael Joseph
1995	*Mir Hardware Heritage*, David S.F. Portree, NASA JSC 26770
1996	*The New Russian Space Programme*, Brian Harvey, Wiley–Praxis
1997	*Walking to Olympus: An EVA Chronology*, David S.F. Portree and Robert C. Treviño, NASA Monographs in Aerospace History, No.7
1998	*The Space Shuttle: Roles, Missions and Accomplishments*, David M. Harland, Wiley–Praxis
1999	*The Last Man on the Moon*, Eugene Cernan and Don Davis, St. Martin's Press

	Exploring The Moon: The Apollo Expeditions, David M. Harland, Springer–Praxis
2000	*Challenge to Apollo: The Soviet Union and the Space Race, 1945–1974*, Asif A. Siddiqi, NASA SP-2000-4408
	The History of Mir, 1986–2000, Rex D. Hall (*ed.*), British Interplanetary Society
	Apollo By The Numbers: A Statistical Reference, Richard W. Orloff, NASA SP-2000-4029
2001	*Taking Science to the Moon,: Lunar Experiments and the Apollo Program*, Donald A. Beattie, Johns Hopkins University Press{?}
	Russia In Space: The Failed Frontier?, Brian Harvey, Springer–Praxis
	Mir: The Final Year, Rex D. Hall (*ed.*), British Interplanetary Society
	The Rocket Men, David J. Shayler and Rex D. Hall, Springer–Praxis
	Skylab: America's Space Station, David J. Shayler, Springer–Praxis
	Gemini: Steps to the Moon, David J. Shayler, Springer–Praxis
2002	*Apollo: The Lost and Forgotten Missions*, David J. Shayler, Springer–Praxis
	The International Space Station: From Imagination to Reality, Rex D. Hall (*ed.*), British Interplanetary Society
	Creating the International Space Station, David M. Harland and John E. Catchpole, Springer–Praxis
	The Continuing Story of the International Space Station, Peter Bond, Springer–Praxis
2003	*Soyuz: a Universal Spacecraft*, David J. Shayler and Rex D. Hall, Springer–Praxis
	Russian Spacesuits, Isaak P. Abramov and Å. Ingemaar Skoog, Springer–Praxis

WEB SITES

In addition to the above, the following are excellent sources of information.

NASA Human Spaceflight, http://spacefilght.nasa.gov/home/index.html
Apollo Spaceflight Journal, http://hq.nasa.gov/office/pao/hitory/alsj/frame.html

Index